Essential Maths

► INTERMEDIATE ◄

MARK BINDLEY
RENIE VERITY

Stanley Thornes (Publishers) Ltd

First published in 1996 by

Stanley Thornes (Publishers) Ltd
Ellenborough House
Wellington Street
CHELTENHAM
GL50 1YW

A catalogue record of this book is available from the British Library.

ISBN 0 7487 2431 1

96 97 98 99 00 / 10 9 8 7 6 5 4 3 2 1

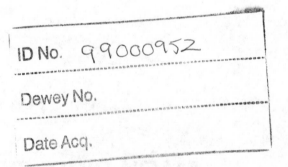

Artwork by Phil Ford, Hardlines

Typeset by Wyvern Typesetting, Bristol
Printed and bound in Great Britain at T J Press, Padstow, Cornwall

Contents

Introduction

About Essential Maths for GCSE

This book is designed to help students pass GCSE at Intermediate Level.

It can be used:

- as a two year course for students in Years 10 and 11.
- as a one year course for students in Further Education.
- as a course for an Adult Education class.
- as a teaching and revision aid for parents or private tutors helping students to prepare for GCSE mathematics.

It features:

- a contents organisation carefully matched to the new GCSE syllabuses in mathematics.
- very extensive explanations and worked examples.
- PAUSE exercises. These follow explanations and worked examples and are designed to test understanding of key points and provide practice in skills and techniques.
- REWIND exercises. These provide a systematic revision of skills and techniques at the end of major sections of the book.
- FAST FORWARD exercises. These provide a wide range of actual past paper questions to ensure students are familiar with the question format used in examinations.

This book assumes that:

- students will own and use a scientific calculator (unless it is specifically indicated otherwise, all exercises should be completed with a calculator);
- Students will obtain a copy of the individual GCSE syllabus they are studying. This will indicate the exact coverage required and dictate the sections of the book that must be studied.

Some advice to students

There is no 'correct' way to succeed in examinations and everybody has their own special ideas and study methods.

I will however offer the following general advice which I hope you may find helpful in developing your own failsafe route to success.

- As you work through the PAUSE exercises, try to keep your solutions in an organised folder or exercise book. These can form a valuable revision aid during the final weeks, days and hours before the examination.
- Working through the REWIND exercises during the final weeks of your course is a very good way to revise. You can make a note of all the questions that cause difficulty and repeat them after a few days to make certain that you have mastered the necessary techniques.
- Most GCSE examinations include a page giving formulae etc. Make sure you obtain a copy of this so you do not waste time memorising unnecessary information.
- Obtain as many past papers for your particular syllabus as you possibly can and work through them using the teaching notes in this book to help you.
- During the final revision period, try to work through as many of the worked examples in this book as possible. Don't just read the examples, cover the solution with a piece of paper and try to solve the problem. If you cannot solve the problem study the solution and then try again. If you are really serious about passing with a good grade, you need to master every worked example in this book!
- Go into the examination feeling prepared. Make sure you have calculator, spare calculator, pencils, pens, geometrical instruments, sweets, tissues, mascots, cushions and everything else you need.

Good luck! Remember, everything is possible with enough effort.

Mark Bindley

Number

People in earlier times *needed* mathematics.

- They needed to *count* their possessions.
- They needed to *measure time* and, since the sun and stars were our first clocks, this meant *measuring angles*.
- They needed to *measure* the *length* of the land that they farmed.
- As trade developed, they needed to *measure weight* and *volume*.

At first, people only used whole numbers.

'I own 7 cows, 5 goats and 24 sheep.'

'There are 200 people living in our settlement.'

You cannot, however, solve all problems with whole numbers and so *fractions* and *decimals* were invented and used.

This is the multiplication square.

x	1	2	3	4	5	6	7	8	9	10	11	12
1	1	2	3	4	5	6	7	8	9	10	11	12
2	2	4	6	8	10	12	14	16	18	20	22	24
3	3	6	9	12	15	18	21	24	27	30	33	36
4	4	8	12	16	20	24	28	32	36	40	44	48
5	5	10	15	20	25	30	35	40	45	50	55	60
6	6	12	18	24	30	36	42	48	54	60	66	72
7	7	14	21	28	35	42	49	56	63	70	77	84
8	8	16	24	32	40	48	56	64	72	80	88	96
9	9	18	27	36	45	54	63	72	81	90	99	108
10	10	20	30	40	50	60	70	80	90	100	110	120
11	11	22	33	44	55	66	77	88	99	110	121	132
12	12	24	36	48	60	72	84	96	108	120	132	144

12 can be divided exactly by 3: $12 \div 3 = 4$

12 cannot be divided exactly by 5: $12 \div 5 = 2$ remainder 2

This is a list of all the numbers which divide exactly into 12: 1, 2, 3, 4, 6 and 12.

Any number which divides exactly into 12 is called a *factor* of 12. So, the factors of 12 are, 1, 2, 3, 4, 6 and 12.

If a number has 2 as one of its factors, it is called an *even* number.

14 is an even number because it divides exactly by 2

If a number does not have 2 as one of its factors, it is called an *odd* number.

9 is an odd number because it does not divide exactly by 2

If a number has exactly two factors it is called a *prime* number. A prime number is a number which can only be divided exactly by itself and the number 1.

> 7 is a prime number because it has only two factors,
> the numbers 7 and 1

> 9 is not a prime number because it has three factors,
> the numbers 9, 3 and 1

When we multiply 12 by the numbers 1, 2, 3, 4, and 5, we get 12, 24, 36, 48 and 60. These numbers all divide exactly by 12. Any number which divides exactly by 12 is called a *multiple* of 12.

> The first 5 multiples of 12 are, 12, 24, 36, 48 and 60.

PAUSE

1 Write down the factors of 8.

2 Write down the factors of 25.

3 Write down the factors of 13.

4 a) Write down all the factors of these numbers:
   i) 18 ii) 21 iii) 23 iv) 16
   b) i) Write down the condition for a number to be prime.
   ii) Which number in a) is prime?

5 Which of 42, 170, 901, 1325 and 1500 are even numbers?

6 The first 10 prime numbers are 2, 3, 5, 7, 11, …

7 The first 7 multiples of 6 are, …

8 The first 6 multiples of 7 are, …

9 Look at the numbers 11, 12, 13, 14, 15, 16, 17, 18.
   a) Which of the numbers are multiples of 3?
   b) Which of the numbers are factors of 36?

10 Look at the numbers 7, 10, 13, 16, 19, 21, 24.
   a) Which of the numbers are factors of 21?
   b) Which of the numbers are multiples of 7?
   c) Which of the numbers are prime?

Examples

1. Find the **highest common factor** (HCF) of 28 and 42.

The factors of 28 are 1, 2, 4, 7, (14), 28.
The factors of 42 are 1, 2, 3, 6, 7, (14), 21, 42.
The largest number in both lists is 14.
14 is the HCF of 28 and 42.

2. Find the **lowest common multiple** (LCM) of 16 and 24.

The multiples of 16 are 16, 32, (48), 64, 80, 96.
The multiples of 24 are 24, (48), 72, 96.
The smallest number to appear on both lists is 48.
48 is the LCM of 16 and 24.

PAUSE

1 Find the HCF of each pair of numbers.
   a) 8, 12   b) 15, 20   c) 18, 30   d) 24, 36   e) 15, 30   f) 25, 35

2 Find the LCM of each pair of numbers.
   a) 6, 8   b) 10, 15   c) 13, 26   d) 4, 7   e) 14, 35   f) 27, 36

Prime factors

1, 2, 3, 4, 6 and 12 are all factors of the number 12.

2 and 3 are called the **prime factors** of 12 because they are **factors** of 12 and also **prime numbers**.

Any number which is not prime can be written as a chain multiplication of its prime factors. We can write:

$$12 = 2 \times 2 \times 3$$

Here is a quick method you can use with large numbers, like 360.

First, we break 360 into *any* multiplication.

$$360 = 10 \times 36$$

Now, we keep breaking each part of this multiplication until only prime numbers are left.

$$360 = 10 \times 36$$
$$360 = 2 \times 5 \times 36$$
$$360 = 2 \times 5 \times 6 \times 6$$
$$360 = 2 \times 5 \times 2 \times 3 \times 6$$
$$360 = 2 \times 5 \times 2 \times 3 \times 2 \times 3$$

Or, more neatly, we can write:

$$360 = 2 \times 2 \times 2 \times 3 \times 3 \times 5$$

There is a special way to write chain multiplications of the same number.

$2 \times 2 \times 2$ can be written as 2^3

3×3 can be written as 3^2

So, we can write:

$$360 = 2^3 \times 3^2 \times 5$$

PAUSE

Copy and complete.

1 The factors of 50 are ...

2 The prime factors of 50 are ...

3 50 can be written as a chain multiplication of its prime factors in this way ...

4 400 can be broken down into a chain multiplication of prime factors like this:

$$400 = 10 \times 40$$
$$400 = 2 \times 5 \times 40$$

Copy and complete the chain.

5 Write each of these numbers as a multiplication of prime factors:

a) 36   d) 25   g) 2700

b) 100   e) 125

c) 48   f) 162

Square, rectangular and triangular numbers

These are the first five *square numbers*.

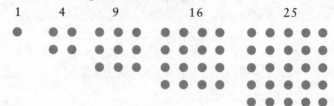

These are the first six *rectangular numbers*.

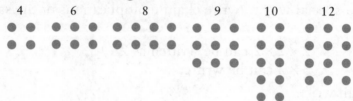

The pattern of dots must have at least two rows and two columns. Squares count as special rectangles.

These are the first five *triangular numbers*.

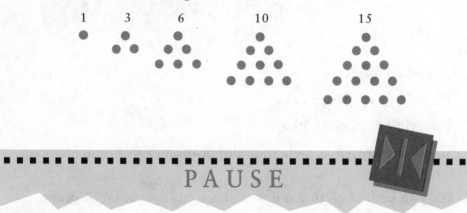

PAUSE

1 Write down the first 12 square numbers.

2 Write down the first 20 rectangular numbers.

3 Write down the first 10 triangular numbers.

4 What could these numbers be called? Draw diagrams of the next three numbers in the set.

Square roots

The numbers 1, 4, 9, 16, 25, ... are the square numbers.
The numbers 1, 2, 3, 4, 5, ... are the **square roots** of these numbers.
This is the special sign for the square root of a number $\sqrt{}$ or $\sqrt{}$, so

$$\sqrt{36} = 6$$

■■■■■■■■■■■■■■■■■■■■■■■■■■■■■■■■■■■■ ■■■
PAUSE

1 Copy and complete.

a) $\sqrt{81}$ =   d) $\sqrt{100}$ =   g) $\sqrt{400}$ =   j) $\sqrt{2500}$ =

b) $\sqrt{49}$ =   e) $\sqrt{144}$ =   h) $\sqrt{900}$ =   k) $\sqrt{6400}$ =

c) $\sqrt{121}$ =   f) $\sqrt{64}$ =   i) $\sqrt{1600}$ =   l) $\sqrt{10\,000}$ =

■■■■■■■■■■■■■■■■■■■■■■■■■■■■■■■■■■■■■■■

SECTION 2 Estimation and Calculation

Multiplying whole numbers without a calculator

An estimate is made first by rounding numbers between 10 and 100 to the nearest 10 and numbers between 100 and 1000 to the nearest 100. The calculations are then worked out like this:

Examples

1. 385×7

   Estimate: $400 \times 7 = 2800$

   Step 3
   7×3 is 21 plus 5 (carried)
   makes 26

$$
\begin{array}{r}
385 \\
\times \quad 7 \\
\hline
2695 \\
{\scriptstyle 5\ 3}
\end{array}
$$

   Step 1
   7×5 is 35
   5 down, 3 carried

   Step 2
   7×8 is 56 plus 3 (carried) makes 59
   9 down, 5 carried

7

2. 264 × 30

Estimate: 300 × 30 = 9000

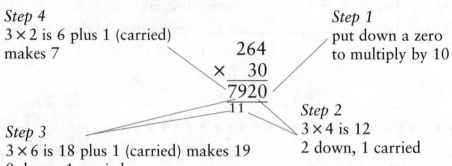

Step 4
3 × 2 is 6 plus 1 (carried)
makes 7

Step 1
put down a zero
to multiply by 10

$$
\begin{array}{r}
264 \\
\times\ \ 30 \\
\hline
7920 \\
\end{array}
$$
11

Step 3
3 × 6 is 18 plus 1 (carried) makes 19
9 down, 1 carried

Step 2
3 × 4 is 12
2 down, 1 carried

PAUSE

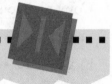

Do not use a calculator in this exercise. Show an estimate and an accurate answer for every calculation.

1 18 × 6	6 573 × 9	11 79 × 800	16 374 × 60
2 34 × 9	7 57 × 90	12 23 × 300	17 990 × 90
3 7 × 59	8 5 × 602	13 450 × 70	18 832 × 400
4 2 × 92	9 63 × 50	14 675 × 3	19 400 × 567
5 8 × 76	10 85 × 400	15 9 × 556	20 2341 × 500

Long multiplication is used with larger numbers. It is a combination of the two techniques you have already used.

Examples

1. 385 × 37

Estimate: 40 × 40 = 16 000

$$
\begin{array}{r}
385 \\
\times\ \ 37 \\
\hline
11550 \\
\end{array}
$$
— *Step 1* 385 × 30
2 1

$$
2695
$$
— *Step 2* 385 × 7
5 3

$$
14245
$$
— *Step 3* 11550 + 2695
1 1

8

2. 426×258

Estimate: $400 \times 300 = 120\,000$

$$
\begin{array}{r}
426 \\
\times\ \ 258 \\
\hline
85200 \\
{\scriptstyle 1} \\
21300 \\
{\scriptstyle 1\ 3} \\
3408 \\
{\scriptstyle 2\ 4} \\
\hline
109908 \\
{\scriptstyle 1}
\end{array}
$$

85200 —— *Step 1* 426×200

21300 —— *Step 2* 426×50

3408 —— *Step 3* 426×8

109908 —— *Step 4* $85\,200 + 21\,300 + 3408$

PAUSE

Do not use a calculator in this exercise. Show an estimate and an accurate answer for every calculation.

1 234×45	6 762×41	11 367×231	16 361×194
2 571×26	7 208×53	12 333×421	17 282×828
3 672×44	8 1102×22	13 406×203	18 327×109
4 980×53	9 675×25	14 292×315	19 447×774
5 747×29	10 337×89	15 928×273	20 621×234

Division of whole numbers by 10, 100 or 1000

Examples

1. $670 \div 10 = 67$

Hundreds	Tens	Units
6	7	0
	6	7

2. $3500 \div 100 = 35$

Thousands	Hundreds	Tens	Units
3	5	0	0
		3	5

9

3. $24\,000 \div 1000 = 24$

Ten Thousands	Thousands	Hundreds	Tens	Units
2	4	0	0	0
			2	4

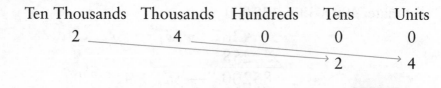

PAUSE

1 A supermarket sells bags of ten oranges.

How many bags of oranges could be made with:
a) 60 oranges, b) 230 oranges, c) 790 oranges, d) 100 oranges,
e) 140 oranges?

2 There are 100 centimes in a French franc.

How many francs are worth:
a) 400 centimes, b) 2000 centimes, c) 4200 centimes,
d) 10 100 centimes, e) 27 800 centimes?

3 Angela was doing a project on newspapers. She collected this information about the daily sales of some popular newspapers.

Paper	Circulation
Daily Mail	1 784 000
Guardian	400 000
Daily Mirror	2 484 000
Daily Star	746 000
Independent	284 000
Today	580 000
Sun	4 008 000
Times	472 000
Financial Times	297 000

Angela decided to sort the papers into order of sales. She also decided to write the numbers divided by 1000 to make her table simpler. This is the start of her new table.

Paper	Circulation (thousands)
Sun	4008

a) How would Angela write the sales for the Guardian?
b) How would Angela write the sales for the Daily Mirror?
c) Which paper has a circulation of 297 thousand?
d) Which paper has a circulation of 1784 thousand?
e) i) Which paper has the smallest circulation?
   ii) How would Angela write this number?

■■■■■■■■■■■■■■■■■■■■■■■■■■■■■■■■■■■■■

Using division by 10, 100 or 1000 to extend multiplication tables

We can extend our use of the multiplication table by dividing by 10, 100 or 1000.

Examples

1. $720 \div 80$

$$720 \div 80 = (720 \div 10) \div 8 = 72 \div 8 = 9$$

11

2. $45\,000 \div 900$

$$45\,000 \div 900 = (45\,000 \div 100) \div 9 = 450 \div 8 = 50$$

3. $24\,000 \div 6000$

$$24\,000 \div 6\,000 = (24\,000 \div 1000) \div 6 = 24 \div 6 = 4$$

PAUSE

Do not use a calculator in this exercise.

1 $120 \div 40$	11 $3600 \div 60$	21 $27\,000 \div 900$
2 $100 \div 20$	12 $6300 \div 70$	22 $27\,000 \div 3000$
3 80×40	13 $3200 \div 80$	23 $28\,000 \div 4000$
4 $720 \div 90$	14 $1200 \div 60$	24 $120\,000 \div 6000$
5 $420 \div 70$	15 $2400 \div 30$	25 $200\,000 \div 5000$
6 $490 \div 70$	16 $1600 \div 400$	26 $450\,000 \div 9000$
7 $540 \div 90$	17 $4000 \div 800$	27 $1\,600\,000 \div 8000$
8 $180 \div 30$	18 $4800 \div 600$	28 $21\,000\,000 \div 7000$
9 $630 \div 90$	19 $2400 \div 300$	29 $3\,000\,000 \div 5000$
10 $640 \div 80$	20 $15\,000 \div 300$	30 $810\,000 \div 9000$

Dividing whole numbers without a calculator

An estimate is made first by rounding numbers between 10 and 100 to the nearest 10 and numbers between 100 and 1000 to the nearest 100. The calculations are then worked out like this:

Examples

1. $216 \div 8$

   Estimate: $200 \div 8 = 25$

   Step 1 8 into 21 goes 2 times

   Step 2 8×2 is 16

   Step 3 $21 - 16$ is 5

   Step 4 Bring down the 6

   Step 5 8 into 56 goes 7 times

   Step 6 8×7 is 56

   Step 7 $56 - 56 = 0$

   $$\begin{array}{r} 27 \\ 8{\overline{)216}} \\ 16 \\ \hline 56 \\ 56 \\ \hline 0 \end{array}$$

   Answer: $216 \div 8 = 27$

12

2. $651 \div 15$
Estimate: $700 \div 20 = 35$

Step 1 15 into 65 goes 4 times

Step 5 15 into 51 goes 3 times

$$\begin{array}{r} 43 \\ 15\overline{)651} \\ \end{array}$$

Step 2 15×4 is 60 — 60

Step 3 $65 - 60$ is 5 — 51 —— *Step 4* Bring down the 1

Step 6 15×3 is 45 — 45

6 —— *Step 7* $51 - 45 = 6$

Answer: $651 \div 15 = 43$ remainder 6

PAUSE

Do not use a calculator in this exercise. Show an estimate and an accurate answer for every calculation.

1 715×13	6 $418 \div 38$	11 $923 \div 34$	16 $99 \div 21$
2 275×25	7 $377 \div 29$	12 $798 \div 37$	17 $302 \div 21$
3 $858 \div 39$	8 $378 \div 42$	13 $721 \div 22$	18 $602 \div 39$
4 $817 \div 19$	9 $416 \div 26$	14 $587 \div 33$	19 $998 \div 46$
5 $528 \div 24$	10 $784 \div 14$	15 $826 \div 13$	20 $957 \div 36$

When we are using a calculator, the answer is usually checked with an *inverse operation*.

Here a calculator is used to find remainders.

Example

A baker starts filling bags of 24 bread rolls from a pile of 453 rolls. How many bags can she fill and how many rolls will she have left over?

$\boxed{4}\,\boxed{5}\,\boxed{3}\,\boxed{\div}\,\boxed{2}\,\boxed{4}\,\boxed{=}$ 18.875

Check $\boxed{1}\,\boxed{8}\,\boxed{.}\,\boxed{8}\,\boxed{7}\,\boxed{5}\,\boxed{\times}\,\boxed{2}\,\boxed{4}\,\boxed{=}$ 453

So, the baker can fill 18 bags.

13

$$\boxed{1}\ \boxed{8}\ \boxed{\times}\ \boxed{2}\ \boxed{4}\ \boxed{=}\ 432$$

Check $\quad\boxed{4}\ \boxed{3}\ \boxed{2}\ \boxed{\div}\ \boxed{2}\ \boxed{4}\ \boxed{=}\ 18$

$$\boxed{4}\ \boxed{5}\ \boxed{3}\ \boxed{-}\ \boxed{4}\ \boxed{3}\ \boxed{2}\ \boxed{=}\ 21$$

Check $\quad\boxed{4}\ \boxed{3}\ \boxed{2}\ \boxed{+}\ \boxed{2}\ \boxed{1}\ \boxed{=}\ 453$

So, the baker has 21 rolls left over.

The next exercise is designed to give you practice in solving problems which require a mixture of calculations.

■ ■

PAUSE

1 In each part, write down how much is received for each equal share and how much is left over.

a) £28 shared by 7 people.
b) 38 sweets shared by 9 children.
c) 6 dozen roses shared by 5 brides.
d) 2 dozen mince pies shared by 10 carol singers.
e) 3000 exercise books shared by 12 school departments.
f) £200 000 shared by 8 lottery winners.
g) 153 marbles shared by 8 children.
h) 327 mice shared into 33 cages.
i) 250 sandwiches shared by two rugby teams and a referee (1 team has 15 members).
j) 280 sandwiches shared by 4 hockey teams and two referees (1 team has 11 members).

■ ■

SECTION 3 Decimals

Place value in decimals

Remember, whole numbers are written under these column headings:

..... 1000s 100s 10s 1s

So, a whole number like 2743 represents:

.....	1000s	100s	10s	1s
	2	7	4	3

Or, $2000 + 700 + 40 + 3$

Decimals are written under these column headings:

| ..... | 1000s | 100s | 10s | 1s | $\frac{1}{10}$s | $\frac{1}{100}$s | $\frac{1}{1000}$s | ..... |

The *decimal point* is used to separate the whole number columns from the fraction columns.

So, the decimal number 2743.823 represents:

.....	1000s	100s	10s	1s	$\frac{1}{10}$s	$\frac{1}{100}$s	$\frac{1}{1000}$s	.....
	2	7	4	3	8	2	3	

Or, $2000 + 700 + 40 + 3 + \frac{8}{10} + \frac{1}{100} + \frac{3}{1000}$

Examples

1. Write 74.13 as an addition of separate column values.

$$74.13 = 70 + 4 + \frac{1}{10} + \frac{3}{100}$$

2. What decimal number represents $\frac{9}{10} + 5 + 70 + \frac{3}{100}$?

$$\frac{9}{10} + 5 + 70 + \frac{3}{100} = 75.93$$

3. What are the largest and the smallest numbers you can make with the digits 0, 3 and 6 and a decimal point?

The largest is 63.0, the smallest is 0.36.

PAUSE

1 Write each number as an addition of separate column values.

a) 2.7
b) 3.9
c) 75.2
d) 7.52
e) 138

f) 13.8
g) 1.38
h) 0.138
i) 8.75
j) 0.875

k) 87.5
l) 34.105
m) 0.001
n) 1.01
o) 101.101

2 What decimal number represents each of these additions?

a) $4 + \frac{5}{10}$

b) $10 + 3 + \frac{2}{10}$

c) $300 + 40 + 8 + \frac{3}{10}$

d) $\frac{9}{10} + 1$

e) $7 + \frac{6}{10} + 10$

f) $3 + 30 + \frac{1}{10}$

g) $6 + \frac{1}{10} + \frac{5}{100}$

h) $7 + \frac{3}{10} + \frac{7}{100}$

i) $7 + \frac{6}{10} + \frac{5}{100} + \frac{3}{1000}$

j) $700 + 50 + \frac{1}{100}$

k) $600 + 4 + \frac{5}{10}$

l) $\frac{3}{10} + 30 + 500 + \frac{1}{1000} + \frac{7}{100}$

3 What are the largest and the smallest numbers you can make with each set of digits and a decimal point?

a) 0, 7 and 4

b) 0, 9 and 1

c) 4, 5 and 6

d) 7, 2 and 8

■ ■

Example

Mrs Stockley was in charge of the long jump on school sports day. This is her record of the lengths of the last ten jumps.

Length (in m) 3.05, 3.78, 3.55, 3.31, 3.10, 3.01, 3.48, 3.82, 3.66, 3.93

a) Draw a number line from 3.0 to 4.0 and show these lengths.
b) Write a list of the lengths sorted into order.

a)

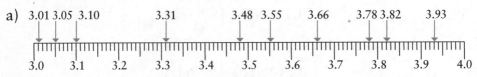

b) This is the ordered list of lengths.

3.01, 3.05, 3.10, 3.31, 3.48, 3.55, 3.66, 3.78, 3.82, 3.93

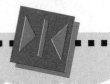

■ ■

PAUSE

Use graph paper to draw the number lines.

1 Paul measured the heights of ten of his relations for a maths project. These are his results.

Heights (in m) 1.63, 1.97, 1.15, 1.48, 1.70, 1.53, 1.08, 1.32, 1.85, 1.23

a) Draw a number line from 1 metre to 2 metres and show these heights.
b) Write a list of the heights sorted into order.

2 a) Draw a number line from 5.0 to 6.0 and show these numbers:

5.23, 5.57, 5.51, 5.8, 5.06, 5.33, 5.4, 5.61, 5.9, 5.17

   b) Write a list of the numbers sorted into order.

3 a) Draw a number line from 40.0 to 41.0 and show these numbers:

40.5, 40.07, 40.7, 40.91, 40.19, 40.68, 40.86, 40.23, 40.32, 40.4

   b) Write a list of the numbers sorted into order.

■ ■

Example

A plant researcher weighs the ten tubers produced by a new variety of potato. These are her results.

Weight (in kg) 0.437, 0.454, 0.489, 0.467, 0.47, 0.405, 0.424, 0.496, 0.443, 0.418

a) Draw a number line from 0.40 to 0.50 and show these weights.
b) Write a list of the weights sorted into order.

a)

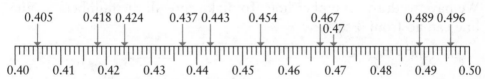

b) This is the ordered list of weights:

0.405, 0.418, 0.424, 0.437, 0.443, 0.454, 0.467, 0.47, 0.489, 0.496

■ ■

PAUSE

Use graph paper to draw the number lines.

1 A quality control inspector measures the actual volume of liquid inside ten bottles marked 'contents 1 litre'. These are her results.

Volume (in litres) 1.013, 1.07, 1.032, 1.047, 1.052, 1.068, 1.02, 1.009, 1.041, 1.037

a) Draw a number line from 1.0 to 1.1. Label the line 1.00, 1.01, 1.02 ..... 1.10. Show these volumes.

b) Write a list of the volumes sorted into order.

2 A trading standards officer checks the actual weight of ten blocks of cheese, each labelled 'weight 5 kg'. These are his results:

Weight (in kg) 5.007, 4.962, 4.981, 5.027, 5.048, 4.956, 4.972, 4.99, 5.015, 5.03

a) Draw a number line from 4.95 to 5.05 and show these weights.

b) Write a list of the weights sorted into order.

3 a) Draw a number line from 15.30 to 15.40 and show these numbers:

15.392, 15.308, 15.38, 15.319, 15.374, 15.328, 15.363, 15.339, 15.35, 15.340

b) Write a list of the numbers sorted into order.

■ ■

Example

Write the numbers 3.71, 3.07, 3.7, 3.17, 3.107 in order of size, smallest first.

We have to draw a number line. To make sure all the numbers fit, the line can go from 3 to 4.

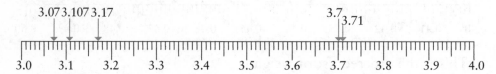

3.107 is hard to place on this line. It lies between 3.10(0) and 3.11(0).

The numbers in order are: 3.07, 3.107, 3.17, 3.7, 3.71.

■ ■

PAUSE

Use graph paper to draw your own number lines. List the numbers in order, smallest first.

1 5.84, 5.8, 5.48, 5.408, 5.08   2 2.1, 2.61, 2.16, 2.06, 2.106

3 8.57, 8.517, 8.75, 8.5, 8.705

4 10.306, 10.6, 10.613, 10.31, 10.36

5 7.21, 7.12, 7.2, 7.102, 7.82

6 9.165, 9.65, 9.56, 9.5, 9.605

7 0.7, 0.3, 0.73, 0.07, 0.873

8 17.503, 17.05, 17.305, 17.5, 17.35

Approximating decimals

Decimals are often rounded.

For example, if a carpenter measured a length as 5.7 cm, he might say, 'it's about 6 cm long.'

Or, if a shopkeeper weighted a piece of cheese as 2.1 kg, she might say, 'it weighs about 2 kg'.

This is called rounding a decimal *to the nearest whole number*.

We can use a number line to help us round decimals to the nearest whole number. This diagram shows how.

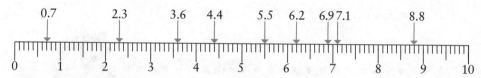

The sign ≈ means 'approximately equal to'.

From the diagram we can see:

$$0.7 \approx 1 \qquad 2.3 \approx 2 \qquad 3.6 \approx 4 \qquad 4.4 \approx 4 \qquad 8.8 \approx 9$$
$$5.5 \approx 6 \qquad 6.2 \approx 6 \qquad 6.9 \approx 7 \qquad 7.1 \approx 7$$

PAUSE

In questions 1 to 5 use a number line. Write out each list of numbers rounded to the nearest whole number.

1 6.1, 2.2, 8.3, 7.4, 3.5, 4.6, 5.7, 1.8, 9.9, 1.3

2 3.1, 5.2, 9.3, 1.4, 4.5, 2.6, 6.7, 7.8, 7.0, 8.9

3 8.1, 1.2, 6.3, 3.4, 7.5, 9.6, 4.7, 5.8, 2.7, 1.9.

4 1.6, 0.3, 2.3, 3.6, 4.2, 5.9, 6.5, 7.7, 8.4, 9.7.

5 1.5, 2.4, 0.5, 3.9, 5.1, 6.4, 7.3, 8.6, 4.9, 9.8.

Decimals are often rounded to a set number of decimal places.
For example, if a plan researcher measured the height of a shrub as
1.23 m, she might record the height as 1.2 m.

Or, if she measured the height of another shrub as 2.37 m, she might
record the height as 2.4 m.

This is called rounding a decimal *to one decimal place*.

We can use a number line to help us round decimals to one decimal
place. This diagram shows how:

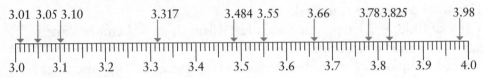

3.01 3.05 3.10   3.317   3.484 3.55   3.66   3.78 3.825   3.98

3.0   3.1   3.2   3.3   3.4   3.5   3.6   3.7   3.8   3.9   4.0

From the picture we can see:

$3.01 \approx 3.0$ $3.05 \approx 3.1$ $3.10 \approx 3.1$ $3.78 \approx 3.8$

$3.484 \approx 3.5$ $3.55 \approx 3.6$ $3.66 \approx 3.7$

$3.825 \approx 3.8$ $3.98 \approx 4.0$ $3.317 \approx 3.3$

PAUSE

1 Paul measured the heights of ten of his relations for a maths
   project. These are his results.

   Heights (in metres) 1.63, 1.97, 1.15, 1.48, 1.70, 1.53, 1.08, 1.32,
                       1.85, 1.23

   Use a number line to write out a list of these heights rounded to
   the nearest 10 centimetres (one decimal place).

   In questions 2 to 7, use a number line to write each list of numbers
   correct to one decimal place.

2 5.23, 5.57, 5.51, 5.80, 5.06, 5.33, 5.40, 5.61, 5.9, 5.17

3 5.275, 5.537, 5.598, 5.892, 5.041, 5.373, 5.408, 5.695, 5.98,
   5.133

4 8.07, 8.90, 8.73, 8.65, 8.15, 8.20, 8.47, 8.33, 8.50, 8.82

5 8.035, 8.917, 8.771, 8.654, 8.155, 8.296, 8.429, 8.363, 8.547,
   8.888

6 15.73, 15.09, 15.42, 15.94, 15.26, 15.18, 15.61, 15.83, 15.35, 15.50

7 15.7556, 15.0673, 15.4892, 15.9956, 15.2373, 15.1329, 15.6923, 15.8613, 15.3474, 15.5534

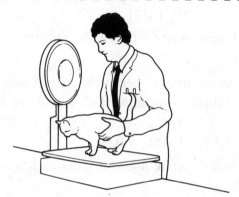

A vet weighs a cat and finds it weighs 1.237 kg. He records this weight as 1.24 kg, this is called rounding a decimal *to two decimal places*.

We can use a number line to help us round decimals to two decimal places. This diagram shows how:

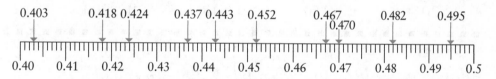

From the diagram we can see:

$0.403 \approx 0.40$	$0.418 \approx 0.42$	$0.424 \approx 0.42$	$0.470 \approx 0.47$
$0.443 \approx 0.44$	$0.452 \approx 0.45$	$0.467 \approx 0.47$	
$0.482 \approx 0.48$	$0.495 \approx 0.50$	$0.437 \approx 0.44$	

PAUSE

Use a number line drawn on graph paper to write each list of numbers correct to two decimal places.

1 5.591, 5.509, 5.582, 5.518, 5.573, 5.527, 5.562, 5.538, 5.551, 5.549

2 15.3984, 15.3025, 15.3872, 15.3161, 15.3764, 15.3228, 15.3641, 15.3315, 15.3587, 15.3413

3 A quality control inspector measures the actual volume of liquid inside ten bottles marked 'contents 1 litre'. These are her results.

Volume (in litres) 1.013, 1.073, 1.032, 1.047, 1.052, 1.068, 1.021, 1.009, 1.041, 10.37

Use a number line to write a list of the volumes correct to two decimal places.

4 A trading standards officer checks the actual weight of ten blocks of cheese, each labelled 'weight 5 kg'. These are his results:

Weight (in kg) 5.007, 4.962, 4.981, 5.027, 5.048, 4.956, 4.972, 4.994, 5.015, 5.030

Use a number line to write a list of the weights correct to two decimal places.

5 Use a calculator to work these out. Give each answer correct to two decimal places.

a) $\sqrt{3}$ d) $\sqrt{8.4}$ g) $\sqrt{3.25}$ j) $\sqrt{19.1}$

b) $\sqrt{7.5}$ e) $\sqrt{42}$ h) $\sqrt{90}$ k) $\sqrt{22.7}$

c) $\sqrt{20}$ f) $\sqrt{150}$ i) $\sqrt{18}$ l) $\sqrt{200}$

■ ■

Adding and subtracting decimals without a calculator

The decimal points go underneath each other. Extra 0s can be added to fill in the gaps.

Examples

1. 3.57+18.173

$$\begin{array}{r} 3.57 \\ +18.173 \\ \hline 21.743 \\ \hline {\scriptstyle 1 \quad 1} \end{array}$$

2. 17−14.84

$$\begin{array}{r} {\scriptstyle 6 \; 9} \\ 17.\!\!\!\not{0}0 \\ -14.84 \\ \hline 2.16 \end{array}$$

PAUSE

Work these out without using a calculator.

1 a) 4.65+7.8   c) 0.968+2.17   e) 19+15.7+12.35
   b) 16.705+10.53   d) 0.0432+0.608   f) 18.14+12+14.9

2 a) 16.73−14.84   d) 10.6−7.135   g) 14−7.4
   b) 11.3−8.14   e) 0.56−0.162   h) 4−3.61
   c) 19.6−15.43   f) 0.07−0.056   i) 10−8.715

Multiplying and dividing decimals without a calculator

Examples

1. 3.741×6

$$\begin{array}{r} 3.741 \\ \times \quad\ 6 \\ \hline 22.446 \\ \hline {\scriptstyle 2\ 4\ 2} \end{array}$$

2. $14.5 \div 4$

$$\begin{array}{r} 3.625 \\ 4\overline{)14.{}^25{}^10{}^20} \end{array}$$

3. 0.12×0.3

Multiply the numbers ignoring the decimal points. $12 \times 3 = 36$.

Count the decimal places in the original question. $2+1=3$, there will be 3 decimal places in the answer.

Answer 0.036

4. $0.9 \div 0.03$

$\dfrac{0.9}{0.03}$ the same as $\dfrac{9}{0.3}$ is the same as $\dfrac{90}{3} = 30$

5. $0.12 \div 0.3$

$\dfrac{0.12}{0.3}$ the same as $\dfrac{1.2}{3}$, so $3\overline{)1.2}$ gives 0.4

PAUSE

Work these out without using a calculator.

1. a) 12.4×4 c) 6.034×6 e) 12.506×9
   b) 8.24×7 d) 10.28×5 f) 0.243×8

2. a) $14.7 \div 2$ c) $18.85 \div 4$ e) $11.73 \div 5$
   b) $17.4 \div 3$ d) $0.127 \div 4$ f) $9.007 \div 6$

3. a) 0.4×0.5 c) 0.15×4 e) 0.24×0.02
   b) 1.2×0.4 d) 0.07×0.8 f) 0.031×0.03

4. a) $0.8 \div 0.2$ d) $0.15 \div 0.5$ g) $16 \div 0.8$ j) $0.8 \div 0.04$
   b) $0.6 \div 0.3$ e) $0.18 \div 0.6$ h) $14 \div 0.7$ k) $2.8 \div 0.07$
   c) $1.2 \div 0.6$ f) $6 \div 0.2$ i) $0.025 \div 0.05$ l) $0.24 \div 0.03$

Multiplication of decimals by 10, 100 and 1000

These are the important metric tables:

■ Length

> 10 millimetres (mm) = 1 centimetre (cm)
> 100 cm = 1 metre (m)
> 1000 m = 1 kilometre (km)

■ Weight

> 1000 grams (g) = 1 kilogram (kg)

■ Volume

> 1000 cubic centimetres (cm^3) = 1 litre (L)
> or
> 1000 millilitres (ml) = 1 litre (L)

Note: millilitres and cubic centimetres are different names for the same unit.

Examples

1. Change 8.67 centimetres into millimetres.

   $8.67 \times 10 = 86.7$ millimetres

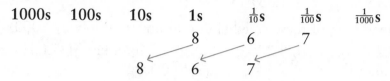

2. Change 2.3 metres into centimetres.

   $2.3 \times 100 = 230$ centimetres

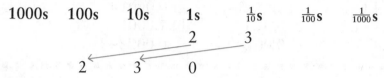

3. Change 17.893 metres into centimetres.

   $17.893 \times 100 = 1789.3$

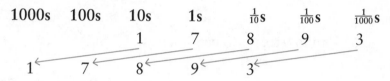

4. Change 17.893 grams into kilograms.

   $17.893 \times 1000 = 17893$

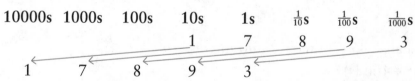

PAUSE

Try to do as many as you can without a calculator. Use your calculator to check your work.

1 Change each of these lengths into millimetres (multiply by 10).

   a) 17 cm
   b) 1.7 cm
   c) 0.17 cm
   d) 8.9 cm
   e) 9.32 cm
   f) 7.85 cm
   g) 8.93 cm
   h) 1.243 cm
   i) 1.47 cm
   j) 1.047 cm

2 Change each of these lengths into centimetres (multiply by 100).

a) 2.45 m   d) 12.375 m   g) 1.2 m   j) 0.765 m
b) 2 m   e) 0.895 m   h) 5.7 m
c) 0.45 m   f) 1.557 m   i) 7.3 m

3 Change each of these weights into grams (multiply by 1000).

a) 1.237 kg   d) 8.23 kg   g) 16.6666 kg   j) 7.0314 kg
b) 1.657 kg   e) 7.6 kg   h) 0.1 kg
c) 7.045 kg   f) 7.8923 kg   i) 0.456 kg

4 Change each of these volumes into millilitres (multiply by 1000).

a) 6.4 L   d) 6.4735 L   g) 0.0001 L   j) 0.00025 L
b) 6.47 L   e) 0.01 L   h) 1.0001 L
c) 6.473 L   f) 0.001 L   i) 0.00234 L

■ ■

Division of decimals by 10, 100 and 1000

Examples

1. Change 654 millimetres into centimetres.

$654 \div 10 = 65.4$

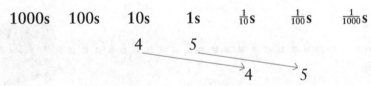

2. Change 45 centimetres into metres.

$45 \div 100 = 0.45$

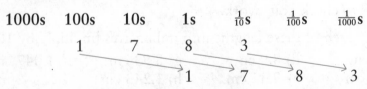

3. Change 178.3 centimetres into metres.

$178.3 \div 100 = 1.783$

1000s	100s	10s	1s	$\frac{1}{10}$s	$\frac{1}{100}$s	$\frac{1}{1000}$s
	1	7	8	3		
		1	7	8	3	

4. Change 1.4 grams into kilograms.

$1.4 \div 1000 = 0.0014$ kilograms

1000s	100s	10s	1s	$\frac{1}{10}$s	$\frac{1}{100}$s	$\frac{1}{1000}$s	$\frac{1}{10000}$s
			1	4			
						1	4

PAUSE

Try to do as many as you can without a calculator. Use a calculator to check your work.

1 Change each of these lengths into centimetres (divide by 10).

  a) 17 mm   d) 8.9 mm   g) 78.5 mm   j) 0.6 mm
  b) 1.7 mm   e) 93.2 mm   h) 785 mm
  c) 0.17 mm   f) 7.85 mm   i) 0.785 mm

2 Change each of these lengths into metres (divide by 100).

  a) 245 cm   d) 12.3 cm   g) 1.2 cm   j) 8652 cm
  b) 200 cm   e) 8.95 cm   h) 5.7 cm
  c) 45 cm   f) 1.557 cm   i) 7355 cm

3 Change each of these weights into kilograms (divide by 1000).

  a) 1237 g   d) 823 g   g) 16.6 g   j) 7.3 g
  b) 1657 g   e) 76 g   h) 100 g
  c) 7045 g   f) 78 923 g   i) 10 g

4 Change each of these volumes into litres (divide by 1000).

  a) 6.4 cm^3 d) 6473.5 cm^3 g) 300 cm^3 j) 650 cm^3
  b) 6.47 cm^3 e) 0.01 cm^3 h) 70 cm^3
  c) 6 cm^3 f) 0.001 cm^3 i) 5000 cm^3

Weekly wages

People who earn a *weekly wage* are paid a fixed rate for each hour they work.

Example

Claire works 34 hours at a rate of £4.20 per hour. Calculate her weekly wage.

$$\text{weekly wage} = 34 \times £4.20 = £142.80$$

PAUSE

Calculate the weekly wage in each case:

1. 28 hours at £4.15 per hour
2. 35 hours at £4.80 per hour
3. 16 hours at £3.50 per hour
4. 18 hours at £2.90 per hour
5. 40 hours at £5.31 per hour
6. 36 hours at £2.75 per hour
7. 32 hours at £5.60 per hour
8. 18 hours at £4.50 per hour
9. 22 hours at £7.22 per hour
10. 27 hours at £4.23 per hour

Overtime

Jemma Hill works for a catalogue company and is paid £3.80 per hour. If she works more than 40 hours in a week she is paid for each extra hour at a higher *overtime rate*. This rate is calculated by multiplying her normal hourly rate by 1.5. This is how her wages are calculated in a week when she works 44 hours:

$$40 \times £3.80 = £152$$
$$4 \times 1.5 \times £3.80 = £22.80$$
$$\text{weekly wage} = £152 + £22.80 = £174.80$$

An overtime rate like this is often called 'time and a half'. Other overtime rates may be 'time and a quarter' (basic rate × 1.25) or 'double time' (basic rate × 2). People are also paid overtime rates for working at weekends or during public holidays.

Example

Surbajit Singh works as a driver. He is paid £4.50 per hour for each hour he works from Monday to Saturday. If he works more than 40 hours he is paid overtime at a rate of 'time and a half'. If he works on Sunday he is paid overtime at a rate of 'double time'. Calculate his wage in a week when he works 46 hours from Monday to Saturday and 5 hours on Sunday.

$$40 \times £4.50 = £180$$
$$6 \times 1.5 \times £4.50 = £40.50$$
$$5 \times 2 \times £4.50 = £45$$
$$\text{weekly wage} = £180 + £40.50 + £45 = £265.50$$

PAUSE

A company pays its workers £4.80 per hour for a basic 40 hour week from Monday to Saturday. Overtime is paid at a rate of 'time and a half'. Any work on Sunday is paid at an overtime rate of 'double time'.

Calculate each persons' wage:

	Name	Hours worked Monday to Saturday	Hours worked Sunday
1	Mr Smith	38	2
2	Ms Brown	40	3
3	Ms Jones	43	0
4	Mr Black	48	1
5	Mr Patel	42	3
6	Ms Hodgson	35	6
7	Ms Dearman	48	4
8	Mr Williams	46	5
9	Ms Toms	51	2
10	Mr Baines	34	10

Significant figures

Correcting a value to one *significant figure* is a way to approximate both whole numbers and decimals. The technique is very similar to that used to correct to a given number of decimal places, but you start at *the first significant figure, not at the decimal point.*

The *first significant figure* is the first digit from the left of the number which is *not* a zero.

Examples

1. Write 8.75 correct to one significant figure.
$$8.75 \approx 9$$

2. Write 17.34 correct to one significant figure.
$$17.34 \approx 20$$

3. Write 236 correct to one significant figure.
$$236 \approx 200$$

4. Write 0.56 correct to one significant figure.
$$0.56 \approx 0.6$$

5. Write 0.0053 correct to one significant figure.
$$0.0053 \approx 0.005$$

PAUSE

Write each of the following numbers correct to one significant figure.

1 13	6 3.7	11 12.9	16 0.35	21 0.0309
2 17	7 2.2	12 14.7	17 0.72	22 19.23
3 85	8 3.5	13 25.9	18 0.0037	23 3425
4 256	9 9.9	14 46.8	19 0.0031	24 27.31
5 841	10 6.3	15 52.893	20 0.00239	25 0.0098

Numbers are rounded to one significant figure to estimate calculations.

Examples

1. Estimate the value of 39.9×48
$$\text{Estimate } 40 \times 50 = 2000$$

2. Estimate the value of 435×0.172

$$\text{Estimate } 400 \times 0.2 = 80$$

3. Estimate the value of $117 \div 3.14$

$$\text{Estimate } 100 \div 3 = 33$$

4. Estimate the value of $19.3 \div 0.0437$

$$\text{Estimate } 20 \div 0.04 = 500$$

PAUSE

Estimate the value of each calculation.

1 13×17	6 14.7×0.35	11 $17 \div 13$	16 $0.0309 \div 0.002\,39$
2 85×3.5	7 0.35×0.72	12 $25.9 \div 2.2$	17 $14.7 \div 0.35$
3 256×9.9	8 0.0037×17	13 $46.8 \div 9.9$	18 $27.31 \div 256$
4 6.3×2.2	9 0.0031×9.9	14 $3245 \div 14.7$	19 $256 \div 841$
5 0.72×46.8	10 19.23×0.0098	15 $0.72 \div 0.35$	20 $0.72 \div 17$

Mixed problems

PAUSE

1 Megan paints plates in a factory.

She earns 16p each for the first 300 plates she paints in one week.

a) Calculate the total amount she earns for the first 300 plates.

She earns 22p each for every plate more than 300 she paints in the same week One week she earned £74.84.

b) Calculate the total number of plates she painted in that week.

(ULEAC)

2

> **ASSISTANT COOK**
> **£3.22 per hour**
> To work 28 hours per week at the
> Chuter Ede Primary School,
> Wolfit Avenue.

a) Work out the Assistant Cook's full weekly wage.

b) Work out the Assistant Cook's yearly wage if she works 38 weeks each year.

(ULEAC)

3 Sally and Charles are asked to work out 2.175^2.

Their calculator display shows:

Sally gives the answer as 4.7.

Charles gives the number correct to three decimal places.

Write down his answer.

(SEG)

4 During October, a central heating boiler was used, on average, for 4.5 hours per day.

During November, the same boiler was used, on average, for 5.8 hours per day.

a) Calculate the total time for which the boiler was used during October and November.

While it is being used the boiler uses 2.44 litres of oil per hour.

b) Calculate the amount of oil used during the two months, correct to the nearest litre.

(NICCEA)

■■■

Imperial units

Imperial units were used in the UK until the 1970s when the country started to change to metric units. Many people still think and work in imperial units.

These are the main Imperial tables:

	Imperial measures	Approximate metric equivalents
Length	12 inches = 1 foot 3 feet = 1 yard 1760 yards = 1 mile	1 inch = 2.54 centimetres 1 foot = 30.48 centimetres, 1 yard = 0.91 metres 5 miles = 8 kilometres
Weight	16 ounces = 1 pound	2.2 pounds = 1 kilogram
Volume	8 pints = 1 gallon	1.76 pints = 1 litre, 1 gallon = 4.55 litres

Examples

1. Mr Brown has just bought a 10 kg bag of weedkiller for his lawn. How many pounds of weedkiller has he bought?

$$10 \times 2.2 = 22 \text{ pounds of weedkiller}$$

2. Mrs Brown has just bought a 100 pint cask of beer. How many litres of beer has she bought?

$$100 \div 1.76 = 56.82 \text{ litres (correct to two decimal places)}$$

3. Jenny Jones is driving in France when she sees this sign:

$$\text{PARIS } 272 \text{ km}$$

How far is she from Paris?

$$272 \div 8 \times 5 = 170 \text{ miles}$$

4. Jack is an old carpenter who prefers to work in feet and inches. He is asked to build a framework 80 cm wide by 150 cm high. What are these measurements in inches?

$$80 \div 2.54 = 31.5 \text{ inches (to one decimal place)}$$
$$150 \div 2.54 = 59.1 \text{ inches (to one decimal place)}$$

1 Mr Smith is buying lawn seed. The type he is interested in comes in 1 kg, 2 kg, 5 kg, 15 kg and 20 kg sizes. Mr Smith can't think in kilograms and wants to know what these weights are in pounds. Convert all these sizes into pounds for Mr Smith.

2 Mrs Baker did a milk round and had always delivered milk in pint bottles. She wanted to check her prices against those in a supermarket which sold milk in litre containers. To help she asked her daughter to complete this table for her.

Pints	1	2	3	4	5	6	7	8	9	10
Litres										

Pints	10	20	30	40	50	60	70	80	90	100
Litres										

Copy and complete the table, giving all your answers correct to two decimal places.

3 Convert all these French road signs into miles.

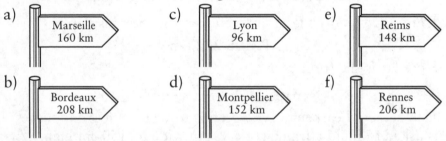

a) Marseille 160 km

b) Bordeaux 208 km

c) Lyon 96 km

d) Montpellier 152 km

e) Reims 148 km

f) Rennes 206 km

4 Jack the carpenter uses these tables to help him convert metric measurements into imperial units but they are covered with grease.

Centimetres	5	10	15	20	25	30
Inches	2.0	3.9				

Centimetres	30.5	61.0				
Feet	1	2	3	4	5	6

Write out some new tables for Jack.

Using the memory and brackets on a calculator

Calculations like:

$$\frac{3.45 + 6.73}{7.22 - 3.56}$$

mean you must work out the top value and the bottom value. Then you divide the top value by the bottom value.

Calculations like this can be done on a calculator without writing anything down.

There are two ways to do this:

Method 1: If your calculator has a memory.

If a calculator has a memory there will be at least two keys to use it with.

One of these keys will be used to place the display number in the memory. The key may be marked:

M   **Min**   **STO**   or in some other way.

One of these keys will be used to return the number in the memory to the display. The key may be marked:

MR   **RCL**   or in some other way.

If we assume the keys are marked M and MR, the calculation above would be done like this:

7 . 2 2 − 3 . 5 6 = M 3 . 4 5 + 6 . 7 3 = ÷ MR = 2.7814207

Notice that the calculation was started from the *bottom*.

Method 2: If your calculator has brackets.

If a calculator can use brackets there will be two keys marked **(** and **)**

The calculation is equivalent to $(3.45 + 6.73) \div (7.22 - 3.56)$ and it is entered in this way.

(3 . 4 5 + 6 . 7 3) ÷ (7 . 2 2 − 3 . 5 6) = 2.7814207

Use either memory keys or bracket keys to complete these calculations, only writing down your final answer.

Give your answers to 3 decimal places where necessary.

1 $\dfrac{36 + 74}{23 - 5}$ 4 $\dfrac{5.6 \div 1.2}{2.7 \times 1.5}$ 7 $\dfrac{3.1 \times 5.5}{6.5 \div 3.7}$ 10 $\dfrac{5.67 \times 1.76}{67.3 \div 1.85}$

2 $\dfrac{56 \times 19}{27 + 57}$ 5 $\dfrac{3.6 + 5.4}{2.7 - 0.4}$ 8 $\dfrac{3.1 \times 2.5}{1.7 \div 0.5}$

3 $\dfrac{4.5 \times 4.5}{1.3 - 0.9}$ 6 $\dfrac{8.1 \div 0.7}{8.5 \times 2.3}$ 9 $\dfrac{3.67 + 5.78}{1.34 - 0.72}$

Using the constant function on a calculator

We often need to repeat the same arithmetic with a list of numbers. For example, a plumber may need to complete this table.

Cost of labour + parts	Call out fee (£25)	VAT (17.5%)
£47		
£85		
£59		
£128		
£250		

To complete the middle column, the plumber needs to add £25 to *every* number in the first column, so he uses the *constant function* on his calculator.

Some calculators are different, but on his the plumber enters:

The plumber then enters all the other numbers like this:

| 4 | 7 | = | 72 |

| 8 | 5 | = | 110 |

| 5 | 9 | = | 84 |

| 1 | 2 | 8 | = | 153 |

| 2 | 5 | 0 | = | 275 |

The table now looks like this:

Cost of labour + parts	Call out fee (£25)	VAT (17.5%)
£47	£72	
£85	£110	
£59	£84	
£128	£153	
£250	£275	

The plumber now needs to add the VAT. He knows that to do this he needs to multiply every number in the second column by 1.175. He enters:

| 1 | . | 1 | 7 | 5 | × | = |

He then enters all the other numbers like this:

| 7 | 2 | = | 84.6 |

| 1 | 1 | 0 | = | 129.25 |

| 8 | 4 | = | 98.7 |

| 1 | 5 | 3 | = | 179.775 |

| 2 | 7 | 5 | = | 323.125 |

The table now looks like this:

Cost of labour + parts	Call out fee (£25)	VAT (17.5%)
£47	£72	£84.60
£85	£110	£129.25
£59	£84	£98.70
£128	£153	£179.78
£250	£275	£323.13

The plumber sometimes makes special offers. Once, he offered to reduce all bills by £12. To enter a constant subtraction of 12 on his calculator, the plumber enters:

$$\boxed{-}\ \boxed{-}\ \boxed{1}\ \boxed{2}\ \boxed{=}$$

Once, the plumber offered to supply all parts at half price. To enter a constant division by 2 on his calculator, the plumber enters:

$$\boxed{\div}\ \boxed{\div}\ \boxed{2}\ \boxed{=}$$

PAUSE

Use the constant function on your calculator in every question.

1 This is a list of prices for 5 bouquets of flowers that a florist is to deliver.

£12.90, £15.34, £13.80, £24.55, £17.58

The florist makes a £3.72 charge for delivery. Write out the list of prices with the delivery charge added.

2 These are the banking charges that a bank asked 5 customers to pay.

£45.27, £8.95, £32.51, £43.99, £89.62

The bank discovered that, due to a computer error, £5.31 too much had been added to these charges. Write out a list of charges reduced by £5.31.

3 These are the bills that a shop has prepared for 5 customers.

£132, £140, £155, £178.60, £198.45

The shop is running a special offer to reduce all bills by 10%. This means each bill must be multiplied by 0.9. Write out a list of bills reduced by 10%.

4 These are a professional gardener's notes of how much fertilizer she usually adds to five different flower beds.

1.5 kg, 1.8 kg, 2.7 kg, 3.9 kg, 2.5 kg

She is using a new concentrated fertilizer which has these instructions:

'To calculate the new amount to use, divide the old amount you used by 3.'

Write out a list of the new amounts of fertilizer she needs to use on the five beds.

5 Copy and complete this table for the plumber.

Parts	Parts ÷ 2	Labour	Labour+ parts ÷ 2	Call out fee (£25)	VAT (17.5%)	Special reduction (£12)
£32	£16	£80	£96	£121	£142.18	£130.18
£48		£89				
£64		£32				
£12		£50				
£86		£65				
£25		£37				
£59		£95				
£102		£86				
£155		£43				
£99		£99				
£128		£102				

■ ■

SECTION 4 Fractions

Fractions (sometimes called common fractions) were once an important method of calculation.

Calculators, computers and metric measurements have made decimal arithmetic so easy that fraction arithmetic is now rarely needed.

It is important to have a basic understanding of fractions.

'*Half* our students own a computer.'

'*Nine tenths* of the crowd were well behaved.'

'She is entitled to *one fifth* of the profits.'

Understanding fractions

$\frac{3}{4}$ shaded, $\frac{1}{4}$ unshaded

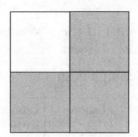

$\frac{5}{6}$ are one type of fish,

$\frac{1}{6}$ is another type of fish

PAUSE

1 This is an abstract painting called 'shapes awake' by the artist Kram Yeldnib.

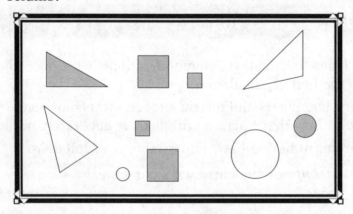

In the painting what fraction of the shapes

a) are triangles
b) are squares
c) are shaded

d) have straight sides
e) are not shaded?

40

2 Estimate the fraction shaded and the fraction unshaded in each of these shapes.

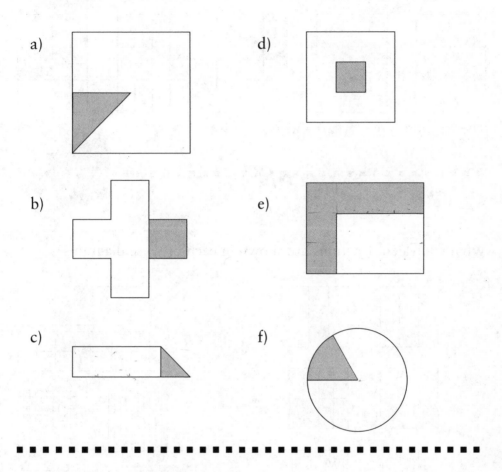

a)

b)

c)

d)

e)

f)

■ ■

This diagram shows that $\frac{1}{2} = \frac{2}{4}$. Pairs of equal fractions like this are called *equivalent fractions*.

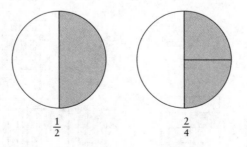

$$\frac{1}{2} \qquad\qquad \frac{2}{4}$$

Example

What equivalent fractions are shown in this diagram?

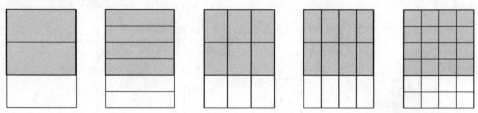

The fractions are $\frac{2}{3}$, $\frac{4}{6}$, $\frac{6}{9}$, $\frac{8}{12}$ and $\frac{16}{24}$.

PAUSE

What equivalent fractions are shown in each of these diagrams?

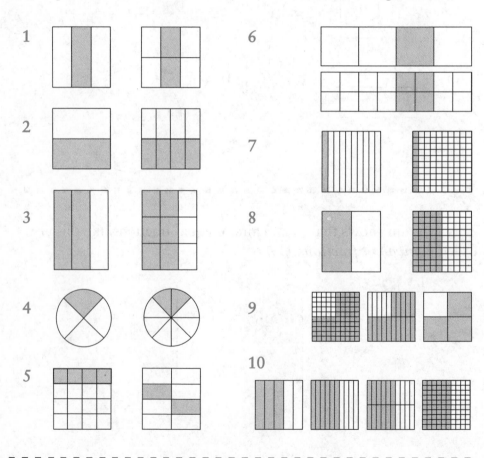

1

2

3

4

5

6

7

8

9

10

Improper fractions

An *improper fraction* is one in which the top number is larger than the bottom number.

A number like $1\frac{5}{7}$, which has both a whole number and fractional part is called a *mixed number*.

Examples

1. Change the improper fraction $\frac{23}{6}$ into a mixed number.

$$1 = \frac{6}{6} \quad 2 = \frac{12}{6} \quad 3 = \frac{18}{6} \quad 4 = \frac{24}{6}$$

so,

$$\frac{23}{6} = 3\frac{5}{6}$$

2. Change the mixed number $2\frac{2}{3}$ into an improper fraction.

$$2 = \frac{6}{3}$$

so

$$2\frac{2}{3} = \frac{8}{3}$$

PAUSE

1 Change each of these improper fractions into a mixed number.

a) $\frac{23}{5}$ c) $\frac{15}{2}$ e) $\frac{19}{4}$ g) $\frac{19}{11}$ i) $\frac{33}{12}$
b) $\frac{9}{7}$ d) $\frac{17}{3}$ f) $\frac{27}{5}$ h) $\frac{27}{8}$ j) $\frac{49}{5}$

2 Change each of the following mixed numbers into an improper fraction.

a) $3\frac{3}{4}$ c) $8\frac{2}{3}$ e) $7\frac{3}{4}$ g) $2\frac{13}{16}$ i) $11\frac{2}{3}$
b) $9\frac{1}{2}$ d) $1\frac{1}{4}$ f) $9\frac{5}{8}$ h) $5\frac{5}{32}$ j) $10\frac{3}{4}$

Equivalent fractions

Remember, two fractions which represent the same part of a quantity are called *equivalent fractions*. An example of a pair of equivalent fractions is:

$$\frac{2}{3} \qquad \text{and} \qquad \frac{10}{15}$$

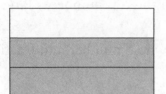 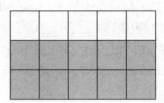

Example

Complete these pairs of equivalent fractions.

$$\frac{3}{4}=\frac{?}{20} \qquad \frac{5}{6}=\frac{?}{42}$$

We replace the question marks by spotting that the bottom numbers have been multiplied by 5 and by 7. The top numbers must be also be multiplied by 5 and by 7.

$$\overset{\times 5}{\frac{3}{4}} = \underset{\times 5}{\frac{15}{20}} \qquad\qquad \overset{\times 7}{\frac{5}{6}} = \underset{\times 7}{\frac{35}{42}}$$

PAUSE

1 Copy and complete.

a) $\frac{2}{5}=\frac{?}{10}$

b) $\frac{5}{8}=\frac{?}{24}$

c) $\frac{7}{6}=\frac{?}{18}$

d) $\frac{3}{7}=\frac{?}{21}$

e) $\frac{12}{17}=\frac{?}{51}$

f) $\frac{1}{9}=\frac{?}{27}$

g) $\frac{11}{12}=\frac{?}{48}$

h) $\frac{4}{5}=\frac{?}{35}$

i) $\frac{5}{9}=\frac{?}{45}$

j) $\frac{8}{11}=\frac{?}{55}$

k) $\frac{7}{10}=\frac{?}{100}$

l) $\frac{1}{9}=\frac{?}{99}$

Simplifying fractions

Example

Simplify $\frac{24}{48}$ and $\frac{45}{85}$.

We simplify fractions by spotting common factors of the top and bottom numbers. Dividing by these common factors simplifies the fraction.

$$\frac{24}{48} \overset{\div 24}{\underset{\div 24}{=}} \frac{1}{2} \qquad \frac{45}{85} \overset{\div 5}{\underset{\div 5}{=}} \frac{9}{17}$$

Fractions can also be simplified by using a calculator with a fraction key $a^{b/c}$.

Keys to press	Calculator display	Answer
2 **4** $a^{b/c}$ **4** **8** **=**	1⌐2	$\frac{1}{2}$
4 **5** $a^{b/c}$ **8** **5** **=**	9⌐17	$\frac{9}{17}$

PAUSE

1 Simplify:

a) $\frac{50}{70}$ c) $\frac{6}{9}$ e) $\frac{24}{30}$ g) $\frac{65}{90}$ i) $\frac{27}{81}$

b) $\frac{25}{60}$ d) $\frac{12}{15}$ f) $\frac{35}{65}$ h) $\frac{45}{63}$ j) $\frac{44}{66}$

Comparing fractions

Which is the bigger fraction, $\frac{2}{3}$ or $\frac{3}{4}$?

We first find a number that both bottom numbers will divide into. We then change each fraction into an equivalent one with this new number at the bottom. In this case, both 3 and 4 divide exactly into 12.

$$\frac{2}{3} \overset{\times 4}{\underset{\times 4}{=}} \frac{8}{12} \qquad \frac{3}{4} \overset{\times 3}{\underset{\times 3}{=}} \frac{9}{12}$$

It is now obvious that $\frac{3}{4}$ is the larger fraction.

1 Which is the larger of each of the following pairs of fractions?

a) $\frac{1}{3}$ and $\frac{1}{4}$ d) $\frac{3}{7}$ and $\frac{5}{9}$ g) $\frac{4}{7}$ and $\frac{7}{10}$ j) $\frac{3}{7}$ and $\frac{5}{11}$

b) $\frac{3}{4}$ and $\frac{4}{5}$ e) $\frac{1}{2}$ and $\frac{4}{6}$ h) $\frac{1}{2}$ and $\frac{3}{7}$ k) $\frac{4}{5}$ and $\frac{9}{11}$

c) $\frac{3}{5}$ and $\frac{5}{6}$ f) $\frac{2}{3}$ and $\frac{5}{7}$ i) $\frac{5}{8}$ and $\frac{2}{3}$ l) $\frac{3}{5}$ and $\frac{7}{12}$

2 Bill scored 8 out of 25 in a Maths test and 6 out of 20 in an English test. In which test did he do better?

3 Jane scored 93 out of 100 in an Art exam and 19 out of 20 in a Biology exam. In which exam did she do better?

4 In one bag of sweets, there are 4 red sweets out of a total of 25. In another bag there are 5 red sweets out of a total of 30. Which bag has the biggest fraction of red sweets?

5 Susan takes 8 cards from a pack and 5 of these are picture cards. Siloben takes 12 cards and gets 7 picture cards. Which girl got the biggest fraction of picture cards?

6 In class 5A, 3 students out of 20 are absent. In class 5B, 5 students out of 24 are absent. Which class has the bigger fraction of absent students?

7 Liverpool have played 10 games and won 7 of them. Arsenal have played 12 games and won 9 of them. Which team has won the bigger fraction of their games?

8 In one batch of 10 television sets, 3 are faulty. In another batch of 24 sets, 7 are faulty. Which is the worse batch?

9 When shooting at a target, John hit the bull with 19 arrows out of 25. Bill hit the bull with 15 arrows out of 20 and Shafiq hit the bull with 37 arrows out of 50. Who is the best shot?

One quantity as a fraction of another

Examples

1 What fraction of 1 hour is 25 minutes?

   25 minutes is $\frac{25}{60}$ or $\frac{5}{12}$ of one hour

2 What fraction of £1 is 64p?

   64p is $\frac{64}{100}$ or $\frac{16}{25}$ of £1

Keys to press	Calculator display	Answer
2 **5** **a**$^{b/c}$ **6** **0** **=**	5⌐12	$\frac{5}{12}$
6 **4** **a**$^{b/c}$ **1** **0** **0** **=**	16⌐25	$\frac{16}{25}$

PAUSE

Give all the answers in this exercise as fractions in their lowest terms.

1 What fraction of 1 minute is:

   a) 10 seconds,   c) 12 seconds,   e) 30 seconds,
   b) 25 seconds,   d) 40 seconds,   f) 15 seconds?

2 What fraction of 1 hour is:

   a) 6 minutes,   c) 42 minutes,   e) 45 minutes,
   b) 50 minutes,   d) 35 minutes,   f) 20 minutes?

3 What fraction of a 24 hour day is:

   a) 1 hour,   f) 12 hours,
   b) 4 hours,   g) 16 hours,
   c) 6 hours,   h) 18 hours,
   d) 8 hours,   i) 13 and a half hours,
   e) 2 hours,   j) 5 and a quarter hours?

4 What fraction of £1 is:

   a) 25p,   c) 10p,   e) 85p,   g) 48p,   i) 2p,
   b) 40p,   d) 50p,   f) 75p,   h) 4p,   j) 27p?

Fractions of quantities

Example

If $\frac{3}{5}$ of the 240 apples in a barrel are bad, how many good apples are there?

To calculate $\frac{3}{5}$ of 240, we first divide by 5 to find $\frac{1}{5}$ and then multiply by 3 to find $\frac{3}{5}$.

$$\frac{1}{5} \text{ of } 240 = 240 \div 5 = 48$$
$$\frac{3}{5} \text{ of } 240 = 48 \times 3 = 144$$

So, 144 apples are bad and therefore, 96 are good (240−144).

On a calculator:

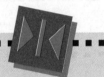

3 $a^{b/c}$ **5** × **2** **4** **0** = 144

144 apples are bad, 96 are good.

PAUSE

1 Find:
  a) $\frac{1}{2}$ of 90
  b) i) $\frac{1}{4}$ of 48
     ii) $\frac{3}{4}$ of 48
  c) i) $\frac{1}{5}$ of 250
     ii) $\frac{2}{5}$ of 250
  d) i) $\frac{1}{7}$ of 91
     ii) $\frac{3}{7}$ of 91
  e) $\frac{4}{5}$ of 100
  f) $\frac{2}{3}$ of 24
  g) $\frac{3}{8}$ of 24
  h) $\frac{7}{9}$ of 81
  i) $\frac{35}{40}$ of 160
  j) $\frac{27}{80}$ of 360

2 There are 25 000 football supporters at a game and the police estimate that $\frac{7}{8}$ of them support the home team. Estimate how many support the away team.

3 Jean-Paul rents 2 videos a week and estimates that $\frac{2}{5}$ of the videos he rents are boring. How many boring videos does he hire in one year?

Adding and subtracting fractions

Examples

1. $\dfrac{1}{3} + \dfrac{3}{4}$

48

We need a number that 3 and 4 both divide into, the LCM of 3 and 4.

Multiples of 3: 3, 6, 9, ⑫ 15 .....

Multiples of 4: 4, 8, ⑫ 16 .....

We need to change both fractions to twelfths:

$$\overset{\times 4}{\underset{\times 4}{\frac{1}{3}}} = \frac{4}{12} \qquad\qquad \overset{\times 3}{\underset{\times 3}{\frac{3}{4}}} = \frac{9}{12}$$

$$\frac{1}{3}+\frac{3}{4} \quad = \quad \frac{4}{12}+\frac{9}{12} \quad = \quad \frac{13}{12} \quad = \quad 1\frac{1}{12}$$

Calculator sequence	Display	Answer
1 $a^{b/c}$ **3** **+** **3** $a^{b/c}$ **4** **=**	⊔ 1 ⊔12	$1\frac{1}{12}$

2. $\dfrac{7}{8}-\dfrac{5}{6}$

Multiples of 8: 8, 16, ㉔ 32, ...

Multiples of 6: 6, 12, 18, ㉔ 30, ...

We need to change both numbers to 24ths:

$$\overset{\times 3}{\underset{\times 3}{\frac{7}{8}}} = \frac{21}{24} \qquad\qquad \overset{\times 4}{\underset{\times 4}{\frac{5}{6}}} = \frac{20}{24}$$

$$\frac{7}{8}-\frac{5}{6} \quad = \quad \frac{21}{24}-\frac{20}{24} \quad = \quad \frac{1}{24}$$

Calculator sequence	Display	Answer
7 $a^{b/c}$ **8** **–** **5** $a^{b/c}$ **6** **=**	1⊔ 24	$\frac{1}{24}$

3. $1 -\dfrac{2}{5} = \dfrac{5}{5}-\dfrac{2}{5} = \dfrac{3}{5}$

Calculator sequence	Display	Answer
1 **–** **2** $a^{b/c}$ **5** **=**	3⊔5	$\frac{3}{5}$

PAUSE

1 a) $\frac{2}{5}+\frac{1}{2}$ c) $\frac{1}{4}+\frac{3}{8}$ e) $\frac{1}{2}+\frac{2}{5}$ g) $\frac{1}{6}+\frac{3}{4}$

   b) $\frac{1}{4}+\frac{1}{3}$ d) $\frac{2}{3}+\frac{1}{4}$ f) $\frac{3}{7}+\frac{2}{5}$ h) $\frac{7}{8}+\frac{5}{6}$

2 a) $\frac{1}{2}-\frac{1}{3}$ c) $\frac{3}{4}-\frac{2}{3}$ e) $\frac{3}{8}-\frac{1}{4}$ g) $1-\frac{5}{6}$

   b) $\frac{1}{3}-\frac{1}{4}$ d) $\frac{5}{6}-\frac{1}{3}$ f) $\frac{5}{7}-\frac{3}{5}$ h) $2-\frac{5}{8}$

Multiplying and dividing fractions

Examples

1. $\frac{3}{4}\times\frac{1}{3} = \frac{3}{12} = \frac{1}{4}$

Calculator sequence	Display	Answer
3 $a^{b/c}$ **4** × **1** $a^{b/c}$ **3** =	1⌐4	$\frac{1}{4}$

2. $\frac{2}{3}\times1\frac{1}{4} = \frac{2}{3}\times\frac{5}{4} = \frac{10}{12} = \frac{5}{6}$

Calculator sequence	Display	Answer
2 $a^{b/c}$ **3** × **1** $a^{b/c}$ **1** $a^{b/c}$ **4** =	5⌐6	$\frac{5}{6}$

3. $1\frac{2}{3}\div\frac{1}{4} = \frac{5}{3}\div\frac{1}{4} = \frac{5}{3}\times\frac{4}{1} = \frac{20}{3} = 6\frac{2}{3}$

*turn the fraction you are dividing
by upside down and multiply*

Calculator sequence	Display	Answer
1 $a^{b/c}$ **2** $a^{b/c}$ **3** ÷ **1** $a^{b/c}$ **4** =	6⌐2 ⌐3	$6\frac{2}{3}$

50

1 a) $\frac{1}{2}\times\frac{1}{3}$ c) $\frac{2}{3}\times\frac{1}{6}$ e) $1\frac{1}{2}\times\frac{1}{3}$ g) $\frac{3}{5}\times\frac{5}{6}$
   b) $\frac{1}{2}\times\frac{3}{4}$ d) $\frac{5}{6}\times\frac{2}{5}$ f) $\frac{1}{4}\times2\frac{1}{2}$ h) $\frac{7}{8}\times1\frac{1}{3}$

2 a) $\frac{1}{2}\div\frac{1}{4}$ c) $\frac{2}{3}\div\frac{1}{3}$ e) $\frac{3}{4}\div4\frac{1}{2}$ g) $2\div\frac{1}{3}$
   b) $\frac{1}{4}\div\frac{1}{8}$ d) $1\frac{1}{2}\div\frac{1}{2}$ f) $6\div1\frac{1}{2}$ h) $7\div\frac{1}{4}$

SECTION 5 Percentages

Percentages and fractions

Percentages are a special way of writing fractions. Each percentage is a fraction with 100 as the bottom number.
For example:

$$25\% \text{ is the same as } \frac{25}{100} \quad \text{and } 66\% \text{ is the same as } \frac{66}{100}$$

Examples
1. Write 66% and 20% as fractions.

$$66\% = \frac{66}{100} = \frac{33}{50}$$
$$20\% = \frac{20}{100} = \frac{1}{5}$$

2. Write $\frac{4}{5}$ and $\frac{3}{4}$ as percentages.

To do this, we must change the bottom numbers of the fraction to 100.

$$\overset{\times 20}{\frac{4}{5}} = \underset{\times 20}{\frac{80}{100}} \quad \text{and} \quad \overset{\times 25}{\frac{3}{4}} = \underset{\times 25}{\frac{75}{100}}$$

So, $\frac{4}{5} = 80\%$ and $\frac{3}{4} = 75\%$

1 Write each of the following percentages as a fraction in its simplest form.

a) 25%   c) 68%   e) 64%   g) 40%   i) 32%
b) 60%   d) 90%   f) 99%   h) 15%   j) 95%

2 Write each fraction as a percentage.

a) $\frac{1}{2}$ c) $\frac{7}{10}$ e) $\frac{12}{25}$ g) $\frac{6}{25}$ i) $\frac{5}{10}$
b) $\frac{17}{20}$ d) $\frac{11}{20}$ f) $\frac{1}{10}$ h) $\frac{3}{25}$ j) $\frac{11}{25}$

3 By what fraction has the price been reduced in this special offer?

> GIFT SETS
> COTY, YARDLEY, LENTHÉRIC   **UP TO 30% OFF**
> SELECTED LINES

Calculating a percentage of a quantity

Examples

1. There were 160 questions in a Maths test and Mary Jones obtained a mark of 45%. How many questions did she get right?

We must calculate 45% of 160, to do this, we remember that 45% is the same as $\frac{45}{100}$.

$$45\% \text{ of } 160 = \frac{45}{100} \text{ of } 160$$

$$45\% \text{ of } 160 = 160 \div 100 \times 45 = 72$$

Step 1: 160 ÷ 100
This finds $\frac{1}{100}$ or
1% of 160.

Step 2: × 45
This finds $\frac{45}{100}$ or
45% of 160.

So, Mary got 72 answers correct out of 160.

2. A garage adds 17.5% VAT to the marked price on all the tyres that it sells. How much will a tyre marked with a price of £25.70 actually cost?

$$17.5\% \text{ of } 25.7 = \frac{17.5}{100} \text{ of } 25.7$$

$$17.5\% \text{ of } 25.7 = 25.7 \div 100 \times 17.5 = \pounds 4.4975$$

£4.4975 to the nearest penny is £4.50, so the tyre costs
£25.70 + £4.50 = £30.20

PAUSE

1 Calculate:

a) 25% of 156	d) 90% of 855	g) 40% of £56.80
b) 60% of 145	e) 64% of 64	h) 15% of £45
c) 68% of 275	f) 99% of 2500	i) 32% of 45 kg

2 28% of Ms Mutton's monthly income of £1240 goes on mortgage repayments. How much does she pay each month?

3 Mr Asquith, the Labour candidate received 44% of the 28 000 votes cast. How many people voted for Mr Asquith?

4 A year ago, Ms Fox's house was worth £85 500. During the last year house prices have fallen by an average of 12%. Estimate the decrease in the value of Ms Fox's house.

5 Find the amount of VAT (at 17.5%) which will be added to the cost of the replacement window in this advertisement.

> ### REPLACEMENT WINDOWS
> **EXAMPLE:**
> 4ft x 3ft 6in Double Glazed Window with single top
> opening in attractive maintenance-free uPVC.
> FULLY FITTED (including sill)
> ## ONLY £184 PLUS VAT
> ★ With Full 10 Year Guarantee.

Writing one number as a percentage of another

Rule: To write one number as a percentage of another:
   ■ form a fraction with the two numbers,

- divide the top number by the bottom number,
- multiply by 100.

Examples

1. Kim Ashton scored 72 correct marks out of a possible 80 in a Science test. What was Kim's percentage score?

$$\text{As a fraction, Kim's score} = \frac{72}{80}.$$

$$\text{As a percentage, Kim's score} = 72 \div 80 \times 100 = 90\%.$$

2. Write 45 as a percentage of 190.

Percentage $= 45 \div 190 \times 100 = 23.7\%$ (correct to one decimal place).

3. In one box of 240 apples, 36 are bad. In another box of 200 apples, 26 are bad. Which box has the better apples?

$$\text{In the first box, the percentage of bad apples} =$$
$$36 \div 240 \times 100 = 15\%.$$

$$\text{In the second box the percentage of bad apples} =$$
$$26 \div 200 \times 100 = 13\%.$$

The second box has the better apples.

4. What is the percentage reduction in price on this toolbox?

BLACK & DECKER
BDK202 Workshop Kit.
Cat. No. 710/4215.
~~£87.95~~
Sale Price
£77.95

5. The box originally cost £87.95 and has been reduced to £77.95, a cash reduction of £10. The percentage reduction is found by writing £10 as a percentage of £87.95.

$$\text{Percentage reduction} = 10 \div 87.95 \times 100 = 11.4\%$$
$$\text{(correct to one decimal place).}$$

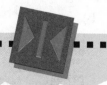

1 Write:

a) 8 as a percentage of 25,
b) 34 as a percentage of 85,
c) £1.25 as a percentage of £50,
d) 42 as a percentage of 120,
e) £5.25 as a percentage of £35,
f) 48p as a percentage of 64p,
g) £8.50 as a percentage of £25,
h) 45 minutes as a percentage of 1 hour,
i) 36p as a percentage of 45p,
j) 14 grams as a percentage of 35 grams,
k) 12 minutes as a percentage of 1 hour 36 minutes,
l) 5.1 cm as a percentage of 6.8 cm.

2 These are the marks that a group of 20 pupils scored in a test with a maximum mark of 120.

| 43 | 56 | 78 | 89 | 94 | 96 | 102 | 112 | 65 | 105 |
| 84 | 22 | 17 | 68 | 72 | 118 | 88 | 71 | 59 | 60 |

Convert them to percentage marks, correct to the nearest 1%.

3 In one bag of potatoes, 3 out of 20 are rotten, in a second bag 5 out of 30 potatoes are rotten and in a third bag 7 out of 40 potatoes are rotten.

a) What is the percentage of rotten potatoes in each bag?
b) Which is the best bag?

4 In one school, 6 out of 10 teachers are female, in a second school 14 out of 25 teachers are female and in a third school 22 out of 50 teachers are male. Which school has the largest percentage of female staff?

5 Calculate the percentage reductions in price on each of these items:

a)

SWAN
Compact
Microwave Oven.
Cat. No. 420/8341.
~~£144.95~~
Sale Price
£134.95

b)

BRITAX
"Babysure" Car Seat.
Cat. No. 375/2582.
~~£27.99~~
Sale Price
£22.99

c)

OLYMPUS
Supertrip Compact.
35 mm Camera.
Cat. No. 560/6100.
~~£54.95~~
Sale Price
£44.95

6 A personal stereo was priced at £48. In a sale it was reduced to £42.

SALE
Was £48
Now £42

a) By what fraction was the original price reduced?
b) Write your fraction as a percentage.

(ULEAC)

7 When 15 oranges are bought individually the total cost is £1.20.
When 15 oranges are bought in a pack the cost is £1.14.
a) What is the percentage saving by buying the pack?

SPECIAL OFFER
20% EXTRA FREE

b) A special offer pack of these oranges has 20% extra free.
How many extra oranges are in the special offer pack?
c) What fraction of the oranges in the special offer pack are free?

(SEG)

8

> **Super Ace Games System**
> Normal Price   £120
> **Sale Price** $\frac{1}{3}$ off

a) Work out the sale price of the Super Ace Games System.

> **Mega Ace Games System**
> Normal Price   £320
> **Sale Price   £272**

b) Find the percentage reduction on the Mega Ace Games System in the sale.
(ULEAC)

9   *Cost breakdown of an £11.49 compact disc*

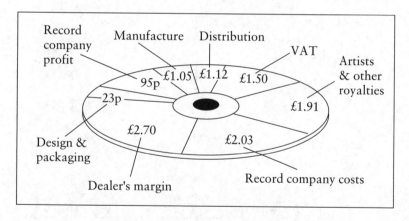

a) How much is the record company profit on each compact disc?

b) What is the total of all the *other* parts of the cost breakdown?

c) Calculate, to the nearest whole number, the record company's profit as a percentage of the total of all the other parts of the cost breakdown.

d) The cost of manufacture rose by 8% and the cost of distribution rose by 5%.

By what percentage, calculated to 1 decimal place, should the dealer's margin fall to keep the cost of a compact disc at £11.49?
(NICCEA)

10 A company bought machines from America for 11 500 US dollars each.

Transport costs added 12% to the cost of a machine.

a) Calculate, in US dollars, the *total* cost of a machine.

The exchange rate was £1 = 1.78 US dollars.

b) Convert your answer to a) into £.

The company sold the machines for £9000 each.

c) Calculate the percentage profit on the total cost price for each machine.

Give your answer correct to 1 decimal place.   (ULEAC)

■ ■

Fractions, percentages and decimals

Fractions, percentages and decimals are all systems to represent part of a quantity.

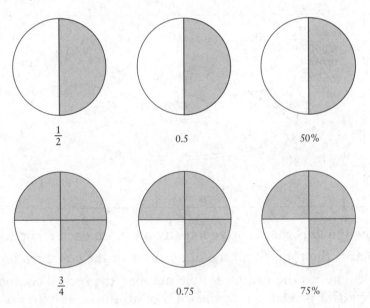

$\frac{1}{2}$ 0.5   50%

$\frac{3}{4}$ 0.75   75%

Rule: To change a fraction to a decimal, divide the top number by the bottom number.

Example

Change $\frac{2}{5}$ to a decimal:

$$2 \div 5 = 0.6$$

58

Rule: To change a decimal to a percentage, multiply by 100.

Example

Change 0.333 to a percentage:

$$0.333 = 33.3\%$$

Rule: To change a percentage to a fraction, write the percentage over 100 and simplify if possible.

Example

Change 65% to a fraction:

$$65\% = \frac{65}{100} = \frac{13}{20} \qquad \div 5$$

Any given fraction, percentage or decimal can always be converted to each of the other systems.

Examples

1. Starting with a fraction.

   Change $\frac{5}{8}$ to a decimal and a percentage.

   $$\frac{5}{8} = 5 \div 8 = 0.635$$
   $$0.625 = 62.5\%$$

2. Starting with a decimal.

   Change 0.85 to a percentage and a fraction.

   $$0.85 = 85\%$$

   $$85\% = \frac{85}{100} = \frac{17}{20} \qquad \div 5$$

3. Starting with a percentage.

   Change 68% to a fraction and a decimal.

   $$68\% = \frac{68}{100} = \frac{17}{25} \qquad \div 4$$

   $$68\% = 0.68$$

PAUSE

1 Change each of these fractions into decimals.

   a) $\frac{3}{4}$ c) $\frac{7}{8}$ e) $\frac{1}{2}$ g) $\frac{1}{4}$ i) $\frac{2}{3}$

   b) $\frac{4}{5}$ d) $\frac{9}{10}$ f) $\frac{1}{5}$ h) $\frac{3}{16}$ j) $\frac{1}{12}$

2 Convert each of the decimals obtained in question 1 into percentages.

3 Change each of these decimals into percentages.

   a) 0.1   c) 0.3   e) 0.64   g) 0.005   i) 0.5

   b) 0.8   d) 0.35   f) 0.08   h) 0.05   j) 0.375

4 Convert each of the percentages obtained in question 3 into fractions.

5 Change each of these percentages into fractions and decimals.

   a) 80%   c) 25%   e) 98%   g) 40%   i) 65%

   b) 32%   d) 70%   f) 55%   h) 48%   j) 50%

6 In a nine carat gold ring $\frac{9}{24}$ of the weight is pure gold.

   What percentage of the weight of the ring is pure gold?   (SEG)

7 A cockroach has a reaction time of 54 thousandths of a second.

   a) Write the cockroach's reaction time as a fraction in figures.

      Give your answer in its lowest terms.

   b) Give your answer to part a) as a decimal fraction.   (ULEAC)

8 A college is offered a discount of $\frac{1}{8}$ if it pays within 1 week of receiving an invoice.

   a) Write $\frac{1}{8}$ as:

     i) a decimal,

     ii) a percentage.

   b) Calculate the amount to be paid by the college if an invoice for £4324.35 is paid within one week.

    Give your answer to the nearest penny.   (NEAB)

Reciprocals

The reciprocal of a number is one divided by that number.

There is a reciprocal key $\boxed{\frac{1}{x}}$ on most scientific calculators.

Examples

1. Find the reciprocal of 5.

   The reciprocal of 5 is $\frac{1}{5}$ or 0.2.

Calculator sequence	Answer
$\boxed{5}$ $\boxed{\frac{1}{x}}$	0.2

2. Find the reciprocal of 9.

   The reciprocal of 9 is $\frac{1}{9}$ or 0.1111.....

Calculator sequence	Answer
$\boxed{9}$ $\boxed{\frac{1}{x}}$	0.1111.....

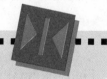

PAUSE

1 Write the reciprocals of these numbers as fractions and then as decimals corrected to 3 significant figures when necessary.

   a) 2   c) 6   e) 16   g) 40
   b) 4   d) 11   f) 20   h) 30

Repeated percentage changes

Examples

1. A bank pays 5% interest each year. Julie leaves £250 saved in the bank for four years. Each year the interest is added to the amount Julie has in the bank. What are her savings worth after 3 years?

   Julie's saving will increase by 5% each year.

   This means that at the end of each year they will be worth 105% of their value at the start of the year.

After 1 year Julie's savings $= 250 \div 100 \times 105 = £262.50$

After 2 years Julie's savings $= 262.5 \div 100 \times 105 = £275.63$
(to the nearest penny)

After 3 years Julie's savings $= 275.63 \div 100 \times 105 = £289.41$
(to the nearest penny)

2. A shopkeeper wants to sell a carpet which costs £400. She offers to reduce the cost by 20% every day until the carpet is sold. What will the carpet cost if it is still unsold after three reductions?

The carpet will be reduced by 20% each day.

This means that each day it will cost 80% of its cost the day before.

Cost after 1 reduction $= 400 \div 100 \times 80 = £320$

Cost after 2 reductions $= 320 \div 100 \times 80 = £256$

Cost after 3 reductions $= 256 \div 100 \times 80 = £204.80$

PAUSE

1 Find the final value of each bank savings account.

a) £500 invested for 3 years at 6%
b) £1000 invested for 2 years at 8%
c) £800 invested for 2 years at 3%
d) £300 invested for 2 years at 8%
e) £2000 invested for 2 years at 3%
f) £750 invested for 3 years at 2%

2 A Department Store is determined to sell off old stock and puts them all on special offer. Find the value of each of these special offers after 3 reductions.

a) A £200 bed reduced by 10% each day until sold.
b) A £500 wardrobe reduced by 20% each day until sold.
c) A £250 dress reduced by 25% each day until sold.
d) A £140 walkman reduced by 50% each day until sold.
e) A £400 mountain bike reduced by 35% each day until sold.
f) A £5000 oriental carpet reduced by 4% each day until sold.

3 a) Pauline paid £80 for a stamp for her collection. This stamp appreciated in value by 5% each year. What was its value 2 years after she bought it?

 b) Stephen bought a motor-cycle for £2500. It depreciated in value by 10% each year. What was its value 3 years after he bought it?   (NICCEA)

4 Nesta invests £508 in a bank account at an interest rate of 8.5% per annum.

 a) Calculate the interest on £508 after 1 year.

 At the end of the first year the interest is added to her bank account. The interest rate remains at 8.5%.

 b) Calculate the total amount of money in Nesta's bank account at the end of the second year.   (ULEAC)

5 £115,000 was invested for three years at 8% per year Compound Interest. Calculate the total interest earned over the three years.

(NICCEA)

■ ■

Reverse percentage calculations

Examples

1. A coat has been reduced by 20% and is on sale for £50. What did it originally cost?

 The £50 represents 100% − 20% = 80% of the original cost

$$1\% \text{ of the original cost} = 50 \div 80 = 0.625$$
$$100\% \text{ of the original cost} = 0.625 \times 100 = £62.50$$

2. An Office Manager received a 5% pay rise and now earns £315 a week. What was she earning before the rise?

 The £315 represents 100% + 5% = 105% of her original wage.

$$1\% \text{ of her original wage} = 315 \div 105 = 3$$
$$100\% \text{ of her original wage} = 3 \times 100 = £300$$

1 A shop has reduced all its prices by 20% in a sale. Find the original cost of items which in the sale cost:

a) £ 160   c) £ 360   e) £ 320   g) £ 45   i) £ 110
b) £ 240   d) £ 64   f) £ 40   h) £ 90   j) £ 36.50

2 All the workers in a factory receive a 5% pay rise. Find the old weekly wage of people who each week now earn:

a) £210   c) £157.50   e) £189   g) £262.5   i) £450
b) £420   d) £147   f) £220.50   h) £300   j) £385

3 A large London Department Store estimates that Christmas Trading is 8% down on last year. If this year they sold goods worth £1 886 000, what value of goods did they sell last year?

4 A self-employed driving instructor is earning 12% more than he did a year ago. If he is earning £480 a week now, what was he earning a year ago?

5

The manager of a carpet shop has made a mistake in writing the advertisement above.

The prices are correct, but the 'percentage off' is NOT correct.

a) Find, correct to one decimal place, the true 'percentage off'.
b) In another shop, where a genuine reduction of 20% had been made, the sale price of a carpet was also £6.39. What was the original price in this shop?   (NICCEA)

Hire purchase

If you buy something on hire purchase you do not pay the full price immediately. You pay a **deposit,** usually based on a percentage of the full price. You then pay a number of weekly or monthly instalments. This is often more expensive than paying the full price immediately.

Examples

A compact disc player can be bought for £299.99 cash or on hire puchase with a 10% deposit and 100 weekly payments of £2.90. Find the difference between the cash and hire purchase prices.

$$10\% \text{ of } £299.99 = 299.99 \div 100 \times 10 = £30.00$$
$$\text{(to the nearest penny)}$$
$$100 \times £2.90 = £290.00$$
$$\text{Hire purchase price} = £290.00 + £30.00 = £320.00$$
$$\text{Difference} = £320.00 - £299.99 = £20.01$$

PAUSE

Calculate in each question.

a) The hire purchase price.

b) The difference between the cash and hire purchase price.

	Item	Cash price	Deposit	Repayments
1	Cooker	£280	20%	50 of £5.70 each
2	Camera	£190	10%	24 of £8.45 each
3	Fridge	£350	10%	12 of £28 each
4	T.V.	£420	15%	24 of £15.20 each
5	Car	£3500	20%	36 of £81.25 each
6	Motor Bike	£1250	25%	24 of £44.21 each
7	Video Recorder	£399	10%	100 of £3.80 each
8	Chair	£180	20%	12 of £13.50 each
9	Computer	£1249	15%	36 of £32.44 each

Establishing and simplifying a ratio

A *ratio* is used to compare two or more numbers. For example, we may say

'the ratio of girls to boys in the class is 5 to 4'.

By this we mean that for every 5 girls in the group there are 4 boys. This is equivalent to saying that the *fraction* of girls in the class is $\frac{5}{9}$. Fractions compare the separate parts into which an object is divided with the *whole object*. Ratios compare the separate parts into which an object is divided *with each other*. Ratios are often written with the word 'to' replaced by the symbol ' : '

The ratio of girls to boys in the class can be written as 5 : 4.

Ratios can be simplified in the same way as fractions.

Examples

1. Simplify the ratio 24 : 90.

    24 and 90 both divide by 6, so we can write:

$$24 : 90 = 4 : 15$$

2. To make shortcrust pastry, we mix 400g of plain flour with 100 grams of butter and 100 grams of fat. What is the ratio of these ingredients in its simplest form?

    The basic ratio is: 400 : 100 : 100

    We can divide by 100 and simplify the ratio to:

$$4 : 1 : 1$$

PAUSE

1 Simplify each of the following ratios

  a) 4:6   g) 54:90   l) £2.50:£3.00

  b) 12:16   h) 24:84   m) £2.50:£3.50

  c) 6:30   i) 20:200   n) 40 days:1 year (not a leap year)

  d) 15:18   j) 96:204   o) 45 minutes:3 hours

  e) 30:35   k) 1 cm to 1 metre p) 60°:300°

  f) 112:64

2 In a school there are 275 boys and 225 girls. What, in its simplest form, is the ratio of boys to girls?

3 Harry Loosealot likes to bet on horses. Last year, out of 250 bets, Harry won 45 times. What is the ratio of Harry's winning bets to his losing bets?

4 Out of 96 students in a school taking GCSE Mathematics, 12 fail. What is the ratio of passes to fails?

■■■■■■■■■■■■■■■■■■■■■■■■■■■■■■■■■■■■■■

Dividing a quantity in a ratio

Ratios are often used to divide a quantity into parts.

Example

Divide 24 sweets between Bill and Belinda in the ratio 5 : 7.

We require 5+7 or 12 'shares':

$$24 \div 12 = 2$$
$$5 \times 2 = 10 \text{ sweets}$$
$$7 \times 2 = 14 \text{ sweets}$$

Bill gets 10 sweets and Belinda gets 14 sweets.

■■■■■■■■■■■■■■■■■■■■■■■■■■■■■■  ■■■

PAUSE

1 Divide 36 kg in the ratios:

a) 1 : 2	d) 5 : 4	g) 1 : 11	j) 5 : 13
b) 3 : 1	e) 2 : 7	h) 5 : 7	k) 1 : 2 : 3
c) 5 : 1	f) 8 : 1	i) 17 : 1	l) 2 : 3 : 4

2 Divide £100 in the ratios:

a) 3 : 1	d) 5 : 3	g) 1 : 9	j) 3 : 17
b) 2 : 3	e) 7 : 1	h) 7 : 3	k) 3 : 3 : 4
c) 1 : 4	f) 1 : 1	i) 19 : 1	l) 2 : 5 : 13

3 Mr Bilson has £300 worth of premium bonds and Mrs Bilson has £500 worth of premium bonds. They agree to share any winnings in the same ratio as the number of bonds they hold. How do they share a win of £25 000?

67

4 One sunny day, Jane and Jessica decide to sell squash from a stall. Jane buys £1.20 worth of ingredients and Jessica buys £1.80 worth of ingredients. They agree to share the takings in the same ratio as their contributions to the costs. How do they share total takings of £15.50? Do you think this is a fair way to share the takings?

■■

Scaling up or down from a ratio

Archie has a recipe for a pizza for 4 people which uses 260 grams of cheese. He wants to adapt the recipe to make a pizza for 6 people, how much cheese should he use?

The ratio between the two recipes is 4 : 6. So, Archie needs to fill in the missing number in this pair of equal ratios.

$$4 : 6$$
$$260 : ?$$

Archie works out 260 ÷ 4, which is 65, and, because the ratios must be equal, multiplies 6 by 65.

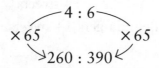

Archie needs to use 390 grams of cheese in the new recipe.

Examples

1. If 5 miles is equal to 8 kilometres, how many kilometres is 250 miles equal to?

   250 ÷ 5 is 50, so we can calculate:

   ```
            ⌒5 : 8⌒
       × 50        × 50
            ↘250 : 400↙
   ```

   250 miles is equal to 400 kilometres.

2. A stack of exercise books 30 cm high contains 36 books.
   a) How many exercise books are there in a pile 45 cm high?
   b) How high is a pile of 90 books?
   a) The heights of the piles are in the ratio 30 : 45 and the numbers of books must be in an equal ratio. So we calculate like this:

Ratio of heights

$$-30 : 45-$$
$$\times 1.2 \qquad \times 1.2$$

Ratio of number
$$\searrow 36 : 54 \nwarrow$$

There are 54 books in the second pile.

b) The numbers of books are in the ratio 36 : 90 and the heights must be in an equal ratio. So we calculate like this:

Ratio of numbers

$$-36 : 90-$$
$$\times 0.83... \qquad \times 0.83...$$

Ratio of heights
$$\searrow 30 : 75 \nwarrow$$

The second pile is 75 cm high.

PAUSE

1 This is a basic recipe for making an apple pie for 4 people.

> 200 grams of flour
> 50 grams of lard
> 50 grams of butter
> 500 grams of apples

Change this recipe into one to make an apple pie for:

a) 8 people,   d) 6 people,   g) 3 people,   j) 1 person.

b) 12 people,   e) 10 people,   h) 14 people,

c) 16 people,   f) 2 people,   i) 18 people,

2 Change to kilometres.

a) 280 miles   c) 450 miles   e) 62 miles

b) 700 miles   d) 35 miles

3 Change to miles.

a) 240 km   c) 40 km   e) 324 km

b) 800 km   d) 96 km

4 A stack of text books 24 cm high contains 10 books. How many books will there be in a stack:

a) 36 cm high,   d) 72 cm high,   g) 57.6 cm high,
b) 48 cm high,   e) 84 cm high,   h) 67.2 cm high?
c) 60 cm high,   f) 43.2 cm high,

5 To make her party fruit drink, Anita mixes 3 litres of orange juice with 1 litre of pineapple juice and 4 litres of lemonade. How much pineapple juice and lemonade would she mix with these quantities of orange juice:

a) 6 litres,   d) 4.5 litres,   g) 5 litres,   j) 4.3 litres?
b) 9 litres,   e) 1.5 litres,   h) 7 litres,
c) 12 litres,   f) 1.8 litres,   i) 4 litres,

■ ■

SECTION 7 Directed Numbers

Understanding negative numbers

On the centigrade temperature scale, the freezing point of water is 0° C. Temperatures on the earth's surface can range from 50° C below freezing to 40° C above freezing.

■ ■
                    PAUSE

1 What temperatures are represented by each arrow on this scale?

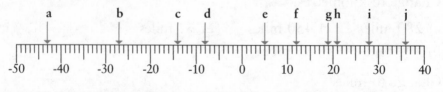

■ ■

Changes in temperature

Examples

1. What is the change in temperature from arrow **c** to arrow **d** on the temperature scale?

   The temperature goes up 8°C between the arrows, so the change is +8°C.

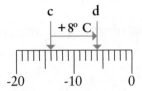

2. What is the change in temperature from arrow **d** to arrow **c** on the temperature scale?

   The temperature goes down 8°C between the arrows, so the change is –8°C.

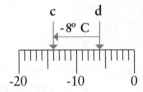

Ordering temperatures

The temperature scale can also be used to order temperatures.

Example

Write these temperatures in order, coldest first.

$$-15°C, 25°C, -20°C, 0°C, 8°C$$

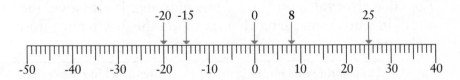

Answer: –20°C, –15°C, 0°C, 8°C, 25°C

What is the change in temperature between arrows:

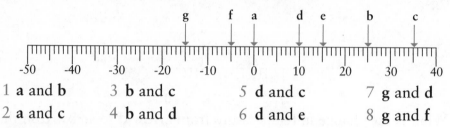

1 **a and b**   3 **b and c**   5 **d and c**   7 **g and d**

2 **a and c**   4 **b and d**   6 **d and e**   8 **g and f**

9 Write the temperatures in order, coldest first.

    a) 14°C, –4°C, –30°C, 25°C, –2°C

    b) –1°C, 18°C, –10°C, 16°C, 7°C

    c) 9°C, –5°C, –12°C, 8°C, 0°C

    d) 4°C, –14°C, 7°C, –8°C, –5°C

    e) –6°C, –13°C, –10°C, –20°C, 0°C

10 The table gives the temperature one night and also the temperature the next day. Find the rise in temperature in each case.

	Temperature at night	Temperature the next day
a)	2°C	14°C
b)	–1°C	7°C
c)	–15°C	–3°C
d)	–5°C	9°C
e)	–3°C	22°C
f)	–18°C	–6°C

11 The table gives the temperature one morning. It also gives the rise or fall in temperature later that day. Find the new temperature in each case.

	Temperature one morning	Rise or fall in temperature
a)	6°C	rise of 12°C
b)	–2°C	rise of 8°C
c)	–13°C	rise of 10°C
d)	5°C	fall of 9°C
e)	0°C	fall of 12°C
f)	–1°C	fall of 6°C

Addition and subtraction of directed numbers

Remember the number line.

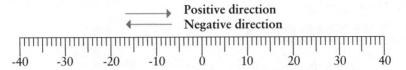

When adding or subtracting negative numbers:

  start with a *position* on the number line,

  make a *change in position* to obtain your answer.

$18 + 7 = 25$ (start at 18, make a change of 7 in the positive direction)

$18 + -7 = 11$ (start at 18, make a change of 7 in the negative direction)

$-18 + 7 = -11$ (start at -18, make a change of 7 in the positive direction)

$-18 + -7 = -25$ (start at -18, make a change of 7 in the negative direction)

$18 - 7 = 11$ (start at 18, make a change of 7 in the negative direction)

$-18 - 7 = -25$ (start at -18, make a change of 7 in the negative direction)

When subtracting a negative number, change this double negative sign to a positive sign:

$$18 - -7 = 18 + 7 = 25$$
$$-18 - -7 = -18 + 7 = -11$$

PAUSE

1 Copy and complete:

$8 + 5 =$	$-8 + 5 =$	$8 - 5 =$	$8 - -5 =$
$8 + -5 =$	$-8 + -5 =$	$-8 - 5 =$	$-8 - -5 =$

2 Copy and complete:

$3 + 7 =$	$-3 + 7 =$	$3 - 7 =$	$-3 - 7 =$
$3 + -7 =$	$-3 + -7 =$	$3 - -7 =$	$-3 - -7 =$

When adding or subtracting negative numbers we can make use of the calculator key ⁺∕₋, which is used to enter negative numbers.

Examples

1. 18 + −7 = ?

   [1] [8] [+] [7] [⁺∕₋] [=]   11

   this is how −7 is entered

2. −18 + 7 = ?

   [1] [8] [⁺∕₋] [+] [7] [=]   −11

   this is how −18 is entered

3. 18 − −7 = ?

   [1] [8] [−] [7] [⁺∕₋] [=]   25

   this is how −7 is entered

P A U S E

Use the ⁺∕₋ key or a number line to answer these.

1 Copy and complete:

13 + 9 =	−13 + 9 =	13 − 9 =	−13 − 9 =
13 + −9 =	−13 + −9 =	13 − −9 =	−13 − −9 =

2 Copy and complete:

5 + 23 =	−5 + 23 =	5 − 23 =	−5 − 23 =
5 + −23 =	−5 + −23 =	5 − −23 =	−5 − −23 =

Multiplication and division of directed numbers

There is just one golden rule when multiplying or dividing directed numbers:

If the signs of the two numbers are the same, your answer will be positive, if the signs of the two numbers are different your answer will be negative.

74

Remember this and you can't go wrong but *don't* try to apply the rule to addition and subtraction!

Examples

Calculator sequences

$18 \times 6 = 108$ `1` `8` `×` `6` `=`   108

$18 \div 6 = 3$ `1` `8` `÷` `6` `=`   3

$-18 \times 6 = -108$ `1` `8` `+/-` `×` `6` `=`   −108

$-18 \div 6 = -3$ `1` `8` `+/-` `÷` `6` `=`   −3

$18 \times -6 = -108$ `1` `8` `×` `6` `+/-` `=`   −108

$18 \div -6 = -3$ `1` `8` `÷` `6` `+/-` `=`   −3

$-18 \times -6 = 108$ `1` `8` `+/-` `×` `6` `+/-` `=`   108

$-18 \div -6 = 3$ `1` `8` `+/-` `÷` `6` `+/-` `=`   3

■

PAUSE

1 Copy and complete:

$10 \times 5 =$ $-10 \times 5 =$ $10 \times -5 =$ $-10 \times -5 =$

$10 \div 5 =$ $-10 \div 5 =$ $10 \div -5 =$ $-10 \div -5 =$

2 Copy and complete:

$12 \times 8 =$ $-12 \times 8 =$ $12 \times -8 =$ $-12 \times -8 =$

$12 \div 8 =$ $-12 \div 8 =$ $12 \div -8 =$ $-12 \div -8 =$

3 Copy and complete:

$6.5 \times 5 =$ $-6.5 \times 5 =$ $6.5 \times -5 =$ $-6.5 \times -5 =$

$6.5 \div 5 =$ $-6.5 \div 5 =$ $6.5 \div -5 =$ $-6.5 \div -5 =$

4 Copy and complete:

$1.2 \times 0.4 =$ $-1.2 \times 0.4 =$ $1.2 \times -0.4 =$ $-1.2 \times -0.4 =$

$1.2 \div 0.4 =$ $-1.2 \div 0.4 =$ $1.2 \div -0.4 =$ $-1.2 \div -0.4 =$

■ ■

Powers of numbers

There is a special way to write chain multiplications of the same number.

3×3 can be written as 3^2

3^2 is read as '3 squared'

$2 \times 2 \times 2$ can be written as 2^3

2^3 is read as '2 to the power 3'
or as '2 cubed'

$4 \times 4 \times 4 \times 4 \times 4 \times 4$ can be written as 4^6

4^6 is read as '4 to the power 6'

Examples

$$3^4 = 3 \times 3 \times 3 \times 3 = 81$$
$$2^5 = 2 \times 2 \times 2 \times 2 \times 2 = 32$$
$$5^2 = 5 \times 5 = 25$$

PAUSE

Calculate the value of:

1 3^5	3 8^2	5 7^4	7 9^3	9 0.9^2
2 5^3	4 2^8	6 4^7	8 2.5^3	10 1.1^5

There are two calculator keys which are useful for finding powers:
$\boxed{x^2}$ and $\boxed{x^y}$ (or $\boxed{y^x}$

Examples

	Calculator sequence	Answer
5^2	$\boxed{5}$ $\boxed{x^2}$	25
2.4^2	$\boxed{2}$ $\boxed{.}$ $\boxed{4}$ $\boxed{x^2}$	5.76
3^4	$\boxed{3}$ $\boxed{x^y}$ $\boxed{4}$ $\boxed{=}$	81
2^5	$\boxed{2}$ $\boxed{x^y}$ $\boxed{5}$ $\boxed{=}$	32

1 Use x^2 to answer these.

a) 9^2 c) 18^2 e) 20^2 g) 7.5^2 i) 0.6^2

b) 12^2 d) 11^2 f) 15^2 h) 8.3^2 j) 0.25^2

2 Use x^y to answer these.

a) 4^3 c) 3^5 e) 10^4 g) 9^4 i) 8^4

b) 2^6 d) 6^3 f) 7^3 h) 11^3 j) 5^6

Zero and negative powers of numbers

The power notation can be extended to zero and negative powers.
The definitions of zero and negative powers can be found from the
patterns.

$$2^5 = 32 \qquad\qquad 3^5 = 243$$
$$2^4 = 16 \qquad\qquad 3^4 = 81$$
$$2^3 = 8 \qquad\qquad 3^3 = 27$$
$$2^2 = 4 \qquad\qquad 3^2 = 9$$
$$2^1 = 2 \qquad\qquad 3^1 = 3$$
$$2^0 = 1 \qquad\qquad 3^0 = 1$$
$$2^{-1} = \tfrac{1}{2} \qquad\qquad 3^{-1} = \tfrac{1}{3}$$
$$2^{-2} = \tfrac{1}{4} \qquad\qquad 3^{-2} = \tfrac{1}{9}$$
$$2^{-3} = \tfrac{1}{8} \qquad\qquad 3^{-3} = \tfrac{1}{27}$$

These patterns suggest these rules for powers:

Any number to the power 1 is equal to itself.

Any number to the power zero is equal to 1.

Any number to a negative power is equal to 1 over the
corresponding positive power.

Examples

1. Evaluate 5^1, 17^0 and 8^{-3}

$$5^1 = 5$$
$$17^0 = 1$$
$$8^{-3} = \frac{1}{8^3} = \frac{1}{512}$$

2. Write as a single power of three, $3^3 \times 3^2$,
   The calculation $3^3 \times 3^2$ means $(3 \times 3 \times 3) \times (3 \times 3)$,
   this can be written as the single power 3^5.

3. Write as a single power of ten, $10^5 \times 10^{-3}$
   The calculation means $(10 \times 10 \times 10 \times 10 \times 10) \times \dfrac{1}{(10 \times 10 \times 10)}$,
   this is equal to 10×10 or 10^2.

PAUSE

1 Find the value of:

a) 5^6	d) 10^3	g) 4^{-2}	j) 12^0	m) 7^0
b) 6^5	e) 10^4	h) 5^{-1}	k) 8^{-2}	n) 35^1
c) 10^2	f) 2^7	i) 10^1	l) 1^{-1}	o) 2^{-8}

2 Write each of the following calculations as a single power of the number.

a) $2^2 \times 2^3$	e) $10^3 \times 10^2$	i) $3^2 \times 3^{-2}$	m) $10^4 \times 10^1$
b) $3^5 \times 3^4$	f) $10^5 \times 10^1$	j) $4^3 \times 4^{-3}$	n) $10^5 \times 10^{-5}$
c) $4^7 \times 4^3$	g) $3^4 \times 3^0$	k) $2^5 \times 2^{-2}$	o) $10^{-4} \times 10^2$
d) $2^9 \times 2^2$	h) $5^0 \times 5^5$	l) $10^3 \times 10^{-2}$	p) $10^7 \times 10^{-2}$

Standard form

Very large and very small numbers are common in Science.
For example, the average distance of the Earth from the Sun is
150 000 000 km and a human blood cell is 0.00001 m long.
Mathematicians have developed a shorthand way to write down very
large and very small numbers called *standard form*.

The numbers are written as a number between 1 and 10, multiplied by a power of 10.

Examples

1. Write 150 000 000 in standard form:
$$150\,000\,000 = 1.5 \times 100\,000\,000$$
So, in standard form 150 000 000 is written 1.5×10^8.

2. Write 0.000 01 in standard form:
$$0.000\,01 = 1.0 \div 100\,000$$
So, in standard form 0.00001 is written 1.0×10^{-5}.

3. Write 1.3×10^5 as a normal number:
$$1.3 \times 10^5 \text{ means } 1.3 \times 100\,000$$
So, $1.3 \times 10^5 = 130\,000$.

4. Write 1.67×10^{-5} as a normal number:
$$1.67 \times 10^{-5} \text{ means } 1.67 \div 100\,000$$
So, $1.67 \times 10^{-5} = 0.000\,016\,7$.

PAUSE

1 Write in standard form:

a) 37 000	d) 150 000	g) 6 500 000	j) 6500
b) 3700	e) 650 000 000	h) 650 000	k) 650
c) 370 000	f) 65 000 000	i) 65 000	l) 65

2 Write as normal numbers:

a) 2×10^5	d) 3.45×10^9	g) 8.3×10^3	j) 1.99×10^{12}
b) 3×10^8	e) 1.6×10^9	h) 4.72×10^5	k) 7.3×10^2
c) 1.7×10^4	f) 3.8×10^3	i) 6.73×10^6	l) 8.09×10^{10}

3 Write in standard form:

a) 0.37	d) 0.000 000 056	g) 0.65	j) 0.00065
b) 0.004	e) 0.000 578	h) 0.065	k) 0.000065
c) 0.000 004 50	f) 0.0000001	i) 0.0065	l) 0.000 006 5

4 Write as normal numbers:
   a) 2×10^{-5} d) 3.45×10^{-9} g) 8.3×10^{-3} j) 1.99×10^{-12}
   b) 3×10^{-8} e) 1.6×10^{-9} h) 4.72×10^{-5} k) 7.3×10^{-2}
   c) 1.7×10^{-4} f) 3.8×10^{-3} i) 6.73×10^{-6} l) 8.09×10^{-10}

5 Rewrite each of the following statements using the standard form.
   a) The lifetime of an omega particle is 0.00000000011 seconds.
   b) The mass of the Earth is 5 967 000 000 000 000 000 000 000 kg.
   c) The average distance of Uranus from the Sun is 2 869 000 000 km.
   d) A large orange contains 0.016 grams of vitamin C.
   e) The moon orbits at an average distance of 384 000 km from the Earth.
   f) A light-year, which is the distance travelled by light in one year is equal to 9 460 500 000 000 000 metres.

■ ■

Standard form on a calculator

Calculator displays can be too small to show the answers to some calculations.

Examples

$$4\,000\,000 \times 3\,000\,000 = 12\,000\,000\,000\,000$$

The calculator display shows   1.2 13   (Very basic calculators may show error). This is the way the calculator shows 1.2×10^{13}.

$$0.0000007 \times 0.000008 = 0.00000000000056$$

The calculator display shows   5.6 -12   (Very basic calculators may show error). This is the way the calculator shows 5.6×10^{-12}.

Scientific calculators have a standard form key. This may be **EXP** or **EE** but look in the instruction book.

These are the key sequences to enter a) 5.6×10^4,   b) 7×10^{-6}.

a) **5** **.** **6** **EXP** **4** Display   5.6 04

b) **7** **EXP** **6** **⁺∕₋**   Display   7 –06

If **=** is pressed after entering the number in standard form, the display may change to a normal number.

80

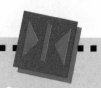

PAUSE

1 Write the numbers shown on these calculator displays as numbers in standard form and as normal numbers.

a) 4 02

b) 7 05

c) 8.4 06

d) 9 –03

e) 1.2 –04

f) 3 –10

g) 6.4 –01

h) 7.2 05

i) 1.9 –02

j) 5.6 02

k) 6.7 01

l) 1.3 –05

2 Use **EXP** to practise entering numbers in standard form. Use the numbers in questions 2 and 4 of the previous exercise.

Calculations with numbers in the standard form

Examples

1. Calculate $8 \times (3.1 \times 10^5)$

$8 \times 3.1 \times 10^5 = 24.8 \times 10^5$

8 ✕ **3** **.** **1** **EXP** **5** **=** 2480000

In standard form this is 2.48×10^6.

2. Calculate $(1.3 \times 10^7) \times (5.6 \times 10^4)$

$1.3 \times 10^7 \times 5.6 \times 10^4 = 1.3 \times 5.6 \times 10^7 \times 10^4 = 7.28 \times 10^{11}$

1 **.** **3** **EXP** **7** ✕ **5** **.** **6** **EXP** **4** **=** 7.28 11

3. Calculate $(1.3 \times 10^5) \div (5.8 \times 10^2)$

$$\frac{1.3 \times 10^5}{5.8 \times 10^2} = \frac{1.3}{5.8} \times \frac{10^5}{10^2} = 0.224 \times 10^3$$

1 **.** **3** **EXP** **5** ÷ **5** **.** **8** **EXP** **2** **=** 224

In standard form this is 2.24×10^2.

4. Calculate $(7.4 \times 10^5) \div (2.8 \times 10^{-4})$

$$\frac{7.4 \times 10^5}{2.8 \times 10^{-4}} = \frac{7.4}{2.8} \times \frac{10^5}{10^{-4}} = 2.6 \times 10^9$$

7 **.** **4** **EXP** **5** ÷ **2** **.** **8** **EXP** **4** **⁺⁄₋** **=** 2.6 09

81

1 Calculate:
   a) $5 \times (1.7 \times 10^4)$
   b) $9 \times (2.5 \times 10^7)$
   c) $15 \times (3.25 \times 10^6)$
   d) $40 \times (2.8 \times 10^{-4})$
   e) $65 \times (7.1 \times 10^{-6})$
   f) $(3.2 \times 10^7) \times (1.9 \times 10^6)$
   g) $(1.78 \times 10^6) \times (1.94 \times 10^2)$
   h) $(2.5 \times 10^5) \times (1.9 \times 10^{-2})$
   i) $(8.3 \times 10^{-3}) \times (1.0 \times 10^8)$
   j) $(1.89 \times 10^{-7}) \times (1.95 \times 10^{-3})$
   k) $(3.2 \times 10^7) \div (1.9 \times 10^6)$
   l) $(1.78 \times 10^6) \div (1.94 \times 10^2)$
   m) $(2.5 \times 10^5) \div (1.9 \times 10^{-2})$
   n) $(8.3 \times 10^{-3}) \div (1.0 \times 10^8)$
   o) $(1.89 \times 10^{-7}) \div (1.95 \times 10^{-3})$

■■■■■■■■■■■■■■■■■■■■■■■■■■■■■■■■■

1 List all the factors of:
   a) 24   b) 51   c) 200

2 List all the prime factors of:
   a) 30   b) 84   c) 105

3 Write these numbers as a chain multiplication of prime factors.
   a) 60   b) 210   c) 2808

4 Write down the first 12 square numbers.

5 Write down the first 12 rectangular numbers.

6 Write down the first 12 triangular numbers.

7 What is the value of:
   a) $\sqrt{81}$ b) $\sqrt{49}$ c) $\sqrt{121}$?

8 Write this list of numbers

   75, 134, 267, 348, 482, 591, 608, 771, 836, 945

   a) correct to the nearest 10,
   b) correct to the nearest 100.

9 Write this list of numbers

897, 1256, 2500, 3789, 4523, 5628, 6171, 7384, 8909, 9173

a) correct to the nearest 10,
b) correct to the nearest 100,
c) correct to the nearest 1000.

10 Estimate the answers to these, then work them out without a calculator.

a) 234+345
b) 689+271
c) 345−268
d) 451−99

e) 23×56
f) 327×87
g) 1458÷27
h) 678÷13

11 Write these lists of numbers in order.

a) 3.4, 6.7, 2.6, 5.4, 7.8, 9.1, 3.7, 2.6, 5.5, 4.0
b) 3.21, 3.25, 2.2, 3.3, 3.27, 3.26, 3.25, 2.28, 3.22, 3.29
c) 1.567, 1.562, 1.567, 1.569, 1.568, 1.563, 1.564, 1.5, 1.56, 1.57

12 Calculate without using a calculator:

a) 89+3.0
b) 8.9+3
c) 8.9+0.3
d) 8.9+30

e) 7.5−4.2
f) 7.9−1.4
g) 11.5−0.6
h) 20−17.5

13 Calculate:

a) 44.67+73.86
b) 523+97.9

c) 114.7+75.93
d) 33.89−13.4

e) 1376−754.56
f) 5431.45−476.56

14 Calculate:

a) 77.5+25.9+703.6
b) 300.1+296.9+456.9

c) (13.956+7834.78)−561.67
d) (19.65+45.78)−(25.65−11.97)

15 Calculate without using a calculator:

a) 3×0.7
b) 0.8×4
c) 0.6×0.6
d) 0.9×2

e) 0.5×0.2
f) 10×0.2
g) 100×0.2
h) 1000×0.5

i) 1000−34.7
j) 10−0.96
k) 1000−803
l) 100−0.012

16 Calculate:

a) 55.5×3
b) 23.4×19

c) 300.5×4.7
d) 36.7×0.34

e) 17.5−8
f) 5−0.347

17 Calculate:

a) 34.67×23 c) 90.98×1.04 e) $23.89 - 7.1$

b) 56.7×45 d) $34.5 - 3$ f) $23.78 - 3.7$

18 a) Calculate correct to 1 decimal place:

     i) $232 \div 7$ ii) 3.4×5.7 iii) $21.5 \div 6$ iv) $18.9 \div 10.4$

b) Write each answer correct to one significant figure.

19 Calculate correct to 2 decimal places:

a) 67.67×23.07 c) $0.127 \div 0.12$

b) $232 \div 7$ d) $2.34 \div 3.2$

20 Change these improper fractions into mixed numbers.

a) $\dfrac{12}{5}$ b) $\dfrac{31}{10}$ c) $\dfrac{34}{7}$ d) $\dfrac{7}{3}$ e) $\dfrac{7}{4}$

21 Change these mixed numbers into improper fractions.

a) $1\frac{3}{4}$ b) $3\frac{2}{3}$ c) $7\frac{1}{2}$ d) $12\frac{1}{11}$ e) $6\frac{5}{6}$

22 Copy and complete.

a) $\dfrac{2}{5} = \dfrac{?}{15}$ c) $\dfrac{7}{6} = \dfrac{?}{24}$ e) $\dfrac{12}{17} = \dfrac{?}{68}$ g) $\dfrac{11}{12} = \dfrac{?}{60}$

b) $\dfrac{5}{8} = \dfrac{?}{16}$ d) $\dfrac{3}{7} = \dfrac{?}{28}$ f) $\dfrac{1}{9} = \dfrac{?}{18}$ h) $\dfrac{4}{5} = \dfrac{?}{65}$

23 Simplify:

a) $\dfrac{50}{60}$ b) $\dfrac{25}{70}$ c) $\dfrac{6}{15}$ d) $\dfrac{12}{18}$ e) $\dfrac{24}{36}$ f) $\dfrac{35}{75}$ g) $\dfrac{65}{75}$

24 Which is the larger of each of the following pairs of fractions:

a) $\dfrac{2}{7}$ and $\dfrac{3}{8}$,   b) $\dfrac{3}{5}$ and $\dfrac{4}{7}$,   c) $\dfrac{3}{50}$ and $\dfrac{5}{60}$?

25 Jane scored 93 out of 120 in an Art exam and 39 out of 50 in a Biology exam. In which exam did she do best?

Write the answers to questions 26 and 27 as fractions in their lowest terms.

26 What fraction of a 24 hour day is:

a) 3 hours,   d) 9 hours,   g) 14 hours,

b) 5 hours,   e) 10 hours,   h) 9 and a half hours,

c) 6 hours,   f) 12 hours,   i) 8 and a quarter hours?

27 What fraction of £1 is:

a) 25p, b) 50p, c) 95p, d) 75p, e) 84p, f) 8p?

28 Calculate:

a) $\frac{3}{5}$ of 250   c) $\frac{3}{4}$ of 144   e) $\frac{4}{5}$ of 250   g) $\frac{3}{8}$ of 96
b) $\frac{1}{2}$ of 70   d) $\frac{3}{7}$ of 21   f) $\frac{2}{3}$ of 48   h) $\frac{7}{9}$ of 108

29 Write each of the following percentages as a fraction in its simplest form.

a) 25%   b) 50%   c) 56%   d) 94%   e) 32%

30 Write each fraction as a percentage:

a) $\frac{1}{2}$ b) $\frac{3}{4}$ c) $\frac{3}{10}$ d) $\frac{17}{20}$ e) $\frac{18}{25}$

31 Calculate:

a) 25% of 144   c) 68% of 325   e) 72% of 72
b) 60% of 230   d) 90% of 600   f) 99% of 10 000

32 Increase:

a) 300 by 20%   c) £297 by 15%   e) £18 493 by 8%
b) £80 by 15%   d) £25 000 by 5%   f) £1500 by 30%

33 Decrease:

a) 380 by 20%   c) £310 by 20%   e) £47 000 by 8%
b) £95 by 15%   d) £39 000 by 5%   f) 1100 by 12%

34 Write:

a) 16 as a percentage of 25
b) 51 as a percentage of 85
c) £12.25 as a percentage of £50
d) 36 as a percentage of 120
e) £5.25 as a percentage of £75
f) 60p as a percentage of 64p
g) £11.30 as a percentage of £25
h) 15 minutes as a percentage of 1 hour
i) 18p as a percentage of 45p
j) 21 grams as a percentage of 35 grams

35 John invests £500 in a savings bank. The bank pays interest at 7% a year which is added to John's account at the end of every year. How much are John's saving worth after three years?

36 A coat is reduced by 25% in a sale and costs £33. What did it cost before it was reduced?

37 Copy and complete this table.

Fraction	Percentage	Decimal
$\frac{1}{2}$	50%	0.5
		0.25
	75%	
$\frac{1}{5}$		
		0.1
	15%	
$\frac{3}{5}$		
		0.52
	4%	
$\frac{1}{20}$		

38 Simplify each of the following ratios:

a) 4 : 10   c) 6 : 36   e) 20 : 35   g) 44 : 84
b) 12 : 18   d) 15 : 21   f) 45 : 90   h) 0.1 : 1

39 Divide 40 kg in the ratios:

a) 1 : 3   b) 4 : 1   c) 5 : 3   d) 7 : 1   e) 7 : 3

40 Divide £120 in the ratios:

a) 3 : 1   b) 2 : 3   c) 1 : 4   d) 15 : 9   e) 5 : 2 : 1

41 A pile of text books 40 cm high contains 25 books.

a) How many books are there in a pile 120 cm high?
b) How high is a pile containing 50 books?

42 Copy and complete.

$8 + 15 =$ $8 + -15 =$ $7 + 12 =$ $7 + -12 =$
$-8 + 15 =$ $-8 + -15 =$ $-7 + 12 =$ $-7 + -12 =$
$8 - 15 =$ $8 - -15 =$ $7 - 12 =$ $7 - -12 =$
$-8 - 15 =$ $-8 - -15 =$ $-7 - 12 =$ $-7 - -12 =$

43 Copy and complete.

$2 \times 5 =$ $2 \div 5 =$ $2.5 \times 10 =$ $2.5 \div 10 =$
$-2 \times 5 =$ $-2 \div 5 =$ $-2.5 \times 10 =$ $-2.5 \div 10 =$
$2 \times -5 =$ $2 \div -5 =$ $2.5 \times -10 =$ $2.5 \div -10 =$
$-2 \times -5 =$ $-2 \div -5 =$ $-2.5 \times -10 =$ $-2.5 \div -10 =$

44 Find the value of:

a) 4^6 c) 10^5 e) 10^{-1} g) 3^{-2} i) 10^1 k) 8^{-2}

b) 2^5 d) 10^4 f) 2^{-7} h) 10^{-4} j) 12^0 l) 1^{-1}

45 Write each of the following calculations as a single power of the number.

a) $2^4 \times 2^3$ c) $2^{-9} \times 2^7$ e) $10^8 \times 10^0$ g) $10^3 \times 10^{-3}$

b) $3^7 \times 3^2$ d) $10^5 \times 10^4$ f) $5^2 \times 5^{-2}$ h) $10^8 \times 10^{-7}$

46 Write in standard form:

a) $3\,000$ b) $45\,000$ c) 346 d) 367.9 e) $23\,400\,000\,000$

47 Write as a normal number:

a) 3.1×10^6 b) 5.67×10^4 c) 1.91×10^8 d) 5.067×10^{10}

48 Write in the standard form:

a) $0.000\,03$ b) 0.017 c) $0.000\,001$ d) $0.009\,009$

49 Write as a normal number:

a) 3.1×10^{-6} b) 5.67×10^{-4} c) 1.91×10^{-8} d) 5.067×10^{-10}

50 Calculate:

a) $50 \times (1.38 \times 10^{-4})$ d) $(5.76 \times 10^{-7}) \times (8.93 \times 10^{-3})$

b) $700 \times (9.03 \times 10^{-6})$ e) $(7.04 \times 10^{-3}) \div (1.21 \times 10^8)$

c) $(3.8 \times 10^{-3}) \times (1.5 \times 10^8)$ f) $(1.5 \times 10^{-7}) \div (8.7 \times 10^{-3})$

FASTFORWARD

1 a) Express as a product of its prime factors:
     i) 126   ii) 420
b) Find the smallest number of which 126 and 420 are factors.

                                                 (SEG)

2 a) Write down all the positive whole numbers which divide exactly into 24. (These are called the factors of 24.)
b) Write down all the factors of 60.
c) Write down the highest common factor of 24 and 60.

                                                 (SEG)

3   8   9   10   11   12

  a) Which of the five numbers is a square number?

  b) Which of the five numbers is a factor of 27?

  c) Which of the five numbers is a multiple of six?

  d) Which of the five numbers is a prime number?   (ULEAC)

4   8,   12,   15,   16,   18,   19,   20,   21,   24,   32

From the ten numbers listed above select:

  a) a prime number,

  b) a perfect square,

  c) a multiple of 7,

  d) the square root of 64,

  e) a factor of 30,

  f) the next term in the sequence 3, 4, 6, 9, 13, ...   (MEG)

5

Daily sales of *The Times* rose in May to 388,196 copies, an increase of 1,938 over April.

Daily sales of *The Independent* were 386,227 in May, a drop of 3,296 from April. *The Guardian* suffered a more severe drop, by 13,636 to 415,426.

Complete the table.

Daily sales

	April	May
The Times		
The Independent		
The Guardian		

(NICCEA)

6 To build an extension to a house, Brickie Builders charge a price of £350 per square metre of floor space plus £1000.

Calculate the cost of an extension of 40 square metres of floor space.

(MEG)

7 A school is planning a disco for 936 pupils.
Each pupil will be given 1 can of drink.
Cans of drink are sold in trays of 24.

88

Work out how many trays of drinks will be needed. Do not use a calculator, show all your working. (ULEAC)

8

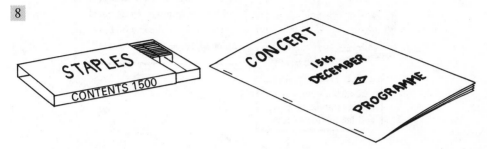

David stapled programmes for the school concert. He needed 3 staples for each programme.

He wasted 2 staples (which got bent) for every 35 programmes he stapled.

How many programmes did he staple from a box containing 1,500 staples? (NICCEA)

9 Flour costs 48p per kilogram. Brett bought 205 kg and shared it equally among 14 people. He calculated that each person should pay £0.72.

Without using a calculator, use a rough estimate to check whether this answer is about the right size. *You must show all your working.* (SEG)

10 a) Copy the diagram which shows a mineral water bottle.
Draw the approximate water level on the bottle, if it is three-quarters full.

b) Bottles of mineral water cost 39p each.
Estimate the cost of 142 bottles.
Show how you obtained your estimate.

c) Without using a calculator, work out the *exact* cost of 142 bottles of mineral water at 39p each.
(You must write down enough working to show that you did not use a calculator).

(MEG)

11

CASH PRICE £275
or Hire Purchase 20% deposit of
cash price followed by 24 equal
monthly instalments of £13
or £7.50 per month rental.

22 inch colour T.V.
Full guarantee on parts
and labour for 4 years.

a) How much deposit is required for the Hire Purchase agreement?

b) What is the total cost of buying the television on Hire Purchase?

c) How much is required for 6 months rental?

d) What is the total cost of renting the television for 4 complete years? (SEG)

12 a) Use your calculator to find the value of

$$\frac{730 \times 8.45 \times 7}{83 \times 9}$$

and write down the full calculator display.

b) Express your answer correct to one place of decimals.

(NEAB)

13 a) A girl earns £3.16 an hour as a part-time waitress. If she works for 5 hours, how much does she earn?

b) A boy works in a shop on one day from 9.30 a.m. to 12.30 p.m. and from 1.15 p.m. to 4.45 p.m. He is paid £18.98.
 i) How long does he work?
 ii) How much is he paid per hour? (MEG)

14 John, Mary and Majid charge for baby-sitting. Their charges are given in the table.

For each hour (or part of an hour) before midnight	For each hour (or part of an hour) after midnight
£1.50	£2.10

Work out the amount earned when:
a) John sat for 3 hours before midnight.
b) Mary sat from 9 o'clock at night until 1 o'clock the following morning.
c) Majid sat from 18.00 to 21.20. (ULEAC)

15 Packing cases weigh 28 kg each. 25 of the cases are loaded onto a lift. The weight limit for the lift is 750 kg. How much short of the limit is the load on the lift? (ULEAC)

16 The picture shows a 5 litre can of oil. 1 litre is about $1\frac{3}{4}$ pints.

a) Write $1\frac{3}{4}$ as a decimal.
b) Find the number of pints of oil in the 5 litre can.

(MEG)

17 $\frac{7}{8}$ of an iceberg is under water.

a) What is this as a decimal?
b) What percentage of an iceberg can be seen?

(MEG)

18 A school has 840 pupils. $\frac{3}{10}$ of them live less than 1 mile from school.
45% of them live between 1 mile and 3 miles from school. The rest live more than 3 miles from school.

a) How many pupils live less than 1 mile from school?
b) What percentage of the pupils live more than 3 miles from school?

(ULEAC)

91

19

The sweets in a box are classed as creams, caramels or toffees. In a box of Keely's Choice, $\frac{3}{8}$ of the contents are creams, $\frac{1}{4}$ of them are caramels and the remainder are toffees.

a) Work out the fraction of the sweets in the box that are either creams or caramels.
b) What fraction of the sweets in the box are toffees?
c) There are 24 sweets in the box. Work out the number of caramels in the box. (ULEAC)

20

A saline drip in a hospital releases 0.1 ml every 3 seconds. How long does it take to empty a 500ml bag?

(WJEC)

21 Margaret and David want to find out how much petrol their car uses in miles per gallon. They fill the car with petrol when the reading on the milometer is:

0	2	8	3	4	0

After a few days they stop at a garage selling petrol at £1.70 per gallon. They fill the car up again and the cost of the petrol is £13.60. By this time the milometer reading is:

0	2	8	6	2	0

How many miles per gallon does the car do over this period?

(WJEC)

 22 Taking 8 kilometres per hour to be 5 miles per hour, find
  a) the speed in kilometres per hour equivalent to the British speed limit of 30 miles per hour,
  b) the speed in miles per hour equivalent to the French speed limit of 60 kilometres per hour.

(MEG)

23

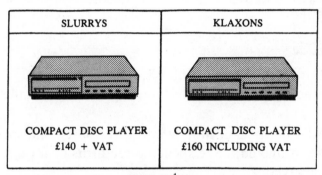

SLURRYS	KLAXONS
COMPACT DISC PLAYER	COMPACT DISC PLAYER
£140 + VAT	£160 INCLUDING VAT

VAT $17\frac{1}{2}$%

Slurrys and Klaxons are selling the same model of compact disc player. Slurry's price is £140 plus $17\frac{1}{2}$% VAT. Klaxon's price is £160 (including VAT).

a) What is Slurry's total price (including VAT)?

At sale time Klaxons cut their prices by 15%.

b) What is Klaxons' sale price?

After the sale the prices return to normal. Slurrys offer special credit arrangements on their disc players.

SLURRYS SPECIAL OFFER

NO DEPOSIT

£4 per week for 52 weeks

c) How much would you pay altogether under Slurrys' offer?

(SEG)

24 A machine always works at the same rate. It produces 150 rods in 5 minutes. How long will it take to produce 375 rods? (SEG)

25 'Alpha Cars' offers a Sierra for hire at £15 a day plus 5p per kilometre. How much would it cost Mr Jones to hire the Sierra for a day and drive 240 kilometres? (SEG)

26 Zenka wants to buy her father a packet of electric drill bits for his birthday. In the first packet that she looks at, there are four different drills and their sizes are marked in inches.

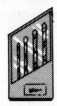 The sizes are $\frac{3}{8}$, $\frac{1}{4}$, $\frac{1}{2}$ and $\frac{1}{8}$

a) Which is the larger size drill bit, $\frac{3}{8}$ or $\frac{1}{4}$?
b) Write the four sizes in order, starting with the smallest.

(NEAB)

27 Joan divides her pocket money into three parts. She spends one third of it on makeup. She spends three-fifths of it on magazines. The rest is saved.

a) What fraction of her pocket money is saved?
b) If she saves 6p each week, how much pocket money does she receive? (SEG)

28 Peter earns £5 per week by working on Saturday mornings. He saves 60% of his wage.

a) How much does he save each week?
b) Given that he spends $\frac{1}{4}$ of the rest on a weekly magazine, how much does this magazine cost him? (SEG)

29 Bill Jones's salary was £950 per month. His boss gave him a salary increase of 8%.

a) Calculate 8% of £950.
b) What was his new salary per month? (SEG)

30 Use your calculator to complete the following calculations:

a) $\dfrac{7.2 \times 2.9}{14.4} = \dfrac{}{14.4} = \ldots\ldots$

94

b) $(8.1)^2 \div 24 = \ldots\ldots\ldots \div 24 = \ldots\ldots\ldots$ (SEG)

31 The rate of exchange is 2.95 German Marks to £1 sterling.

a) A tourist changes £25 into Marks. How many Marks would she receive?

b) She pays 42 Marks for a gift to bring home. What is the cost of the gift in pounds and pence to the nearest penny? (SEG)

32 Mr Ng's telephone bill consists of two parts: a standing charge of £15 and 5p per unit used. Calculate:

a) the total bill if 125 units are used,

b) the number of units used if the total bill is £25.00. (SEG)

33 In an election, the votes cast for the candidates of the various parties were as follows:

SMP Alliance Party	11 997
Calculator Freedom Party	5 126
Euclid Revival Party	2 567
Others	320

a) Write down to the nearest thousand:

 i) the number of votes cast for the winning candidate,
 ii) the total number of votes cast in the election.

b) i) Use your answers to write down the fraction:

$$\frac{\text{number of votes cast for the winning candidate}}{\text{total number of votes cast}}$$

ii) Use this fraction to estimate the percentage share of the votes obtained by the winning candidate. (MEG)

34 A video-recorder sells in the shop for £423. This price is obtained by increasing the net value by 17.5% to account for Value Added Tax. Calculate its net value. (MEG)

35 Mr and Mrs Williams invest £1000 in an investment account which pays 10.5% per annum interest.

a) How much interest do they get in a year?

b) They have to pay tax on this interest at the rate of 27p in the £1. How much tax do they pay? How much of the interest is left after paying tax?

c) What percentage is this 'after tax' interest of their £1000 investment? (WJEC)

36 A trade union negotiates the following rise in wages on behalf of its members:

5% of weekly wage or £6 per week, whichever is greater.

One employee finds that, for him, there is no difference between a rise of 5% and a rise of £6 per week. Calculate this employee's weekly wage before the rise. (NEAB)

37

a) Calculate the reduction in the price of the camera.
b) What is this reduction as a percentage of the normal price?

SALE
35mm Camera
was £39.95
Now £31.96

(NEAB)

38 At the beginning of 1987 a house was valued at £49 500. During 1987 the prices of houses in the area went up by 13%. What was the house worth at the end of 1987? (NEAB)

39 Sarah Jones earns £720 per month.
a) Calculate how much Sarah earns in a year.
b) Her tax allowances for the year are £3400. Calculate her taxable income for the year.
c) The rate of tax is 27% of her taxable income. Calculate the amount of income tax Sarah pays in a year. (WJEC)

40 David wants to buy a television. He can buy a Beovision (the best set there is) for £345 with 12 months free credit or a basic Flan for £295.

SUPPOSE he pays £59 deposit on the Beovision and then 11 equal monthly instalments.
a) How much is left to pay after the deposit?
b) How much is each of the 11 instalments?

IF he wants to buy the Flan, he has to pay 20% deposit.
c) What is 20% of £295?
d) How much is left to pay?

Interest is charged on the balance at $16\frac{1}{2}$%.
e) How much is left to pay including the interest?

This has to be paid in 11 instalments.
f) How much is each instalment?
g) Which television would you advise David to buy, and why?

(MEG)

41 Rachel is to buy a new stereo system, costing £240, on an interest free credit scheme.

a) She must pay a 25% deposit. How much is this?

b) How much is left to pay after she has paid the deposit?

c) How much are her monthly payments, if she pays the remainder in six equal monthly payments? (ULEAC)

42 'Electric Kettle: Special Offer 15% off marked price.' The marked price is £27.95. How much would you pay, to the nearest penny, at the 'special offer' price? (ULEAC)

43 Of the total land area of Belgium, 1.4 million hectares is cultivated and six hundred thousand hectares is forest.

a) Write both these numbers in figures.

b) The total land area of Belgium is 3 million hectares. What area of land is neither cultivated nor forest? (MEG)

44 In 1972 India had 100 251 000 houses and a population of 638 389 000. How many people were there to each house? Give your answer to the nearest whole number. (MEG)

45 Because of bad weather, my journey from Norwich to Nottingham took $5\frac{1}{2}$ hours instead of the usual $3\frac{1}{2}$ hours.

a) Calculate the percentage increase in my travel time, correct to one decimal place.

The distance from Norwich to Nottingham is about 135 miles.

b) Calculate the approximate percentage decrease in my average speed. (ULEAC)

46 Supergrowth Unit Trust claims that the value of its units is likely to grow by 21% compound interest per annum. Assuming that this claim is true, calculate the value, after 3 years, of an investment of £1000 in Supergrowth Unit Trust. (MEG)

47 Each member of Melchester Diners' Club has to pay an annual subscription of £45 and a fee of £5.50 at each meeting attended. The Club has 42 meetings each year.

a) Find the total cost of a year's membership for a member who attends all 42 meetings.

b) To retain membership, a member must attend at least 60% of the year's meetings. Find the minimum number of meetings that a member can attend in a year and still retain membership.

c) Calculate to the nearest penny, the average cost per meeting for a member who attends 32 meetings in the year. (MEG)

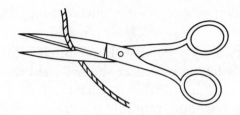

48 A piece of string is 36 cm long. John cuts it into two pieces, making one piece twice as long as the other. What are the lengths of the two pieces? (SEG)

49 Don has 120 records. The ratio of the number of Don's records to the number of Phil's record is 3:5. How many records does Phil have? (ULEAC)

50 There are 180 pupils in the first year at Bronglais Comprehensive School. The ratio of the number of boys to the number of girls in the first year is 5:4. How many girls are in the first year?

 (WJEC)

51 Jill has 40 pence and Dave has 60 pence. They agree to spend all the money on sweets and to share them in the same ratio as their money.

They buy 30 toffee-lumps. How many toffee-lumps should each have? (ULEAC)

52 A gang of workers is digging a trench. When there are six workers they manage to dig a trench 18 m long in one day. All the workers dig at the same rate.

 a) Work out the length of trench that one worker could dig in one day.
 b) A group of workers digs 12 m in one day. How many workers are there in the group? (ULEAC)

53 a) In her will Mrs Hannah Pennypincher left half her money to the SMP, a third to the SDP, a tenth to the SNP, and the remainder to her husband, Simon Alan Pennypincher (SAP). What fraction of her money would he inherit?
 b) If she changed her will to share the money between the SMP and SAP only, in the ratio 4 to 1 respectively, what fraction would her husband now inherit? (MEG)

54 If the following ingredients are mixed together, there will be enough dough to make eight pizzas.

PIZZA RECIPE
(MAKES EIGHT)

16 ounces of flour

4 ounces of butter

3 eggs

¼ pint of milk

1 ounce of yeast

a) How much flour is needed to make one pizza?

b) How much butter would be needed for six pizzas?

c) Ahmed makes twelve pizzas to store in his freezer. What is the total weight in pounds and ounces of all the ingredients, excluding milk and eggs?

(16 ounces = 1 pound.) (NEAB)

55 On Monday at 0600 the temperature was – 6°C. At 0800 it was –1°C and by 1000 it was 12 degrees higher than at 0600.

a) By how much had the temperature risen at 0800?

b) What was the temperature at 1000? (SEG)

56 a) At noon on a January day the temperature was 3°C. By 6 p.m. it had fallen by 5°C. What was the temperature at 6 p.m.?

b) At midnight the temperature was – 7°C. By how many degrees had the temperature fallen between noon and midnight?

(MEG)

57 This table gives the temperature in Sheffield during one week in January 1987.

Day	Sun	Mon	Tues	Wed	Thurs	Fri	Sat
Noon	– 3°C	– 2°C	1°C	– 3°C	2°C	3°C	– 2°C
Midnight	– 8°C	– 8°C	– 6°C	– 10°C	– 6°C	– 3°C	– 5°C

a) What is the lowest temperature in the table?

b) On which day was there the biggest drop in temperature between noon and midnight?

c) How much was this drop?

d) What was the least rise in temperature between midnight one day and noon the following day?

e) On the next Sunday the temperature was 8° higher at noon than at midnight the previous night. What was the temperature at noon? (MEG)

58 A spaceship is approximately 77 million miles from Mars. 1 mile is equivalent to 1.6 kilometres.

a) How far, to the nearest million kilometres, is the spaceship from Mars?

b) Write your answer to part a) in standard form. (NEAB)

59 The number 10^{100} is called a googol.

a) Write the number 50 googols in standard index form.

A nanometre is 10^{-9} metres.

b) Write 50 nanometres, in metres.

Give your answer in standard index form. (ULEAC)

60 The approximate population of the United Kingdom is given in standard form as 5.2×10^7.

Write this as an ordinary number. (SEG)

61 Evaluate 1.2^8. (SEG)

62 The mass of a meteorite is 3.61×10^7 kg.

Write 3.61×10^7 as an ordinary number. (ULEAC)

63 The mean distance of the Earth from the Sun is 149.6 million kilometres.

Write the number 149.6 million in standard index form.

(ULEAC)

64 a) Find the value of n for which

$2^n = 128$

b) Calculate 4^{-3} giving your answer
   i) as a vulgar or common fraction,
   ii) in standard form. (NICCEA)

Algebra

Algebra is a branch of mathematics developed by Hindu and Moslem mathematicians.

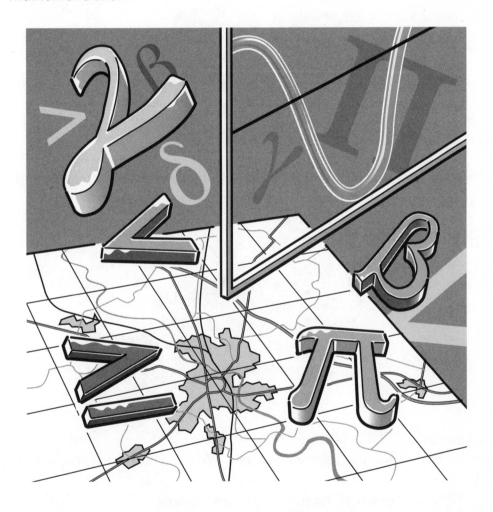

The word algebra was first used by the Arab mathematician Alkarismi in the ninth century.

Using letters to represent numbers

John is playing marbles with Vali, Ben and Hitesh. This picture shows the numbers of marbles in each person's bag.

We can write this information more simply if we let the letter *m* stand for the number of marbles in John's bag.

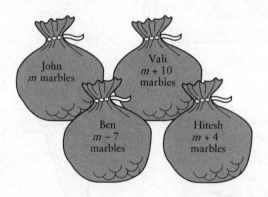

Example

In a Maths test, Susan scored *p* marks. Write each of these people's marks using *p*.

a) Wendy, who scored 6 more marks than Susan.

b) Errol, who scored 5 marks less than Susan.

a) Wendy scored $p+6$ marks.

b) Errol scored $p - 5$ marks.

1 In a Science test, Hitesh scored *y* marks. Write each of these people's marks using *y*.

    a) John, who scored 21 less marks than Hitesh.

    b) Tom, who scored 11 more marks than Hitesh.

    c) Dick, who scored 1 less marks than Hitesh.

    d) Harry, who scored 5 less marks than Hitesh.

2 Sharon has *x* plants in her garden. Write each of these people's number of plants using *x*.

    a) Wendy, who has 7 more plants than Sharon.

    b) Kimberly, who has 13 more plants than Sharon.

    c) Nadia, who has 7 less plants than Sharon.

    d) Shelly, who has 13 less plants than Sharon.

After playing a game, the number of marbles owned by each person changes:

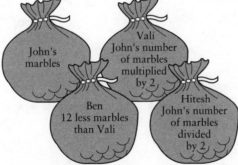

We can write this information more simply if we let the letter *r* stand for the number of marbles in John's bag.

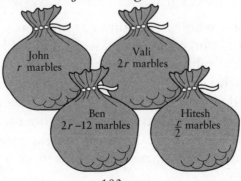

In algebra we usually do not use the multiplication sign.

$$2 \times r \text{ is written as } 2r \text{ (never as } r2).$$

In algebra divisions are usually shown by putting one number over another.

$$r \div 2 \text{ is written as } \frac{r}{2}$$

Examples

1. Sheila, Alan, Sally and Jonathan all collect stamps. Sheila has k stamps. Write each of these people's marks using k.

   a) Alan, who has 2 times as many stamps as Sheila.

   b) Sally, who has Sheila's number of stamps divided by 5.

   c) Jonathan, who has 20 less stamps than Alan.

   a) Alan has $2k$ stamps.

   b) Sally has $\frac{k}{5}$ stamps.

   c) Jonathan has $2k-20$ stamps.

2. Write as simply as possible:

a) 4 times s	c) 6 more than b	e) 20 divided by h
b) t divided by 7	d) 12 less than z	f) v less than 8

a) $4s$	c) $b+6$	e) $\frac{20}{h}$
b) $\frac{t}{7}$	d) $z-12$	f) $8-v$

■  ■ ■

PAUSE

1 Greg Allen earns n pounds each week. Write what each of these people earn each week using n.

   a) Janice Beesley who earns 3 times as much as Greg.
   b) Ruth Nicholl who earns Greg's wage divided by 2.
   c) Daphne Hicks who earns 25 pounds a week more than Janice.
   d) Jo Seddon who earns 15 pounds less a week than Ruth.
   e) Kathy Hobday who earns 30 pounds less a week than Janice.
   f) Chris Whitney who earns 12 pounds more each week than Ruth.
   g) Ian Goodall who earns 20 pounds less each week than Chris.
   h) David Sheppard who earns 45 pounds more a week than Kathy.

2 Write as simply as possible:

    a) 3 more than *e* d) *w* divided by 5 g) 12 times *d*   j) 2 divided by *t*
    b) 18 less than *v* e) 5 divided by *w* h) 13 more than *y* k) 2 less than *m*
    c) 5 times *w*   f) *v* less than 18 i) *t* divided by 2   l) *m* less than 2

Writing rules (simple formulae)

There are 12 pencils in a box.

How many pencils are there in:

a) 2 boxes   b) 5 boxes   c) *b* boxes

a) number of pencils $= 12 \times 2 = 24$ c) number of pencils $= 12 \times b = 12b$
b) number of pencils $= 12 \times 5 = 60$

In words we can write the rule:

      the number of pencils $= 12 \times$ the number of boxes.

This is called a ***formula*** for working out the number of pencils in any number of boxes.

Let *n* represent the number of pencils and *b* represent the number of boxes. We can write the formula *in symbol form* like this:

$$n = 12b$$

Examples

1. A plumber is called out for a lot of emergencies. He charges a call out fee of £20 which he adds to the cost of any repairs he does. A formula for working out the plumber's total charge is:

$$\text{total charge}(t) = \text{cost of repairs}(r) + £20$$

a) Work out the total charge if the cost of repairs is:

   i) £50   ii) £135

b) Write this formula in symbol form, using the letters in the brackets.

a) i) total charge = 50 + 20 = £70 ii) total charge = 135 + 20 = £155

b) $t = r + 20$

2. Jane works in a supermarket. To work out her weekly wage she uses the formula:

$$\text{wages}(w) = \text{number}(n) \text{ of hours worked} \times £5$$

a) Work out her weekly wage if she works:

   i) 20 hours   ii) 35 hours

b) Write this formula in symbol form, using the letters in the brackets.

a) i) wage = 20×5 = £100   ii) wage = 35×5 = £175

b) $w = 5n$

PAUSE

1 When making a pot of tea for a group of people, Mr Mitchell uses the formula:

number of tea bags$(b) = $ number of people$(p) + 1$

a) Work out the number of tea bags used for:

   i) 2 people   ii) 3 people   iii) 4 people   iv) 5 people

b) Write this formula in symbol form, using the letters in the brackets.

2 A baker works out the number of rolls needed to fill bags of two dozen rolls. He uses the formula:

number of rolls(r) needed = number of bags(b)×24

a) Work out the number of rolls needed to fill:
  i) 2 bags   ii) 5 bags   iii) 10 bags   iv) 24 bags
b) Write this formula in symbol form, using the letters in the brackets.

3 A theatre has 500 seats. The manager works out the number of tickets left to sell using this formula:

number(n) of tickets left to sell = 500 − number of tickets sold(s)

a) Work out the number of tickets left to sell if the manager has sold:
  i) 40 tickets   ii) 100 tickets   iii) 231 tickets   iv) 449 tickets
b) Write this formula in symbol form, using the letters in the brackets.

4 A lottery syndicate of 12 people agree to share any winnings using the formula:

each person's share(s) = amount won(w) ÷ 12

a) Work out each person's share if the amount won is:
  i) £144   ii) £6000   iii) £8858.40   iv) £125 783
b) Write this formula in symbol form, using the letters in the brackets.

5 A football club pays £60 to hire a mini-bus each time they play an away game. Some players travel in their own cars. The treasurer works out what each player must pay to use the bus using the formula:

cost(c) of using bus = £60 ÷ number(n) travelling on the bus

a) Work out what each player must pay if:
  i) 12 use the bus   iii) 8 use the bus
  ii) 10 use the bus   iv) 11 use the bus
b) Write this formula in symbol form, using the letters in the brackets.

6 Sam Brown works in a factory. Each week she gives her mother £40 out of her wages towards the housekeeping. She uses this formula to work out how much she has left.

amount left(l) = wages(w) − £40

a) Work out how much Sam has left in a week when she earns:
  i) £280   ii) £230   iii) £229.50   iv) £267.86
b) Write this formula in symbol form, using the letters in the brackets.

Extending the use of letters to represent numbers

Examples

1. A factory supervisor uses this formula to work out each person's wage.

   wage(w) = number of hours(h) worked × rate(r) per hour

   a) Work out the wage of somebody who:

   i) works for 35 hours at £4.50 per hour,   ii) works for 40 hours at £3.80 per hour.

   b) Write this formula in symbol form, using the letters in the brackets.

   a) i) wage = 35 × £4.50 = £157.50   ii) wage = 40 × £3.80 = £152.00

   b) $w = hr$

2. A shop stocks two brands of crisps, Walker's and Smith's. The shopkeeper makes a profit of 2p for each bag of Walker's crisps sold and a 3p profit for each bag of Smith's crisps sold. The shopkeeper uses this formula to work out the total profit on crisps.

   total profit(p) = 2 × number of bags of Walker's(w) sold + 3 × number of bags of Smiths's(s) sold

   a) Work out the total profit if the shopkeeper sells:

   i) 200 bags of Walker's and 300 bags of Smith's,

   ii) 300 bags of Walker's and 200 bags of Smith's.

   b) Write this formula in symbol form, using the letters in the brackets.

   a) i) total profit = 2 × 200 + 3 × 300 × 1300p = £13.00

   ii) total profit = 2 × 300 + 3 × 200 = 1200p = £12.00

   b) $p = 2w + 3s$

3. A student works out the volume of a cube using this formula:

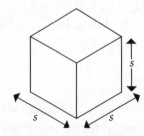

Volume(V) = side length (s) × side length (s) × side length (s)

a) Work out the volume of a cube with a side length of:

   i) 2 cm   ii) 5 cm

b) Write this formula in symbol form, using the letters in the brackets.

a) i) Volume $= 2 \times 2 \times 2 = 8$ cubic centimetres
   ii) Volume $= 5 \times 5 \times 5 = 125$ cubic centimetres

b) $V = s \times s \times s$, which is written as $V = s^3$

PAUSE

1 A student works out the perimeter of a rectangle using this formula:

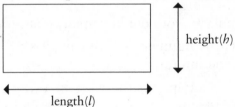

height(h)

length(l)

$$\text{Perimeter}(p) = 2 \times \text{length}(l) + 2 \times \text{height}(h)$$

a) Work out the perimeter of a rectangle with:
   i) length $= 5$ cm, height $= 2$ cm
   ii) length $= 14$ cm, height $= 4$ cm
   iii) length $= 8$ cm, height $= 2.5$ cm
   iv) length $= 22.8$ cm, height $= 8.4$ cm

b) Write this formula in symbol form, using the letters in the brackets.

2 A teacher is organising a trip to the theatre. A 53 seater coach will cost £225 and each ticket will cost £12. She uses this formula to work out the total cost of the trip.

total cost(t) $= 12 \times$ number going(n) $+ 225$

a) Work out the total cost of the trip if:
   i) 30 pupils go   ii) 40 pupils go   iii) 45 pupils go.

b) Write this formula in symbol form, using the letters in the brackets.

3 To work out the cost per pupil for a theatre trip a teacher uses the formula:

cost per person$(p) = 225 \div$ number going$(n) + 12$

a) Work out the cost per person of the trip if:

    i) 30 pupils go   ii) 40 pupils go   iii) 45 pupils to go.

b) Write this formula in symbol form, using the letters in the brackets.

4 A market gardener sells plants in trays of various sizes.

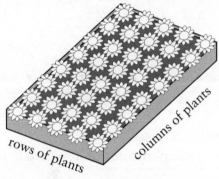

She uses this formula to work out how many plants there are in a tray.

number of plants$(p) =$ number in each row$(r) \times$ number in each column (c)

a) Work out how many plants there are in a tray with:

    i) 3 rows and 4 columns   iii) 5 rows and 8 columns

    ii) 6 rows and 6 columns   iv) 8 columns and 5 rows

b) Write this formula in symbol form, using the letters in the brackets.

5 Another market gardener always sells plants in square trays of various sizes.

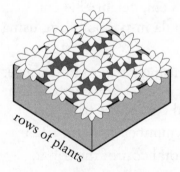

He uses this formula to work out how many plants there are in a tray.

number of plants$(p) =$ number in each row$(r) \times$ number in each row (r)

a) Work out how many plants there are in a tray with:

    i) 3 rows   ii) 5 rows   iii) 8 rows   iv) 4 rows

b) Write this formula in symbol form, using the letters in the brackets.

6 A toy manufacturer sells wooden bricks in packs of various sizes like this:

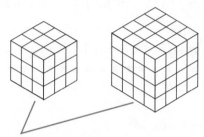

Number of bricks in a row

This is the formula used to work out how many bricks there are in each pack:

number of bricks(b) = number in a row(r) ×
number in a row(r) × number in a row(r)

a) Work out how many bricks there are in a pack with this number of bricks in each row:

    i) 3 bricks   ii) 4 bricks   iii) 5 bricks   iv) 8 bricks

b) Write this formula in symbol form, using the letters in the brackets.

■■■■■■■■■■■■■■■■■■■■■■■■■■■■■■■■■■■■

Substituting values for letters

Example

If $p = 15$, find the value of:

a) $p + 7$ b) $p - 12$ c) $20 - p$

a) $p + 7 = 15 + 7 = 22$ b) $p - 12 = 15 - 12 = 3$ c) $20 - p = 20 - 15 = 5$

■■■■■■■■■■■■■■■■■■■■■■■■■■■■■■■

PAUSE

1 If $p = 25$ find the value of:

  a) $p + 8$ b) $p - 12$ c) $p + 50$ d) $50 - p$ e) $25 - p$

2 If $x = 3$ find the value of:
   a) $x + 10$ b) $10 - x$ c) $x - 2$ d) $x + 2$ e) $x + x$

3 If $y = 19$ find the value of:
   a) $y + 11$ b) $y - 11$ c) $y + 31$ d) $31 - y$ e) $y + y + y$

4 If $n = 30$ find the value of:
   a) $n - 25$ b) $n + 25$ c) $n - 13$ d) $n + 13$ e) $47 - n$

5 If $q = 7$ find the value of:
   a) $q + 7$ b) $q + q$ c) $q + 14$ d) $q + q + q$ e) $28 - q$

■ ■

Example

If $z = 20$, find the value of:

a) $2z$ b) $2z + 5$ c) $\frac{z}{10}$ d) $\frac{z}{5} - 2$

a) $2z = 2 \times 20 = 40$ c) $\frac{z}{10} = \frac{20}{10} = 2$

b) $2z + 5 = 2 \times 20 + 5$ d) $\frac{z}{5} - 2 = \frac{20}{5} - 2$

$\qquad = 40 + 5$ $\qquad = 4 - 2$

$\qquad = 45$ $\qquad = 2$

■ ■

PAUSE

1 If $y = 4$ find the value of:
   a) $3y$ d) $\frac{y}{2} + 7$

   b) $3y - 6$ e) $\frac{5y}{10}$

   c) $\frac{y}{2}$

2 If $m = 12$ find the value of:
   a) $5m$ d) $\frac{m}{3} - 3$

   b) $2m + 1$ e) $\frac{5m}{6}$

   c) $\frac{m}{4}$

3 If $t = 20$ find the value of:
   a) $\frac{t}{5}$ d) $\frac{7t}{10}$

   b) $3t$ e) $\frac{t}{4} + 22$

   c) $50 - 2t$

4 If $p = 10$ find the value of:
   a) $5p - 1$ d) $\frac{5}{p}$

   b) $5p + 1$ e) $\frac{p}{2} + 7$

   c) $\frac{p}{5}$

5 If $c=9$ find the value of:

a) $7c+3$ d) $\frac{5c}{3}+2c$

b) $3c-10$ e) $\frac{90}{c}-c$

c) $100-9c$

- -

Examples

If $m=5$ and $n=8$, find the value of:

a) $3m$ b) mn c) m^3 d) $3m^2n$ e) $3m+2n$

a) $3m=3\times5$ c) $m^3=5\times5\times5$ e) $3m+2n=3\times5+2\times8$
$\quad\ =15$ $\quad\quad=125$ $\quad\quad\quad\quad\ =15+16$

b) $mn=5\times8$ d) $3m^2n=3\times5\times5\times8$ $=31$
$\quad\ =40$ $\quad\quad\quad=600$

- -

PAUSE

1 If $x=5$, $y=6$ and $z=3$ find the value of:

a) $3x+2y$ h) $2xy$ o) $3x^2$

b) $2x+3y$ i) $3xz$ p) y^3

c) $5x+17$ j) $4yz$ q) z^4

d) $6x$ k) $\frac{10}{x}$ r) x^2y

e) xy l) $\frac{y}{z}$ s) xy^3

f) xz m) $\frac{18}{yz}$ t) $2xz^3$

g) yz n) x^2

2 If $a=4$, $b=5$ and $c=2$, find the value of:

a) $a+b$ k) a^2+c^2

b) $b-c$ l) $17+2cb^2$

c) $a+b+c$ m) $a(b+c)$

d) $2a+3b$ n) $ab+c$

e) $5c-2b$ o) $ab+ac$

f) ab p) $2ac+b$

g) abc q) $ac+2b$

h) $2ac$ r) ac^2+b

i) ab^2 s) ac^2+b^2

j) a^3b t) $2ab^2c$

- -

Substitution which involves adding and subtracting negative numbers

Example

If $p=-5$, find the value of:

113

a) $p+7$ b) $p-12$ c) $20-p$

a) $p+7=-5+7=2$ b) $p-12=-5-12=-17$ c) $20-p=20--5=25$

PAUSE

1 If $p=-25$ find the value of:
  a) $p+8$ d) $50-p$
  b) $p-12$ e) $25-p$
  c) $p+50$

2 If $x=-3$ find the value of:
  a) $x+10$ d) $x+2$
  b) $10-x$ e) $x+x$
  c) $x-2$

3 If $y=-19$ find the value of:
  a) $y+11$ d) $31-y$
  b) $y-11$ e) $y+y+y$
  c) $y+31$

4 If $t=-40$ find the value of:
  a) $t+40$ d) $t-t$
  b) $t-40$ e) $t+t+40$
  c) $t+t$

5 If $y=-32$ find the value of:
  a) $y+y+y$ d) $40+y+y$
  b) $y+64$ e) $100-y$
  c) $64-y$

Substitution which involves multiplying and dividing negative numbers

Example
If $z=-20$, find the value of:
a) $2z$

b) $2z+5$

c) $\frac{z}{10}$

d) $\frac{z}{5}-2$

a) $2z = 2\times-20 = -40$

b) $2z+5 = 2\times-20 + 5$
     $= -40+5$
     $= -35$

c) $\frac{z}{10} = \frac{-20}{10} = -2$

d) $\frac{z}{5}-2 = \frac{-20}{5}-2$
     $= -4-2$
     $= -6$

1 If $y=-4$ find the value of:
   a) $3y$ d) $\frac{y}{2}+7$
   b) $3y-6$ e) $\frac{5y}{10}$
   c) $\frac{y}{2}$

2 If $m=-12$ find the value of:
   a) $5m$ d) $\frac{m}{3}-3$
   b) $2m+1$ e) $\frac{5m}{6}$
   c) $\frac{m}{4}$

3 If $x=-8$ find the value of:
   a) $3x+4$ d) $\frac{x}{4}$
   b) $4x+3$ e) $\frac{5x}{20}$
   c) $\frac{40}{x}$

4 If $a=-25$ find the value of:
   a) $4a-100$ d) $\frac{2a}{10}$
   b) $100-2a$ e) $5a-\frac{a}{5}$
   c) $\frac{a}{5}-3$

5 If $b=-36$ find the value of:
   a) $\frac{b}{6}$ d) $\frac{10b}{90}$
   b) $\frac{b}{2}$ e) $2b-70$
   a) $\frac{b}{2}-\frac{b}{6}$

Example

If $m=-5$ and $n=8$, find the values of:

a) $3m$

b) $3m+2n$

c) mn

d) m^3

e) $3mn^2$

f) $3m^2n$

a) $3m=3\times-5=-15$

b) $3m+2n=3\times-5+2\times8=-15+16=1$

c) $mn=-5\times8=-40$

d) $m^3=-5\times-5\times-5=-125$

e) $3mn^2=3\times-5\times8\times8=-960$

f) $3m^2n=3\times-5\times-5\times8=600$

1 If $x=5$, $y=-6$ and $z=-3$ find the value of:

a) $3x+2y$

b) $2x+3y$

c) $5x+17$

d) $6x$

e) xy

f) xz

g) yz

h) $2xy$

i) $3xz$

j) $4yz$

k) $\dfrac{10}{x}$

l) $\dfrac{y}{z}$

m) $\dfrac{18}{yz}$

n) x^2

o) $3x^2$

p) y^3

q) z^4

r) x^2y

s) xy^3

t) $2xz^3$

2 If $a=-4$, $b=5$ and $c=-2$, find the value of:

a) $a+b$

b) $b-c$

c) $a+b+c$

d) $2a+3b$

e) $5c-2b$

f) ab

g) abc

h) $2ac$

i) ab^2

j) a^3b

k) a^2+c^2

l) $17+2cb^2$

m) $a(b+c)$

n) $ab+c$

o) $ab+ac$

p) $2ac+b$

q) $ac+2b$

r) ac^2+b

s) ac^2+b^2

t) $2ab^2c$

■ ■

Substitution in formulae

Formulae can be used to solve problems.

Example

A formula used to calculate the correct dose of medicine for a child older than 12 months is:

$$C=\frac{An}{n+12},$$

where, A is the adult dose, n is the child's age in years and C is the child's dose.

What is the correct dose of a medicine for a child aged 8 years if the adult's dose is 50 ml?

Substituting these values in the formula, we have:

$$C=\frac{50\times8}{8+12}$$

$$=\frac{400}{20}$$

$$=20 \text{ ml}$$

116

1 A formula to calculate an infant's (less than twelve months old) dose of a medicine is:

$$I = \frac{Am}{150},$$

where I is the infant's dose, A is the adult's dose and m is the infant's age in months.

Copy and complete this table to find an infant's dose when the adult dose is 75 ml:

m	Am	I
1	75	0.5
2	150	1
3		

2 When an object is dropped from a height, the distance it has fallen and its velocity are given by the formulae:

$$s = 5t^2 \text{ and } v = 10t,$$

where s is the distance dropped (in metres), v is the velocity (in metres per second) and t is the time (in seconds) that the object has been falling.

a) The depth of a well can be found by dropping a stone and timing how long it takes to hit the water.

Copy and complete this table.

Time taken for stone to drop(t seconds)	Depth of well ($5t^2$ metres)
1	5 metres
2	
3	
4	
5	

b) Copy and complete this table, showing the velocity with which the dropped stone will hit the water in the well.

Time taken for stone to drop(t seconds)	Velocity of stone ($10t$ metres per second)
1	10 metres per second
2	
3	
4	
5	

   c)   An apple falls from a tree and hits a man on the head travelling at a speed of 10 metres per second. How far has the apple fallen?

   d)   If a flowerpot knocked from a window ledge has fallen 20 metres, how fast is it travelling?

3 If a car is travelling at v miles per hour, its stopping distance d metres, is given by the formula:

$$d = \frac{v^2 + 20v}{60}$$

  a) Copy and complete this table of values showing the stopping distances for speeds from 10 mph to 70 mph in steps of 10 mph.

Speed(v)	v^2	$20v$	$v^2 + 20v$	d
10	100	200	300	5
20	400	400	800	13.3

  b) Does the car travel twice as far when stopping from 60 mph as it does when stopping from 30 mph?

■ ■

Simplification

A piece of mathematical shorthand like $3x + 4y$, or $7m^2 + 9m$, is called an *expression*. The separate parts of an expression, like, $3x$ or $4y$, are called the *terms* of the expression.

So, $5y + 7x + 5$ is an *expression* with three *terms*.

Example

The perimeter of a shape is the distance round the outside of the shape. Write down an expression for the perimeter of this shape.

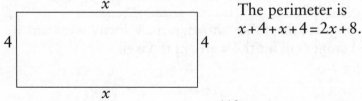

The perimeter is
$x + 4 + x + 4 = 2x + 8$.

PAUSE

Write down an expression for the perimeter of each shape.

1

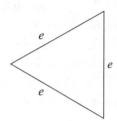

2

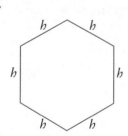

3

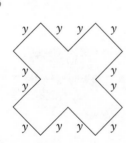

4

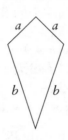

5

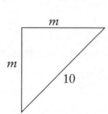

6

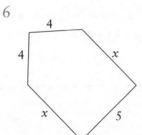

7

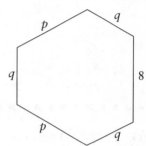

8

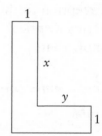

9

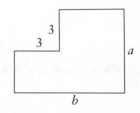

Some expressions can be simplified. For example $7x + 3x$ can be more simply written as $10x$. In the same way:

$$3p + 2p = 5p$$
$$12a - 3a = 9a$$
$$5p + 7q - 2p + 8q = 5p - 2p + 7q + 8q$$
$$= 3p + 15q$$

Simplify each expression.

1 $4m+3m$	7 $8a-4a+2a$	13 $3x+2y+2x+3y$
2 $5t-t$	8 $2w-5w+8w$	14 $2e-5f+7e+6f$
3 $e+8e$	9 $-5x+7x$	15 $6m+n+4m+2n$
4 $8x-6x$	10 $-3m+9m$	16 $3s+2r+s+5r$
5 $2z+6z+3z$	11 $-4d+2d$	17 $5r+2q-3r+2q$
6 $7r+2r-3r$	12 $-3k-2k$	18 $7x+5y+2x-3y$

■ ■

As a general rule, two terms can only be combined if the letter parts are *identical*.

The exception to this rule is when letters can be rearranged to make two terms identical.

We know that:

$$mn = m \times n = n \times m = nm$$

So, $2mn + nm = 2mn + mn = 3mn$.

Examples

$$x^2+x^2=2x^2$$
$$17p^3-12p^3=5p^3$$
$$a^2+3a-5a+7=a^2-2a+7$$
$$x^2+xy+xy+y^2=x^2+2xy+y^2$$

Simplify each of the following expressions.

1 $11x+12y+7x+5y$	7 $m^2+3m+5m+15$
2 $11x+12y-7x+5y$	8 $b^2+5b-2b-10$
3 $11x+12y+7x-5y$	9 $2a^2-12a-2a+12$
4 $11x-12y-7x-5y$	10 $3v^3+3v^2+2v^2+2v$
5 $5x-3xy+6y+2x+2yx$	11 $12ab-8ab+15ba-10ab$
6 $7p+5q+6pq-5p-2q-3qp$	12 $5st-5st+10ts-7st+8ts$

■ ■

Multiplying terms

Multiplying terms together is done by writing out in full exactly what each term means.

Examples

$$3 \times 2e = 3 \times 2 \times e = 6 \times e = 6e$$
$$6 \times 3y^2 = 6 \times 3 \times y \times y = 18y^2$$
$$-5 \times 4y^2 = -5 \times 4 \times y \times y = -20y^2$$

PAUSE

Multiply:

1 $3 \times 4v$	6 $2 \times 9z^2$	11 $-3 \times 2v$	16 $-2 \times -2z^2$
2 $7 \times 2m$	7 $12 \times z^3$	12 $-7 \times 6m$	17 $20x - z^3$
3 $8 \times 5t$	8 $3 \times 5m^4$	13 $6x - 2t$	18 $-3 \times 4m^4$
4 $7u \times 5$	9 $4r^2 \times 6$	14 $3ux - 5$	19 $-2r^2 \times -5$
5 $4c \times 4$	10 $7n^3 \times 3$	15 $-4cx - 2$	20 $-4n^3 \times 9$

Expanding brackets

Brackets are expanded by multiplying all the terms inside the bracket by the term outside the bracket.

Examples

$$4(2m+6) = 4 \times 2m + 4 \times 6 = 8m + 24$$
$$3(5 - 2x^2) = 3 \times 5 - 3 \times 2x^2 = 15 - 6x^2$$
$$-2(3e - 5f) = -2 \times 3e - -2 \times 5f = -6e + 10f$$

PAUSE

Expand each bracket.

1 $3(2w+7)$	3 $6(3+2v)$	5 $6(2x-1)$	7 $4(x^2+1)$	9 $3(5+2m^2)$
2 $5(4m+2)$	4 $5(3r-4)$	6 $4(4-2y)$	8 $6(w^3+5)$	10 $5(3e+2f)$

11 $7(3u-2v)$ 13 $-3(w+7)$ 15 $-6(2+3v)$ 17 $-4(3r-4)$ 19 $-7(2-m)$
12 $3(2s-3t)$ 14 $-5(2m+4)$ 16 $-8(3+2s)$ 18 $-2(2x-5)$ 20 $-4(x^2+3)$

Simplifying expressions with brackets

Examples

$$4(x+1)+5(2x+3) = 4x+4+10x+15 = 14x+19$$
$$5(a+b)-2(a-b) = 5a+5b-2a+2b = 3a+7b$$
$$y(2y-1)+4(2y-1) = 2y^2-y+8y-4 = 2y^2+7y-4$$
$$x(x+y)-y(x+y) = x^2+xy-yx-y^2 = x^2-y^2$$

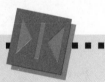

PAUSE

Expand the brackets and then simplify.

1 $3(4r+7)+2(2r-6)$

2 $2(x+1)+5\ (x+3)$

3 $6(3+2r)+3(1-r)$

4 $7(y-1)+2(3y+5)$

5 $8(2x+3)+6(3-x)$

6 $4(x+5)+5(x+5)$

7 $x(x+3)+2(x+3)$

8 $5x(x+1)-2(x+1)$

9 $2x(x-1)+3(x-1)$

10 $5(6g-9)-4(g+7)$

11 $-6(6r+6)-6(6-6r)$

12 $-6(6r-6)-6(6+6r)$

13 $(4m-3n)-(3n-4m)$

14 $-(4m-3n)+4m+11n$

15 $2s(s-1)+5(s-1)$

16 $2(s-1)-s(s-1)$

17 $2(1-s)-s(1-s)$

18 $5n(3n-2n^2+6)+5(n^2-1)$

19 $5n(2n+1)-3(2n+1)$

20 $x(2x+y)-y(2x+y)$

Multiplying a bracket by a bracket

The general rule for multiplying brackets is to multiply each term in the second bracket by each term in the first bracket. In symbols this rule is expressed like this:

$$(a+b)(c+d) = a(c+d)+b(c+d) = ac+ad+bc+bd$$

Examples

1. $(2x+1)(x+3)$ $=2x(x+3)+1(x+3)$
   $=2x^2+6x+x+3$
   $=2x^2+7x+3$

2. $(2x-4)(x+7)$ $=2x(x+7)-4(x+7)$
   $=2x^2+14x-4x-28$
   $=2x^2+10x-28$

PAUSE

Expand:

1 $(x+3)(x+5)$	6 $(2w+3)^2$	11 $(2x-1)(x+1)$	16 $(x+3)(3-x)$
2 $(x+5)(x+7)$	7 $(3x+7)(x-1)$	12 $(3x-2)(x+5)$	17 $(2x+3)(3-2x)$
3 $(2x+3)(x+2)$	8 $(m+5)(2m-5)$	13 $(2e-4)(e+7)$	18 $(w+9)(5-w)$
4 $(2e+4)(e+5)$	9 $(6n+1)(n-3)$	14 $(x-5)(x+7)$	19 $(5-w)^2$
5 $(a+3)(2a+5)$	10 $(5q+3)(2q-3)$	15 $(2x-3)(x+2)$	20 $(2-5x)(2x+1)$

Finding the factors of an expression

To *factorise* an expression we must spot a common factor and then extract this factor outside a bracket.
For example, when we look at the expression:

$$3x+12,$$

we see that each term has 3 as a factor.
Therefore, we can extract 3 outside a bracket and write:

$$3x+12 = 3(x+4)$$

Examples

$$2x+8 = 2(x+4)$$
$$14e-21 = 7(2e-3)$$
$$x-x^2 = x(1-x)$$
$$ab+b = b(a+1)$$

Factorise

1 $5x+10$	6 $5x-15$	11 $10m+15$	16 x^2+x
2 $3y+9$	7 $3g-21$	12 $8y+10$	17 $2x^2+3x$
3 $2t+14$	8 $14-7x$	13 $30+25e^2$	18 $w-2w^2$
4 $7t+14$	9 $16-2z$	14 $10u-35$	19 $xy+y$
5 $6x+15$	10 $11w-22$	15 $24-15y^2$	20 $3yz-z$

If we have extracted the **highest common factor** of the terms in an expression, we say we have *factorised the expression completely*.

Examples

Each of the following expressions has been factorised completely.

$$12w-18 = 6(2w-3)$$
$$21x-7x^2 = 7x(3-x)$$
$$8y^3-6y^2+10y = 2y(4y^2-3y+5)$$
$$ab^2+a^2b = ab(b+a)$$

P A U S E

Factorise completely.

1 $20x+10$	6 $15x^2-30$	11 $2x^2+10x$	16 $8ef-20f$
2 $12y+18$	7 $16-8z$	12 $3y^2-15y$	17 $6yz-15z$
3 $12t+24$	8 $44w^2-22$	13 $8w-12w^2$	18 $12yz+18y$
4 $42t+14$	9 $3x^2+9x$	14 $5xy+15x$	19 a^2b+ab
5 $36x+12$	10 $4y^2-6y$	15 $6xy+12y$	20 a^2b+a^2

Finding the factors of a quadratic expression

An expression like $x^2+7x+10$ which contains an x^2 term is called a **quadratic expression**.

It may be possible to factorise the expression into a bracket multiplied by another bracket.

x^2 can only be the result of multiplying x by x. This tells us that both brackets must contain a single x.

$$x^2+7x+10=(x\quad)(x\quad)$$

We need to fill in a number term in each bracket. These number terms must meet two conditions:

1. They must multiply to produce 10.
2. They must produce two x terms which combine to a total of $7x$.

To complete the brackets we use trial and error.
We could, for example, try 10 and 1. They meet the first condition, but when we try them in the brackets we get:

$$(x+1)(x+10)=x^2+10x+x+10$$
$$=x^2+11x+10$$

so the second condition has not been met.

If, on the other hand, we try 5 and 2, they meet both conditions.

$$(x+5)(x+2)=x^2+5x+2x+10$$
$$=x^2+7x+10$$

So, the quadratic expression can be factorised like this:

$$x^2+7x+10=(x+5)(x+2)$$

Examples
1. Factorise $x^2+10x+24$.

$$x^2+10x+24=(x\quad)(x\quad)$$

Several pairs of numbers multiply to produce 24. The only pair that also combine to produce $10x$ is 4 and 6. So,

$$x^2+10x+24=(x+4)\,(x+6)$$

2. Factorise $2x^2+x-6$.

$$2x^2+x-6=(2x\quad)\,(x\quad)$$

Four pairs of whole numbers multiply to produce −6.

$$-2\times3=-6,\quad 2\times-3=6,\quad -1\times6=-6,\quad 1\times-6=-6$$

That makes this problem more difficult, both because we are dealing with a negative and because there is a $2x$ term in one bracket. There are no tricks here, we must just slog away with trial and error until we come up with the correct numbers. The solution is:

$$2x^2+x-6=(2x-3)\,(x+2)$$

To help you get the feel of this kind of factorisation, you are given one factor of the expression in each part of question 1.

1 Copy and complete.
   a) $x^2+8x+15=(x+3)($ $)$ f) $x^2-7x+10=(x-2)($ $)$
   b) $x^2+7x+12=(x+3)($ $)$ g) $2x^2+7x+3=(2x+1)($ $)$
   c) $x^2+x-12=(x-3)($ $)$ h) $2x^2+19x+9=(x+9)($ $)$
   d) $x^2+3x-10=(x+5)($ $)$ i) $2x^2+5x-12=(x+4)($ $)$
   e) $x^2-10x+21=(x-3)($ $)$ j) $3x^2-26x+35=(x-7)($ $)$

Factorise:

2 a) $x^2+9x+18$ 3 a) x^2-5x+6 4 a) x^2-x-6
   b) x^2+6x+5 b) x^2-6x+5 b) $x^2+3x-10$
   c) x^2+6x+8 c) x^2-6x+8 c) x^2+x-6
   d) $x^2+8x+15$ d) $x^2-8x+15$ d) $x^2-3x-10$
   e) x^2+8x+7 e) x^2-7x+6 e) x^2-3x-4
   f) $x^2+11x+18$ f) $x^2-10x+21$ f) x^2+3x-4
   g) $x^2+19x+18$ g) $x^2-9x+20$ g) x^2-x+30
   h) $x^2+10x+24$ h) $x^2-11x+28$ h) $x^2+4x-21$
   i) $2x^2+7x+3$ i) $2x^2-3x+1$ i) $2x^2+5x-3$
   j) $2x^2+5x+3$ j) $2x^2-11x+12$ j) $2x^2+2x-12$

SECTION 2 Solving Equations

Equations

Examples

1. Solve the equation:

$$y+12=34$$

   Subtract 12 from both sides:

$$y+12-12=34-12$$
$$y=22$$

2. Solve the equation:
$$x+5=2$$

Subtract 5 from both sides:
$$x+5-5=2-5$$
$$x=-3$$

PAUSE

Solve:

1 $d+10=11$	6 $13+m=20$	11 $d+11=5$	16 $13+m=12$
2 $e+4=8$	7 $2+g=2$	12 $e+8=2$	17 $2+g=0$
3 $x+7=9$	8 $s+25=27$	13 $x+9=2$	18 $s+25=20$
4 $x+6=13$	9 $s+2=27$	14 $x+13=3$	19 $s+2=-4$
5 $y+7=14$	10 $x+5=15$	15 $y+3=1$	20 $s+5=-2$

The left hand side and right hand side of an equation are equal. The left hand and right hand sides will therefore still be equal if we choose to *add any number to both sides of the equation.*

Examples
1. Solve the equation:
$$x-22=5$$

Add 22 to both sides:
$$x-22+22=5+22$$
$$x=27$$

2. Solve the equation:
$$x-5=-2$$

Add 5 to both sides:
$$x-5+5=-2+5$$
$$x=3$$

Solve:

1 $d-10=11$	6 $m-13=20$	11 $d-11=5$	16 $m-13=-12$
2 $e-4=8$	7 $g-2=0$	12 $e-8=2$	17 $g-2=-2$
3 $x-7=9$	8 $s-25=27$	13 $x-9=2$	18 $s-25=-20$
4 $x-6=13$	9 $s-2=27$	14 $x-13=3$	19 $s-2=-4$
5 $y-7=14$	10 $x-5=15$	15 $y-3=1$	20 $s-5=-2$

Examples

1. Solve the equation:

$$2w=10$$

Divide both sides by 2:

$$\frac{2w}{2}=\frac{10}{2}$$

$$w=5$$

2. Solve the equation:

$$-3e=15$$

Divide both sides by −3:

$$\frac{-3e}{-3}=\frac{15}{-3}$$

$$e=-5$$

Solve:

1 $2x=12$	3 $5t=25$	5 $8e=16$	7 $6m=24$
2 $4m=16$	4 $7u=49$	6 $5x=10$	8 $9x=90$

9 $5r=40$	12 $-3c=12$	15 $-5z=55$	18 $-5t=-50$
10 $11z=22$	13 $-4m=40$	16 $-2x=-4$	19 $-4x=56$
11 $-2s=22$	14 $-4m=-40$	17 $-8u=40$	20 $-2x=-100$

■■■■■■■■■■■■■■■■■■■■■■■■■■■■■■■■■■■

Examples

1. Solve the equation:

$$\frac{x}{2}=5$$

Multiply both sides by 2:

$$2\times\frac{x}{2}=2\times5$$

$$x=10$$

2. Solve the equation:

$$\frac{x}{4}=-7$$

Multiply both sides by 4:

$$4\times\frac{x}{4}=4\times-7$$

$$x=-28$$

■ ■

PAUSE

Solve:

1 $\frac{x}{5}=2$	6 $\frac{f}{5}=5$	11 $\frac{y}{3}=-4$	16 $\frac{y}{-9}=2$
2 $\frac{z}{3}=4$	7 $\frac{y}{3}=21$	12 $\frac{j}{6}=-2$	17 $\frac{y}{-9}=-2$
3 $\frac{m}{10}=10$	8 $\frac{x}{7}=7$	13 $\frac{y}{2}=-9$	18 $\frac{x}{5}=-10$
4 $\frac{m}{10}=1$	9 $\frac{z}{6}=5$	14 $\frac{y}{9}=-2$	19 $\frac{x}{-5}=-10$
5 $\frac{w}{6}=2$	10 $\frac{z}{5}=6$	15 $\frac{y}{-2}=9$	20 $\frac{m}{-3}=50$

■■■■■■■■■■■■■■■■■■■■■■■■■■■■■■■■■■■

Remember, we can:

- add any number to both sides of an equation,
- subtract any number from both sides of an equation,
- multiply both sides of an equation by any number,
- divide both sides of an equation by any number.

Example

Solve the equation $3e-7=11$

$$3e-7=11$$

add 7 to both sides:

$$3e-7+7=11+7$$
$$3e=18$$

divide both sides by 3:

$$\frac{3e}{3}=\frac{18}{3}$$
$$e=6$$

PAUSE

1 $2x+3=7$ 6 $3x-4=2$ 11 $\dfrac{z}{2}+1=4$ 16 $\dfrac{t}{3}-1=4$

2 $3x+1=16$ 7 $2x-1=9$ 12 $\dfrac{x}{4}+3=5$ 17 $\dfrac{y}{2}-3=-1$

3 $2x+6=20$ 8 $3x-10=2$ 13 $\dfrac{x}{3}+4=7$ 18 $\dfrac{y}{3}+2=-1$

4 $5x+8=18$ 9 $2x+5=1$ 14 $\dfrac{x}{10}+1=2$ 19 $\dfrac{r}{2}-6=0$

5 $5m+3=8$ 10 $4x+14=2$ 15 $\dfrac{y}{4}+5=7$ 20 $\dfrac{x}{2}-7=3$

21 Write down the equation which represents each picture and find the value of the letter.

a) The perimeter
is 18cm.

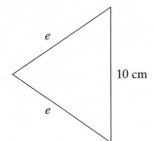

c) The perimeter
is 84cm.

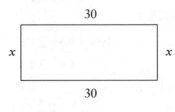

e) The perimeter
is 32cm.

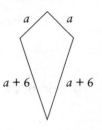

b) The perimeter is
is 64cm.

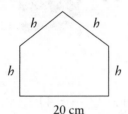

d) The perimeter is
is 90cm.

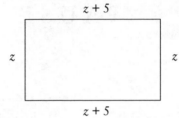

f) The perimeter
is 32cm.

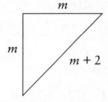

■■■■■■■■■■■■■■■■■■■■■■■■■■■■■■■■■■■■

Equations with brackets

Examples

1. Solve $3(x+5)=12$

    expand the bracket:

    $$3x+15=12$$

    subtract 15 from both sides:

    $$3x+15-15=12-15$$
    $$3x=-3$$

    divide both sides by 3:

    $$\frac{3x}{3}=-\frac{3}{3}$$
    $$x=-1$$

2. Solve $3(2y+1)+2(y-3)=21$

    expand the brackets:

    $$6y+3+2y-6=21$$

simplify:

$$8y-3=21$$

add 3 to both sides:

$$8y-3+3=21+3$$
$$8y=24$$

divide both sides by 8:

$$\frac{8y}{8}=\frac{24}{8}$$
$$y=3$$

PAUSE

Solve:

1 $3(2w+7)=45$
2 $5(4m+2)=50$
3 $6(3+2v)=54$
4 $9(3+s)=36$
5 $5(3r-4)=10$
6 $6(2x-1)=54$
7 $4(4-2y)=0$
8 $6(2-m)=-6$

9 $3(4r+7)+2(2r-6)=41$
10 $2(x+1)+5(x+3)=66$
11 $6(3+2r)+3(1-r)=30$
12 $7(y-1)+2(3y+5)=29$
13 $8(2x+3)+6(3-x)=62$
14 $4(x+5)+2(x+5)=54$
15 $5(2n+1)-3(2n+1)=-6$
16 $3(2x+1)-2(2x-1)=-1$

Equations with letter terms on both sides

Some equations have letter terms on both side of the equals sign. In this case, we must first simplify to a single letter term.

Examples

1. $4t-5=3t+12$

subtract $3t$ from both sides:

$$4t-5-3t=3t+12-3t$$
$$t-5=12$$

add 5 to both sides:

$$t-5+5=12+5$$
$$t=17$$

132

2. $2x-4=3(x-3)-8$

expand the bracket:

$$2x-4=3x-9-8$$

simplify:

$$2x-4=3x-17$$

subtract $2x$ from both sides:

$$2x-4-2x=3x-17-2x$$
$$-4=x-17$$

add 17 to both sides:

$$-4+17=x-17+17$$
$$13=x$$
$$\text{or } x=13$$

PAUSE

Solve:

1 $8p+4=7p-3$

2 $x-15=9-2x$

3 $4-p=15-2p$

4 $3e=18-6e$

5 $5w-17=7-3w$

6 $3m+3-m=3m-8$

7 $\dfrac{x}{2}+3=x+5$

8 $7+\dfrac{a}{3}=a-6$

9 $3(x-5)+2=x-1$

10 $12-x=5(x+3)-3(x+5)$

11 $3(2p-1)=5(p+1)$

12 $3(x+7)-1=7x$

13 $5(2s-1)+4=6s+1$

14 $4+5(2x+3)=8(3-x)$

Equations with decimal numbers

Most of the equations we have dealt with so far have been designed to have whole number solutions.
Most examination problems have been of this kind.
But with some examination questions, you may need to use a calculator.

Example

Solve correct to 2 decimal places:

$$2.4x - 3.7 = 15.4$$

add 3.7 to both sides:

$$2.4x - 3.7 + 3.7 = 15.4 + 3.7$$
$$2.4x = 19.1$$

divide both sides by 2.4:

$$\frac{2.4x}{2.4} = \frac{19.1}{2.4}$$

$$x = 7.96 \text{ (correct to 2 decimal places)}$$

PAUSE

Solve, giving answers correct to 2 decimal places.

1 $2.5x + 3 = 7.2$
2 $3.8x + 1.4 = 1.6$
3 $2.5x + 6.4 = 20$
4 $5.2x + 8.1 = 9.8$
5 $5.4m + 3.2 = 8.6$

6 $3.1x - 4.6 = 12.2$

7 $2.2x - 1.7 = 9.5$
8 $3.4x - 10 = 2.6$
9 $2.4x + 5.2 = 1.7$
10 $4.5x + 1.4 = 2.5$
11 $\frac{z}{2} + 1.3 = 4.7$

12 $\frac{x}{4.2} + 3.6 = 5.4$

Inequalities

There are four inequality signs:

less than	$<$
greater than	$>$
less than or equal to	$\leq$
greater than or equal to	$\geq$

Are these statements true or false?

1 $4 \times 5 > 19$	5 $25 \leq 4 \times 5$	9 $17 - 4 \leq 15$	13 $8 \times 6 \geq 40$
2 $7 - 3 \leq 4$	6 $32 \geq 4 \times 8$	10 $20 \geq 8 \times 3$	14 $7 \times 5 > 8 \times 4$
3 $60 \div 5 \geq 14$	7 $2 + 9 < 11$	11 $8 \times 3 < 25$	15 $9 \times 5 < 10 \times 4$
4 $7 + 5 < 13$	8 $17 - 4 \leq 12$	12 $80 \div 10 \leq 9$	16 $3 \times 8 \geq 6 \times 4$

Example

1. The number x makes the inequality $x < 5$ true. Show the inequality on a number line.

   An inequality like $x < 5$ means x can be any number less than 5. A number line drawn to illustrate $x < 5$ would look like this:
   (the hollow 'blob' indicates that the last number is not included in the solution)

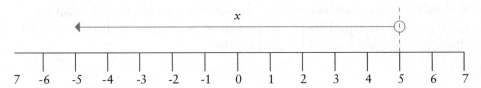

2. The number x makes the inequality $x \geq -2$ true. Show the inequality on a number line.

   A number line drawn to illustrate $x \geq -2$ would look like this:

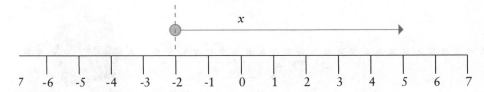

(The solid 'blob' indicates that the last number *is* included in the solution.)

In every question draw number lines to show the inequalities.

1 $x>4$ 6 $x<-1$
2 $x\leq3$ 7 $x\geq-4$
3 $x>0$ 8 $x<-3$
4 $x<0$ 9 $x>2$
5 $x\geq-1$ 10 $x\leq-4$

The negative and positive whole numbers, together with zero are called *the integers*.

Example

a) Show the inequality $-2\leq x<5$ on a number line.

b) Make a list of the integer values of x which make this inequality true.

a) A double inequality like $-2\leq x<5$ means x can be any number greater than or equal to -2 but less than 5. A number line drawn to illustrate $-2\leq x<5$ would look like this:

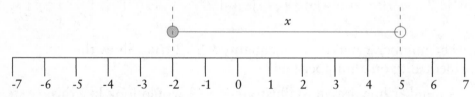

b) Integer values for x are $-2, -1, 0, 1, 2, 3, 4$

In every question:
a) draw a number line to show the inequality,
b) if x is an integer, list the values of x.

1 $2<x\leq5$	6 $-4\leq x\leq-1$
2 $-1<x<3$	7 $-1\geq x>-3$
3 $0\leq x<7$	8 $-5<x<3$
4 $-3<x\leq0$	9 $5\geq x\geq2$
5 $-1<x\leq1$	10 $-1\leq x<4$

■ ■

We can apply the rules of algebra to simplify an inequality.

Example

Write down and then solve the inequality shown in this picture.

The perimeter is > 23 cm.

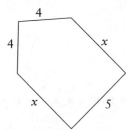

The inequality is:

$$2x+13>23$$

subtract 13 from both sides:

$$2x+13-13>23-13$$
$$2x>10$$

divide both sides by 2:

$$\frac{2x}{2}>\frac{10}{2}$$
$$x>5$$

■ ■

PAUSE

In questions 1 to 12, solve each inequality and illustrate your solution with a number line.

1 $x+4<7$	3 $3e>-21$	5 $5x+2>32$	7 $2x-1<9$
2 $y-7\geq-10$	4 $\frac{y}{4}\geq1.5$	6 $3x-4\leq2$	8 $3x-10\geq2$

9 $2x+5<1$ 12 $\frac{x}{4}+3>4$ 15 $6(3+2v)\geq18$ 18 $15-p\leq10-2p$

10 $4x+14>2$ 13 $3(2w+7)\geq57$ 16 $8p+2<7p-3$ 19 $3e>27-6e$

11 $\frac{z}{2}+1\leq4$ 14 $3(4m+2)<54$ 17 $x+15\geq9-2x$

20 Write down and then solve the inequality which represents each
   picture.

a) The perimeter
   is >26 cm.

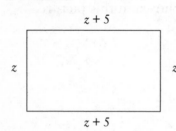

$z+5$

z z

$z+5$

b) The perimeter
   is ≤36 cm.

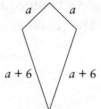

a a

$a+6$ $a+6$

c) The perimeter
   is ≥8 cm.

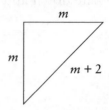

m

m

$m+2$

When there are two inequalities, we can solve each inequality
separately.

Example

$$4\leq2x<11$$

First inequality
divide both sides by 2:

$$4\leq2x$$

$$\frac{4}{2}\leq\frac{2x}{2}$$

$$2\leq x$$

Second inequality
divide both sides by 2:

$$2x<11$$

$$\frac{2x}{2}<\frac{11}{2}$$

$$x<5.5$$

Combined solution is $2\leq x<5.5$

On a number line:

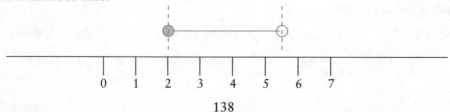

0   1   2   3   4   5   6   7

PAUSE

a) Solve each inequality.

b) Illustrate each solution with a number line.

1 $6 \leq 3x < 15$ 3 $5 < x + 2 \leq 9$ 5 $-5 \leq x + 1 \leq 4$ 7 $3 \leq 2x < 4$

2 $6 \leq 2x \leq 13$ 4 $0 < x - 3 \leq 4$ 6 $-2 < \dfrac{x}{3} \leq 1$ 8 $1.5 < \dfrac{x}{2} < 2$

Inequalities solutions which require division or multiplication by a negative number

We need to add one extra rule for the simplification of inequalities. To simplify an inequality you:

- can add any number to both sides of the inequality,
- can subtract any number from both sides of the inequality,
- can multiply both sides of the inequality by any positive number,
- can divide both sides of the inequality by any positive number.
- cannot multiply or divide both sides by a *negative* number.

Examples

1. $13 - 2x > 21$

    add $2x$ to both sides:

    $$13 - 2x + 2x > 21 + 2x$$
    $$13 > 21 + 2x$$

    subtract 21 from both sides:

    $$13 - 21 > 21 - 21 + 2x$$
    $$-8 > 2x$$

    divide both sides by 2:

    $$-\frac{8}{2} > \frac{2x}{2}$$
    $$-4 > x$$
    $$\text{i.e. } x > -4$$

2. $18 - x \leq 12$

    add x to both sides:

139

$$18-x+x\leq12+x$$
$$18\leq12+x$$
$$6\leq x$$
i.e. $x\leq6$

■ ■

PAUSE

Solve these inequalities and illustrate the solution with a number line.

1 $9-x<15$ 3 $8-2x<14$ 5 $2(3-x)<4$ 7 $3-4x\leq2x+9$

2 $9-x\geq7$ 4 $8-\dfrac{x}{2}\leq9$ 6 $12-5x\geq15-2x$ 8 $5-x>13-3x$

■ ■

Rearranging formulae

A formula is a set of rules for solving a problem.

Example

Make F the subject of the formula:
$$C=\frac{5(F-32)}{9}$$

multiply both sides by 9:
$$9\times C=9\times\frac{5(F-32)}{9}$$
$$9C=5(F-32)$$

divide both sides by 5:
$$\frac{9C}{5}=\frac{5(F-32)}{5}$$
$$\frac{9C}{5}=F-32$$

add 32 to both sides:
$$\frac{9C}{5}+32=F-32+32$$
$$\frac{9C}{5}+32=F$$

i.e. $F=\dfrac{9C}{5}+32$

140

Rearrange each formula to make the letter given in brackets the subject.

1 $t=e+5$ (e)

2 $m=r-4$ (r)

3 $C=3d$ (d)

4 $R=\dfrac{V}{1}$ (V)

5 $y=2x+3$ (x)

6 $y=tx-r$ (r)

7 $s=u+vt$ (v)

8 $y=\dfrac{3x}{2}$ (x)

9 $A=\dfrac{h(a+b)}{2}$ (h)

10 $s=\dfrac{t(u+v)}{2}$ (v)

Simultaneous equations

This is a pair of **simultaneous equations**.

$$x+y=9 \text{ (Equation 1)}$$
$$x-y=3 \text{ (Equation 2)}$$

To solve a pair of simultaneous equations, we must find values of x and y which make both equations true.

A solution can be found by combining the equations in such a way that one letter disappears.

Example

$$x+y=9 \quad ①$$
$$x-y=3 \quad ②$$

add $\quad 2x \quad =12$

$x \quad =6$

substitute $x \quad =6$ into $①$

$6+y=9$

$y=3$

check using $②$

$6-3=3$

The solution is $x=6$, $y=3$.

141

There are many pairs of values for x and y which make *one* of the equations true, for example $x=4$, $y=5$ makes the first equation true and $x=7$, $y=4$ makes the second equation true. The word *simultaneous* means 'at the same time'. Our solution, $x=6$, $y=3$ is the only pair of values that make both equations true *simultaneously*.

Example

$$2a+b=5.7 \quad ①$$
$$a+b=4 \quad ②$$

subtract $\quad a \quad =1.7$

substitute $\quad a \quad =1.7$ into ①

$$3.4+b=5.7$$
$$b=2.3$$

check using ②

$$1.7+2.3=4$$

The solution is $a=1.7$, $b=2.3$.

This is the rule for making one letter disappear:

You need a letter in each equation with the same number in front; *subtract when the signs of these letters are the same, add when they are different.*

PAUSE

Solve these pairs of simultaneous equations.

1 $x+y=8$
   $x-y=3.4$
2 $2x+y=17$
   $x+y=11$
3 $4x+y=9$
   $3x-y=5$
4 $4x+y=19$
   $2x+y=11$
5 $3x+2y=7$
   $3x-2y=-1$

6 $5x+3y=12$
   $2x+3y=12$
7 $3m+4n=14$
   $3m-4n=-2$
8 $3p+2q=10$
   $5p-2q=22$
9 $x-5y=5$
   $x+5y=-5$
10 $3x+2y=16$
   $5x+2y=20$

Sometimes we need to be particularly careful with signs.

Example

$$3u-4w=14 \quad ①$$
$$3u-7w=20 \quad ②$$
subtract $\quad\quad 3w=-6$

The number in front of u in each equation is 3 and both 3's are positive so we subtract.

$$w=-2$$
substitute $\quad w=-2$ into ①
$$3u-4\times-2=14$$
$$3u+8=14$$
$$3u=6$$
$$u=2$$

This term comes from $-4w$ subtract $-7w$ or $-4w--7w$ which gives $-4w+7w=3w$

check using ②
$$3\times2-7\times-2=6+14=20$$
The solution is $u=2$, $w=-2$.

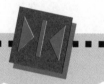

PAUSE

Solve:

1 $p-q=1$
   $q+2p=20$
2 $2m+n=2$
   $2n+2m=0$
3 $7x+3y=14$
   $9x+3y=18$
4 $4a+3b=17$
   $4a-2b=2$
5 $2x-q=3$
   $2x+3q=23$

6 $8x+2y=-15$
   $8x-2y=15$
7 $3e-2f=-17$
   $2e-2f=-16$
8 $r-4s=-30$
   $3r-4s=-18$
9 $6m-2n=-10$
   $8m-2n=-16$
10 $3x-5y=17$
   $8x-5y=37$

If there is not an equal number of one of the letters in both equations we must multiply either one or both of the equations. We multiply by numbers which will produce new equations which *do* have an equal number of one of the letters.

143

Example

Solve:

$$2b+5s=19 \quad \text{(Equation 1)}$$
$$3b+4s=18 \quad \text{(Equation 2)}$$

$$2b+5s=19 \quad ① \quad (\times 3)$$
$$3b+4s=18 \quad ② \quad (\times 2)$$
$$\Rightarrow 6b+15s=57 \quad ③$$
$$\Rightarrow 6b+\ 8s=36 \quad ④$$

subtract $\qquad 7s=21$

$$s=\ 3$$

substitute $\qquad s=\ 3$ into ①

$$2b+5\times3=19$$
$$2b+15=19$$
$$2b=\ 4$$
$$b=\ 2$$

check using ②

$$3\times2+4\times3=\ 6+12$$
$$=18$$

The solution is $s=3$, $b=2$.

PAUSE

Solve:

1 $m+5n=13$
   $3m+n=11$
2 $m+4n=9$
   $6m-3n=27$
3 $4x-3y=30$
   $x+y=4$
4 $8t=4s-40$
   $2t=3m-18$
5 $5e-f=64$
   $2e-5f=54$

6 $4x+4y=26$
   $3x+6y=27$
7 $4m+2n=-2$
   $3m-2n=-5$
8 $8x-3y=9$
   $5x+4y=35$
9 $5t-7s=-5$
   $7t-5s=17$
10 $8a-3b=21$
   $7a-7b=49$

Problems can be solved using simultaneous equations.

Example
x and y are two angles on a
straight line. The difference
between the angles is 50°.
Find x and y.

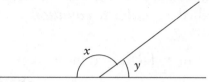

$x+y=180$ ① (angles on a straight line)
$x-y=50$ ②

add $\quad\underline{2x\quad=230}$

$x=115$

substitute in ①

$115+y=180$

$y=65$

check using ②

$115-65=50$

The solution is $x=115°$, $y=65°$.

PAUSE

For each of these problems, write a pair of simultaneous equations.
Solve your equations to find the solution to the problems.

1 The sum of two numbers, x and y is 40. The difference between the
   numbers, where x is larger than y is 8. Find the numbers.

2 A knife costs k pence and a fork f pence. A knife and fork together
   cost £1.70. A knife and three forks cost £3.30. Find the cost of a
   knife and the cost of a fork.

3 Four fish and chips and eight chicken and chips cost £32. Eight fish
   and chips and four chicken and chips cost £31. Find the cost of one
   fish and chips and the cost of one chicken and chips.

4 The perimeter of a rectangle is 36 cm. If the length is 6 cm more
   than the width, find the length and the width of the rectangle.

5 Five coloured pencils and eight felt tips cost £3.40. Seven coloured
   pencils and four felt tips cost £2.60. Find the cost of each.

Quadratic equations

An equation like $x^2-5x+6=0$, which contains a squared term, is called a **quadratic equation**.

$$x^2-5x+6=0$$

It can be factorised into:

$$(x-3)(x-2)=0$$

Since $(x-3)$ multiplied by $(x-2)$ is zero, either $(x-3)$ or $(x-2)$ must be zero. So, either:

$$(x-3)=0$$

or:

$$(x-2)=0$$

This means that either:

$$x=2 \; or \; x=3$$

Example
Solve:

$$x^2+10x=39$$

rearranging gives:

$$x^2+10x-39=0$$

factorising gives:

$$(x+13)(x-3)=0$$

therefore:

$$\text{either } x+13=0 \; or \; x-3=0$$

and:

$$\text{either } x=-13 \; or \; x=3$$

PAUSE

1 Solve by factorisation:
   a) $x^2+9x+18=0$
   b) $x^2+6x+5=0$
   c) $x^2+6x+8=0$
   d) $x^2+8x+15=0$
   e) $x^2+8x+7=0$
   f) $x^2+11x+18=0$
   g) $x^2+12x+35=0$
   h) $x^2+10x+16=0$
   i) $2x^2+13x+15=0$
   j) $2x^2+9x+10=0$

2 Solve by factorisation:
   a) $x^2-5x+6=0$
   b) $x^2-6x+5=0$
   c) $x^2-6x+8=0$
   d) $x^2-8x+15=0$

e) $x^2-7x+6=0$

f) $x^2-10x+21=0$

g) $x^2-11x+28=0$

h) $x^2-11x+18=0$

i) $2x^2-16x+30=0$

j) $2x^2-9x+4=0$

3 Solve by factorisation:

a) $x^2+4x+3=0$

b) $x^2-7x=0$

c) $x^2+9x+18=0$

d) $x^2-4x-5=0$

e) $x^2-6x=-5$

f) $x^2-x=12$

g) $2x^2+7x=30$

h) $x^2=12-4x$

i) $2x^2=5x-3$

j) $m^2-8=2m$

Equations solved by trial and improvement

This is a method which relies on systematic 'guesses', which get closer and closer to the correct solution.

Example

A carpet is 2 metres longer than it is wide. If the width is x, the length will be $x+2$ and the area will be $x(x+2)$ or x^2+2x.

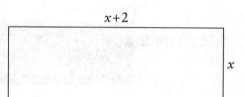

If the area is 19 m²:

a) show that the width of the carpet lies between 3m and 4m.

b) find the width of the carpet correct to 1 d.p.

x	x^2+2x	comment
3	15	too small
4	24	too big

The solution lies between 3m and 4m as $x = 3$ gives a value of $x^2 + 2x$ which is smaller than 19, $x = 4$ gives a value of $x^2 + 2x$ which is larger than 19.

x	x^2+2x	comment
3.2	16.64	too small
3.3	17.49	too small
3.4	18.36	too small
3.5	19.25	too big

The solution lies between 3.4m and 3.5m. 19.25 is closer to 19 than 18.36 so the solution, correct to 1 d.p. is 3.5m.

In each case:

a) show that there is a solution between the two given values;

b) find the solution correct to 1 decimal place.

1   $x^2+x=1$ between 0 and 1   6 $2y^2-8y=7$ between 4 and 5

2   $x^2-7x=5$ between 7 and 8   7 $5x^2-9x+1=0$ between 1 and 2

3   $y^2+5y=3$ between 0 and 1   8 $y+\dfrac{1}{y}=3$ between 2 and 3

4   $2x^2-3x=0$ between 1 and 2   9 $x^3+x=13$ between 2 and 3

5   $x^2-5x+2=0$ between 0 and 1   10 $x^3-2x=0$ between $x=1$ and $x=2$

■ ■

SECTION 3   Number Patterns and Number Sequences

A sequence is a set of numbers which follow a pattern. The simplest example is the set of positive whole numbers.

$$1, 2, 3, 4, 5, 6, 7, 8 \ldots$$

Because this sequence is the most simple and basic of all, these numbers are given a special name. They are called the set of *Natural Numbers*.

Each number in a sequence is called a *term* of the sequence.

The pattern for the whole numbers is that each term is one more than the term before it.

Other sequences have more complicated patterns.

Examples

1. What are the next 2 terms in the sequence: 1, 4, 7, 10, 13, 16 ... ?

   Each term is produced by adding 3 to the one before it so the next two terms are 19 and 22.

2. What are the next 2 terms in the sequence: 1, 2, 4, 8, 16... ?

   Each term is produced by multiplying the one before it by 2 so the next two terms are 32 and 64.

3. What are the next 2 terms in the sequence: 100, 90, 80, 70, ... ?

Each term is produced by subtracting 10 from the one before it so the next two terms are 60 and 50.

4. What are the next 2 terms in the sequence: 1, 3, 7, 15, 31 ..... ?

Each term is produced by multiplying the one before it by 2 and adding 1 so the next two terms are 63 and 127.

PAUSE

Give the next two terms in each sequence. Write down the pattern for each sequence.

1 1, 5, 9, 13, 17 .....
2 23, 21, 19, 17 .....
3 1, 3, 9, 27, 81 .....
4 400, 200, 100, 50 .....
5 2, 5, 10, 17, 26 .....
6 1, 2, 4, 7, 11 .....
7 1, 6, 11, 16, 21 .....
8 3, 5, 8, 12, 17 .....
9 1, 4, 13, 40, 121 .....
10 17, 16, 14, 11 .....
11 1, 1, 2, 3, 5, 8 .....
12 1, 5, 25, 125 .....

Other sequences can be developed from the set of natural numbers. For example, if we multiply all the natural numbers by 2, we produce the set of *even numbers*.

$$2, 4, 6, 8, 10, 12, 14, 16, ...$$

We use the basic sequence of natural numbers to number the terms in other sequences.

The first 8 terms in the sequence of odd numbers are numbered like this:

Term number	1st	2nd	3rd	4th	5th	6th	7th	8th
Term	1	3	5	7	9	11	13	15

Using this notation, we can write:

9 is the 5th term in the sequence of odd numbers

149

PAUSE

1 Copy and complete:
   a) The 10th term of the sequence of even numbers is .....
   b) The 50th term of the sequence of even numbers is .....
   c) The 50th term of the sequence of odd numbers is .....
   d) The 7th term of the sequence of prime numbers is .....
   e) The 5th term in the sequence of square numbers is .....

Describing sequences

We can use mathematical shorthand to describe sequences. If you wanted to know what the 200th even number is, you could write down the first 200 terms of the sequence: 2, 4, 6, 8, 10, 12 ..... You would however probably spot that since the 1st term is 2, the 2nd term is 4 and the 3rd term is 6, it follows that the 200th term must be 400.

In other words, if you want to know the nth term, you would multiply n by 2.
We can summarise this rule for working out even numbers with a formula like this:

> for the even numbers, the nth term $= 2n$.

Example
Find a formula for the nth term of the sequence: 3, 7, 11, 15, 19 .....
It is best to start by constructing a table like this:

T_1	T_2	T_3	T_4	T_5
3	7	11	15	19

(where T_1 is the first term etc)

$$4 \quad 4 \quad 4 \quad 4$$

The terms of the sequence go up in 4s. Compare the sequence with the $4 \times$ table.

150

	T_1	T_2	T_3	T_4	T_5	T_n
Sequence	3	7	11	15	19	4n-1
×Table	4	8	12	16	20	4n

We can see that each term in our sequence is 1 less than the corresponding term of the 4 × table.

The formula for the nth term is $4n-1$.

PAUSE

In questions 1 to 5, find a formula for the nth term of each of the following sequences:

1 1, 3, 5, 7, 9 .....
2 5, 17, 29, 41, 53 .....
3 6, 11, 16, 21, 26, ...

4 4, 7, 10, 13, 16, 19, ...
5 4, 11, 18, 25, 32, ...

6 A gardener always plants flower beds with geraniums, and marigolds. Her planting plans for 1, 2 and 3 geraniums are:

```
M M M          M M M M          M M M M M
M G M          M G G M          M G G G M
M M M          M M M M          M M M M M
1 geranium     2 geraniums      3 geraniums
```

Find a formula for the number of marigolds needed if n geraniums are planted.

Triangular numbers

The sequence of triangular numbers is 1, 3, 6, 10, 16 .... They are called triangular numbers because the terms of the sequence can be represented by triangular patterns of dots like this:

1 Two identical triangular numbers can be placed together to form a rectangle.

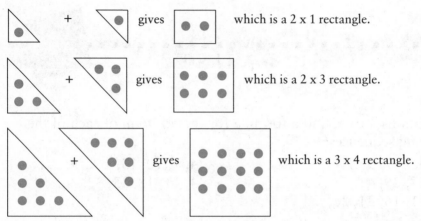

+ gives which is a 2 x 1 rectangle.

+ gives which is a 2 x 3 rectangle.

+ gives which is a 3 x 4 rectangle.

a) Copy this pattern and continue it for 2 more triangular numbers.

b) A rectangle is formed from the 10th triangular number. How many dots high and how many dots wide will this rectangle be?

c) A rectangle is formed from the 50th triangular number. How many dots will this rectangle contain? Use your answer to calculate the 50th triangular number.

d) A rectangle is formed from the nth triangular number. Write down a formula for the number of dots in this rectangle. Use your answer to write down a formula for the nth triangular number.

2 Find a formula for the nth term of the sequences illustrated in these diagrams:

a) The 'pentagonal' numbers:

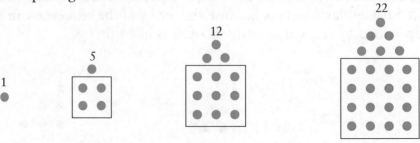

1   5   12   22

b) The 'stellate' numbers:

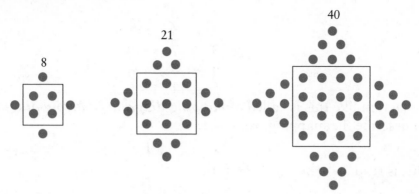

3 Four people meet in a pub.
   a) If each person shakes hands with the other three people, how many handshakes will there be?
   b) How many handshakes are there if 6 people meet?
   c) How many handshakes are there if 3 people meet?
   d) How many handshakes are there if n people meet?

4 Twenty football teams play in the English Premier League. Every season each team plays every other team twice, once at 'home', once 'away'.
   a) How many Premier League games are played each season?
   b) There are ten teams in the Scottish Premier League, how many games do they play in a season?
   c) How many games would be played in a league with n teams?

■■■■■■■■■■■■■■■■■■■■■■■■■■■■■■■■■■■

Number machines

This is a *number machine*:

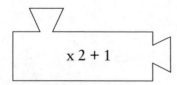

The number machine shows that we feed in numbers they are first multiplied by 2 and then added to 1.
If we feed in the numbers 1, 2, 3, 4, 5, we transform these numbers into 3, 5, 7, 9, 11.

This transformation can be illustrated with a diagram like this:

$$1 \rightarrow 3$$
$$2 \rightarrow 5$$
$$3 \rightarrow 7$$
$$4 \rightarrow 9$$
$$5 \rightarrow 11$$

A diagram like this is called a **mapping diagram** or **arrow diagram**. To describe the mapping, we write:

$$x \rightarrow 2x + 1$$

This is read as 'x becomes $2x + 1$'.

Example

Draw a mapping diagram to illustrate the mapping $x \rightarrow x + 1$ applied to the set of numbers -3, -2, -1, 0, 1, 2, 3.

This is the required diagram:

$$-3 \rightarrow -2$$
$$-2 \rightarrow -1$$
$$-1 \rightarrow 0$$
$$0 \rightarrow 1$$
$$1 \rightarrow 2$$
$$2 \rightarrow 3$$
$$3 \rightarrow 4$$
$$x \rightarrow x+1$$

PAUSE

Draw diagrams to show each of the following mappings applied to the numbers $-3, -2, -1, 0, 1, 2, 3$.

1 $x \rightarrow x+4$

2 $x \rightarrow 2x$

3 $x \rightarrow \dfrac{x}{2}$

4 $x \rightarrow x^2$

5 $x \rightarrow x^2+3x-4$

6 $x \rightarrow \dfrac{12}{x}$

In the sixteenth century, the French mathematician Rene Descartes developed a new form of mathematics in which algebra was applied to geometry. This branch of mathematics has been named in his honour and is called *Cartesian Geometry*.

Descartes developed a grid of numbers on which pictures of mathematical functions could be plotted. This is an example of the basic cartesian graph.

Example

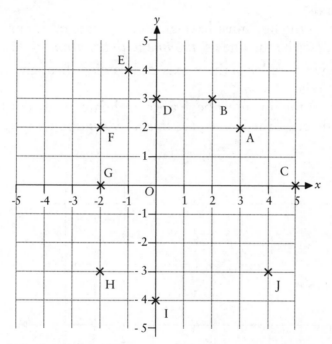

A is the point (3,2)	F is the point (−2,2)
B is the point (2,3)	G is the point (−2,0)
C is the point (5,0)	H is the point (−2,−3)
D is the point (0,3)	I is the point (0,−4)
E is the point (−1,4)	J is the point (4,−3)

■ ■

PAUSE

1 Draw an *x*-axis and a *y*-axis from −5 to 5. Plot each of the

following set of points on your graph, joining them in the order they are plotted.

 a) $(0,4)$, $(2,1)$, $(3,-3)$, $(-1,-2)$, $(-4,0)$, $(-1,-1)$, $(-3,3)$, $(1,1)$, $(0,4)$
 b) $(0,4)$, $(4,0)$, $(4,-4)$, $(0,-4)$, $(-4,0)$, $(-4,1)$, $(-3,1)$, $(-5,2)$, $(-5,0)$,
   $(0.-5)$, $(5,-5)$, $(5,0)$, $(0,5)$, $(-2,5)$, $(-1,3)$, $(-1,4)$, $(0,4)$

■ ■

Equations for lines parallel with the axes

A line is drawn parallel to the y-axis and passing through the number 4 on the x-axis.

Every point on the line must have an x coordinate of 4. For this reason, we describe the line as, *the line with equation $x=4$.*

A line is drawn parallel to the x-axis and passing through the number -1 on the y-axis.

Every point on the line must have a y coordinate of -1. For this reason, we describe the line as, *the line with equation $y=-1$.*

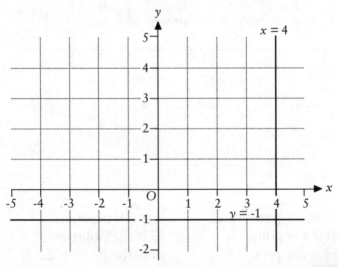

■  ■ ■

PAUSE

1 Draw a pair of axes from -5
   to $+5$. Draw and label the lines
   with equations:

   a) $x=3$ d) $x=4.5$
   b) $x=-2$ e) $x=-3.5$
   c) $x=0$

156

2 On the same diagram draw and label these lines.
   a) y=4   d) y=−2.5
   b) y=−3   e) y=−5
   c) y=0

3 Write down the equation of the line which passes through each of the following pairs of points:
   a) (2,−2) and (5,−2)
   b) (2,−2) and (2,5)
   c) (0,7) and (0,−7)
   d) (0,7) and (−5, 7)
   e) (−5,0) and (5,0)
   f) (−5,0) and (−5,5)

Slopes and gradients

A straight line which is not parallel to either axis is drawn on a graph. There is a common *slope* or *gradient* between any two points on the line.

The slope or gradient of a line measures how 'steep' the line is and is defined as the fraction:

$$\frac{\text{difference between the } y \text{ coordinates of the points}}{\text{difference between the } x \text{ coordinates of the points}}$$

Look at this straight line:

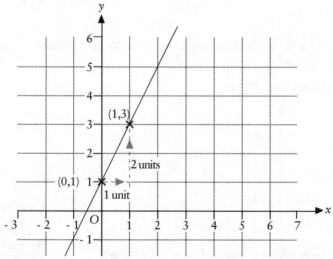

The gradient between the points (0,1) and (1,3) is given by:

$$\text{Gradient} = \frac{3-1}{1-0} = \frac{2}{1}$$

We can write:

The line has a constant gradient of $\frac{2}{1}$.

Or, more simply:

'The line climbs 2 squares up the graph for every 1 square across the graph'.

Examples

1. Find the gradient of the line which connects the points (1,1) and (5,3).

   Drawing a graph we have:

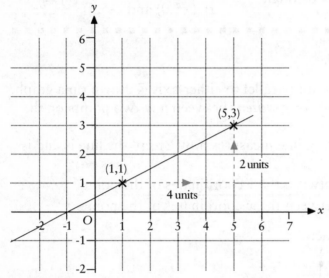

Gradient =
$$\frac{3-1}{5-1} = \frac{2}{4} = \frac{1}{2}$$

2. Find the gradient of the line which connects the points (−2,3) and (4,−2).

   Drawing a graph we have:

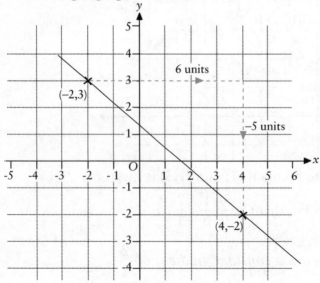

Gradient =
$$\frac{-2-3}{4--2} = \frac{-5}{6}$$

1 Draw a graph and find the gradient of the line which connects each of the following pairs of points:

a) (0,0) and (3,6)
b) (0,0) and (2,6)
c) (0,0) and (4,4)
d) (−2,−1) and (2,1)
e) (−4,−1) and (6,1
f) (−3,1) and (3,5)
g) (−1,1) and (1,4)
h) (−4,4) and (3,−3)
i) (−5,−2) and (−3,−6)
j) (1,6) and (2,4)
k) (−1,5) and (1,−5)
l) (−4,−3) and (4,−5)

Equations for lines

Cartesian graphs can be used to show mathematical relationships.

For example, the equation $y=2x+1$ is a rule for changing the number x into the number y.

If we take a sequence of values for x, the equation produces a sequence of values for y. These values are usually shown in a table.

For example, if we start with a sequence of x values from −3 to 3, we get a table like this:

x	−3	−2	−1	0	1	2	3
y	−5	−3	−1	1	3	5	7

Each pair of x and y values can be shown as coordinates on a graph.

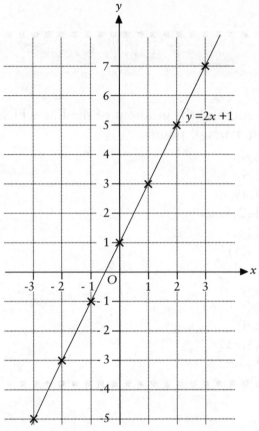

A line has been drawn through the points.
This line represents *all* the pairs of values connected by the equation $y = 2x + 1$.

PAUSE

1 Draw up a table of values for each of the following equations, using all the whole number x values from -3 to 3. Plot the pairs of values you obtain as coordinates on a graph and, if it seems sensible to do so, connect the points with a straight line.

a) $y = x + 4$

b) $y = x - 4$

c) $y = 2x$

d) $y = \dfrac{x}{2}$

e) $y=2x+4$

f) $y=\dfrac{x}{2}-4$

g) $y=4-x$

h) $y=x^2$

■■■■■■■■■■■■■■■■■■■■■■■■■■■■■■■■■■■■■■

Relationships expressed in the form $y=mx+c$

Any equation which can be expressed in the form $y=mx+c$:

■ will be a straight line with a gradient equal to m,

■ will cross the y-axis at the point $(0,c)$.

This rule lets us make predictions.

For example, the graph drawn from the equation $y=4x-5$ will have a gradient equal to 4 and will cross the y-axis at the point $(0,-5)$.

If we know a graph will be a straight line we can reduce the number of values included in a table.

The minimum number of points necessary to fix a straight line is two but usually at least 3 are plotted (the third as a check).

Example

Draw a graph to illustrate the equation $y=11-2x$, using x values from -3 to 3.

We know that the graph will be a straight line, so we use 3 points. It is a good idea to use the two end values of x and one other value. This will tell us what values need to be covered on the y-axis of our graph.

A suitable table of values is:

x	-3	0	3
y	17	11	5

161

Which produces the graph:

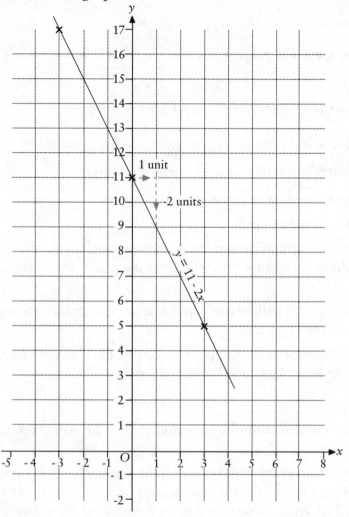

From the graph, we see that the line has a slope of $-\dfrac{2}{1}$ and cuts the y-axis at $(0,11)$.

This checks with the results predicted by comparing $y=11-2x$ with $y=mx+c$.

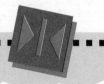

PAUSE

1 Draw a graph to illustrate each of the following equations using x

162

values from –4 to 4 and write down the gradient of the graph and the coordinates of the point where the graph cuts the y-axis.

a) $y=x+3$

b) $y=x-3$

c) $y=3x$

d) $y=\dfrac{x}{4}$

e) $y=3x-3$

f) $y=\dfrac{x}{4}+3$

g) $y=4-2x$

h) $y=\dfrac{3x}{4}+3$

i) $y=2(x-1)$

j) $y=8-\dfrac{x}{2}$

2 Write down the equations of the lines which have the following gradients and cutting points on the y-axis.
a) gradient 2, cutting point (0,3)
b) gradient 3, cutting point (0,2)

c) gradient $\dfrac{1}{2}$, cutting point (0,–1)

d) gradient –1, cutting point (0,0)
e) gradient –2, cutting point (0,12)

■ ■

Relationships not expressed in the form $y = mx + c$

Often, relationships can be rearranged into the form $y=mx+c$.

For example, $4y-3x=12$ can be rearranged into $y=\dfrac{3x}{4}+3$.

This means that the graphs of $4y-3x=12$ will be a straight line.
It is easier to draw the graph of $4y-3x=12$ without rearranging.
A quick way to draw the graph is to let first x and then y take the value 0. A third point can be obtained by selecting one other x value.
We can start with the table:

x	0		4
y		0	

The table is completed like this:

x	0	–4	4
y	3	0	6

163

This is the graph:

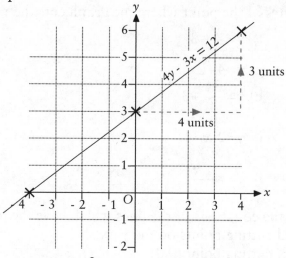

The line has a gradient of $\frac{3}{4}$ and cuts the y-axis at $(0,3)$.

PAUSE

1 Draw a graph to illustrate each of the following equations:
   a) $4y=x+1$ f) $5x+4y=20$
   b) $3y=6x-12$ g) $x-y=7$
   c) $2x+3y=24$ h) $x+y=10$
   d) $2x-3y=24$ i) $3y=21-7x$
   e) $2y-x=8$ j) $40x+60y=240$

Using graphs to solve simultaneous equations

Simultaneous equations can be solved by drawing a graph. For example, suppose we wish to solve the equations:

$$2y+x=8$$
$$4y-x=4$$

We can construct these tables:

x	0	8	2
y	4	0	3

x	0	-4	8
y	1	0	3

164

This is the graph:

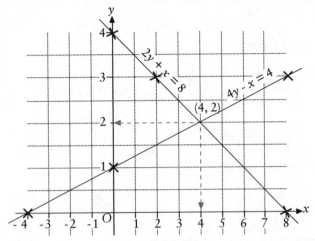

From the graph we can see that the lines intersect at (4,2). This must mean that $x = 4$ and $y = 2$, are values that satisfy *both* equations. So, these values are the solution to our simultaneous equations.

PAUSE

1 Solve each of the following pairs of simultaneous equations by drawing a graph.

a) $y=x-3$
   $x+3y=6$

b) $y=2x+2$
   $2y=x-2$

c) $y+2x=6$
   $2y+x=6$

d) $2y-2x=7$
   $4y+3x=0$

e) $2y=x-8$
   $4y=20-x$

Graphs of quadratics

This is a table of values for the equation $y=x^2$:

x	-3	-2.5	-2	-1.5	-1	-0.5
y	9	6.25	4	2.25	1	0.25

x	0	0.5	1	1.5	2	2.5	3
y	0	0.25	1	2.25	4	6.25	9

This is the graph of $y = x^2$.

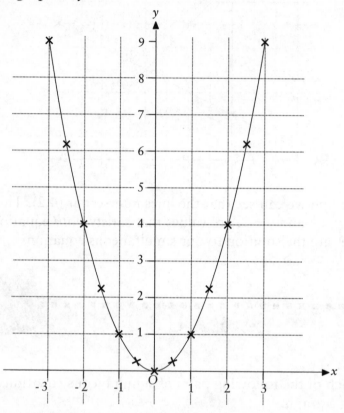

The points should be connected with a smooth curve, *not with a series of straight lines*. Values of x which are not whole numbers can be included to help draw the curve.

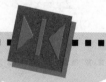

PAUSE

1 Draw an x-axis from -3 to 3 and a y-axis from -5 to 14. It is easier to draw smooth curves if you make the scale on the x-axis 2 cm to 1 unit and the scale on the y-axis 1 cm to 1 unit.

Use the same pair of axes for all the graphs in question 1 and question 2.

a) Use this table to plot a graph of $y=x^2$

x	-3	-2	-1	0	1	2	3
y	9	4	-1	0	1	4	9

Join the points with a smooth curve and label the graph $y=x^2$

b) Copy and complete this table for $y=x^2+1$

x	-3	-2	-1	0	1	2	3
y	10					5	

Plot the points and join them with a smooth curve. Label the graph $y=x^2+1$

c) Make a table like the one in part b) for each of these:

   i) $y=x^2+2$

   ii) $y=x^2+3$ iii) $y=x^2+4$

Plot each set of points and join them with a smooth curve. Label each curve with its equation.

2 Use the same pair of axes as question 1.

a) Copy and complete this table for $y=x^2-1$

x	-3	-2	-1	0	1	2	3
y		3				8	

Plot the points and join them with a smooth curve. Label the graph $y=x^2-1$

b) Make a table like the one in part a) for each of these:

   i) $y=x^2-2$

   ii) $y=x^2-3$

   iii) $y=x^2-4$

Plot each set of points and join them with a smooth curve. Label each curve with its equation.

3 a) Look carefully at the graphs you have drawn in questions 1 and 2 What do you notice about them?

   b) Describe the appearance of these graphs:

   i) $y=x^2+10$

   ii) $y=x^2-15$

This exercise looks at the effect of multiplying x^2 by a number.

PAUSE

1 Draw an x-axis from -3 to 3 and a y-axis from 0 to 28. Make the
  scale on the x-axis 2cm to 1 unit. Make the scale on the y-axis 1cm to
  2 units. Use the same pair of axes for all the graphs in this question.
  a) Use the table from question 1 of the previous exercise to draw
     the graph of $y=x^2$
  b) Copy and complete this table for $y=2x^2$

x	-3	-2	-1	0	1	2	3
y	18					8	

  Plot the points and join them with a smooth curve. Label your
  curve $y=2x^2$
  c) Make a table like the one in part b) for $y=3x^2$. Plot the points,
     join them with a smooth curve and label your graph.
  d) Copy and complete this table for $y=\frac{1}{2}x^2$.

x	-3	-2	-1	0	1	2	3
y		2					4.5

  Plot the points and join them with a smooth curve. Label your
  curve $y=\frac{1}{2}x^2$.
  e) What do you notice about the graphs you have drawn?
     Describe the appearance of these graphs:
     i) $y=5x^2$ ii) $y=\frac{1}{4}x^2$

2 Draw an x-axis from -3 to 3 and a y-axis from -28 to 0. Make the
  scale m the x-axis 2 cm to 1 unit. Make the scale on the y-axis 1 cm
  to 2 units.
  Use the same pair of axes for all the graphs in this question.
  a) Copy and complete this graph for $y=-x^2$

x	-3	-2	-1	0	1	2	3
y	-9				-1		

  Plot the points and join them with a smooth curve. Label your
  graph.
  b) Make a table like the one in part a) for each of these:
     i) $y=-2x^2$ iii) $y=-3x^2$ ii) $y=-\frac{1}{2}x^2$

Plot each set of points and join them with a smooth curve. Label each curve with its equation.

c) What do you notice about the graphs you have drawn? Describe the appearance of these graphs:

i) $y=-4x^2$ ii) $y=-\frac{1}{3}x^2$.

■ ■

The graphs we have drawn are examples of a type of curve called a *parabola*. The parabola always has a 'bowl' shape and is produced by any equation of the form $y=ax^2+bx+c$, where a, b and c are any selected numbers.

Example
Draw a graph to illustrate the equation $y=x^2+2x-8$, as x takes values from -3 to 5.
This is the table:

x	-3	-2	-1	0	1	2	3	4	5
y	7	0	-5	-8	-9	-8	-5	0	7

This is the graph:

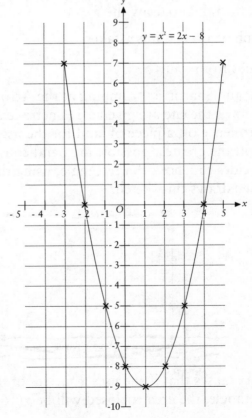

$y = x^2 = 2x - 8$

169

1 a) Copy and complete this table of values for $y=x^2+x-6$.

x	-4	-3	-2	-1	0	1	2	3
y	6		-4	-6				

b) Use the table to draw the graph of $y=x^2+x-6$ as x takes values from -4 to 3.

2 a) Copy and complete this table
of values for $y=\dfrac{x^2}{2}-5$.

x	-5	-4	-3	-2	-1	0	1	2	3	4	5
y	7.5		-0.5				-4.5				

b) Use the table to draw the
graph of $y=\dfrac{x^2}{2}-5$ as x takes
values from -5 to 5.

3 Wayne and Sharon are camping at the Arrow River Campsite.
Campers at the site are given a 16 metre length of rope and four
posts to fence off a piece of land for themselves. Sharon wants to
fence off the greatest possible area and considers the possibilities.
She decides to fence off a rectangle, using the river bank as one
side. She draws this sketch.

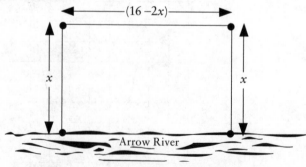

Sharon realises that if she uses x metres of the rope for each side of
the rectangle, the area enclosed will be $x(16-2x)$ square metres.

a) Copy and complete this table of values for the area enclosed as x takes values from 1 to 8 metres.

x	0	0.5	1	1.5	2	2.5	3	3.5	4
$16-2x$	16		14			11			
Area	0	7.5	14		24				31.5

x	4.5	5	5.5	6	6.5	7	7.5	8
$16-2x$								0
Area						14		0

b) Use your table to draw a graph to illustrate the way that the area enclosed varies as x takes values from 0 to 8 metres. (Make the y-axis an 'area' axis and plot the area values against the x values).

c) What advice would you offer to Sharon?

4 When a body is dropped from a height, the distance it has fallen is given by the equation:

$$s=4.9t^2,$$

where s is the distance dropped (in metres) and t is the time (in seconds) that the body has been falling.

a) Copy and complete this table of values for the distance fallen (correct to 1 d.p.) during the first 4 seconds.

t	0	0.5	1	1.5	2	2.5	3	3.5	4
s	0	1.2		11.0			44.1		78.4

b) Use your table to draw a graph to illustrate the distance fallen as the time varies from 0 to 4 seconds (make the x-axis your 'time' axis and the y-axis your 'distance' axis). Use 2 mm graph paper with a scale of 2 cm to represent 1 second on the time axis and 2 cm to represent 10 metres on the distance axis.

c) Wayne drops a stone down a well and it is 2.8 seconds before he hears a splash. Use your graph to estimate the depth of water in the well.

5 If a car is travelling at v miles per hour, its stopping distance d metres, is given by the formula:

$$d=v^2+\frac{20v}{60}$$

(This formula calculates a total stopping distance, including the time taken for a driver to react and apply the brakes)

a) Draw up a table of values showing the stopping distances for speeds from 10 mph to 70 mph in steps of 5 mph.

b) Draw a graph to illustrate your table of values.

c) Sid Pratt likes to prove how clever he is by racing through 30 mph speed limits at speeds of 50 mph and more. He justifies this to his friends, saying, 'Well, you know, its only 20 mph over the limit init, you know, I mean, you know, that's nufin is it, you know'. Write some comments for Sid to consider about speeds and stopping distances (assume that Sid can read).

■ ■

Solving equations with graphs

Having drawn a graph to represent a mathematical relationship, we can use it to solve a range of equations.

For example, this is a table of values for the equation $y = x^2 + 3x - 10$

x	−6	−5	−4	−3	−2	−1	0	1	2	3
y	8	0	−6	−10	−12	−12	−10	−6	0	8

To show the way that a graph can be used to solve a range of equations, we will use our graph to solve:

1. $x^2 + 3x - 10 = 0$
2. $x^2 + 3x - 10 = 6$
3. $x^2 + 3x - 10 = x$

Equation 1

The table of values or the graph can be used immediately to solve $x^2 + 3x - 10 = 0$.

The points where the graph cuts the x-axis (or $y = 0$) give us solutions of $x = -5$ or $x = 2$ for the equation.

Equation 2

In the same way, *Equation 2* can be solved by looking for the crossing points of $y = x^2 + 3x - 10$ and $y = 6$.

Equation 3

Equation 3 can be solved by looking for the crossing points of $y = x^2 + 3x - 10$ and $y = x$.

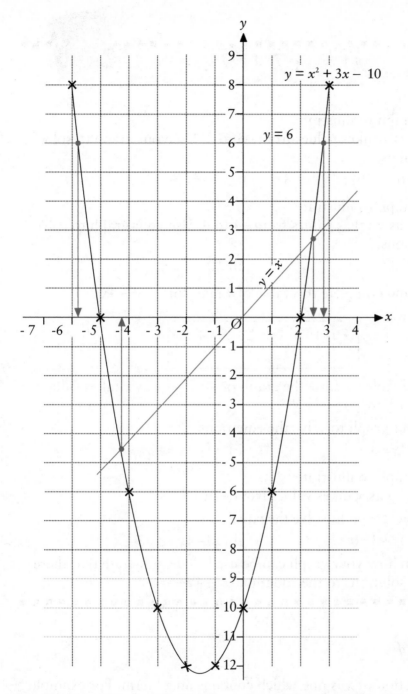

Reading from the graph, our equations have the following solutions:

Equation 2 $x^2+3x-10=6$ $x=2.8$ or -5.8

Equation 3 $x^2+3x-10=x$ $x=2.3$ or -4.3

173

1 Draw a graph to illustrate
$y=x^2-8$, as x takes values from -3 to 3. Use your graph to solve
the equations:

a) $x^2-8=0$ b) $x^2-8=-3$ c) $x^2=7$

2 Draw a graph to illustrate
$y=10-x^2$ as x takes values from -4 to 4. Use your graph to solve
the equations:

a) $10-x^2=1$ b) $10-x^2=x$

3 a) Copy and complete this table of values for $y=x^2-4x$

x	-1	0	1	2	3	4	5
y	5		-3			0	

b) Use the table to draw the graph of $y=x^2-4x$ as x takes values
from -1 to 5.

c) Use your graph to solve the equations:

i) $x^2-4x=0$ ii) $x^2-4x=-2$

4 Draw a graph to illustrate
$y=x^2-4x+3$ as x takes values from -1 to 5.

a) Use your graph to solve the equations:

i) $x^2-4x+3=0$ ii) $x^2-4x+3=3$

b) Explain how your graph can be used to demonstrate that there
are no solutions to the equation, $x^2-4x+3=-2$.

■ ■

Cubic graphs

A *cubic* function of x is one which contains an x^3 term. For example
$y=2x^3+3x^2-x+7$. Questions involving the graphs of cubic functions
are not common in examinations but may sometimes crop up. The
next exercise gives practice in drawing graphs of cubic functions.

1 a) Copy and complete this table for $y=x^3$.

x	−3	−2	−1	0	1	2	3
y	−27					8	

b) Draw an x-axis from −3 to 3 and a y-axis from −30 to 30. Use a scale of 2 cm to 1 unit for x and 2 cm to 5 units for y.

c) Plot the points in your table and join the points with a smooth curve.

2 Can you predict what the graph of $y=x^3+3$ would look like? Draw up a table for $y=x^3+3$ for values of x from −3 to 3. Using the same axes as for question 1, plot the points in your table. Join the points with a smooth curve. Was your prediction correct?

3 Can you predict what the graph of $y=\frac{1}{2}x^3$ would look like? Draw up a table for $y=\frac{1}{2}x^3$ for values of x from −3 to 3. Using the same axes as for questions 1 and 2, plot the points in your table. Join the points with a smooth curve. Was your prediction correct?

■ ■

Reciprocal graphs

A reciprocal function is one which involves a division by an x term. Examples are:

$$y=\frac{12}{x} \ or \ y=3+\frac{3}{x}$$

Example
Draw the graph of $y=\frac{6}{x}$ as x takes values from −6 to 6.
This is the table of values.

x	−6	−5	−4	−3	−2	−1
y	−1	−1.2	−1.5	−2	−3	−6

x	1	2	3	4	5	6
y	6	3	2	1.5	1.2	1

From the table we can draw this graph.

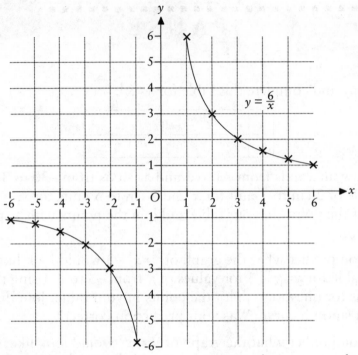

$$y = \frac{6}{x}$$

The two 'halves' of our graph are not connected. As x goes from 1 to 0, taking smaller and smaller values, y becomes a larger and larger *positive* number.

As x goes from −1 to 0, taking smaller and smaller decimal values, y becomes a larger and larger *negative* number.

Division by zero is impossible so, we cannot plot any point on the graph when x is zero.

We are left with a graph which is in two distinct parts, with a boundary line on which no point can be plotted.

PAUSE

1 a) Copy and complete this table for $y = \dfrac{12}{x}$

x	−12	−10	−8	−6	−4	−3	−2	−1	1	2	3	4	6	8	10	12
y			−1.5									3				1

b) Draw a pair of axes from −12 to 12. Use a scale of 1 cm to 2 units on both axes.

c) Plot the points in your table and join with a smooth curve. Label the graph $y=\dfrac{12}{x}$.

d) By drawing a suitable line on your graph, solve the equation $\dfrac{12}{x}=x$.

2 a) In a circuit connected to a 4 volt battery the current (I) is linked to the resistance (R) by the formula $I=\dfrac{4}{R}$. Copy and complete this table for I as R varies from 0 to 8 ohms.

R ohms	1	2	3	4	5	6	7	8
I amps			1.3			0.7		

b) Draw a pair of axes. Make the R axis the horizontal axis with a scale of 1 cm to 1 ohm. The I axis is the vertical axis with a scale of 2 cm to 1 amp.

c) Plot the points in your table and join with a smooth curve.

d) Use your graph to find the resistance (R) when the current (I) is 1.5 amps.

■ ■

Graphs of inequalities

We have used graphs to illustrate many different functions, all of which produced either straight lines or curves. Graphs can also be used to illustrate *inequalities*.

Remember, an inequality is a statement like $x>2$ or $y\le-1$. We have already illustrated these inequalities with simple number lines like this:

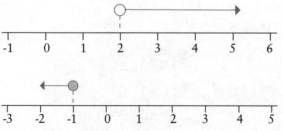

On a coordinate grid, all the points with an x coordinate equal to 2 lie on a straight line. All the points with an x coordinate which is greater than 2 lie to the right of this line. So, using a graph, the inequality $x>2$ is illustrated like this:

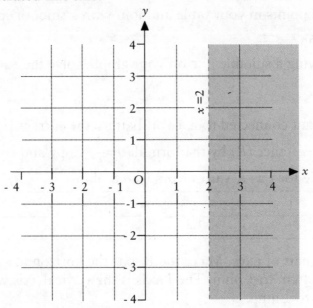

A dotted line is used because points which are on the line $x=2$ are not included in the region.

We can illustrate the inequality $y \leq -1$ with this graph:

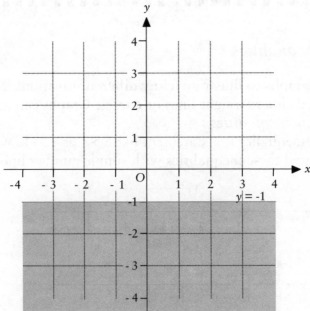

In this case a solid line is used because points which are on the line $y=-1$ are included in the region.

1 Illustrate each on the following inequalities with a graph.
   a) $x<2$ d) $y<0$
   b) $y>-1$ e) $x>-2$
   c) $x>0$ f) $y\leq-3$

2 Write down the inequality illustrated by each of the following graphs:

a)

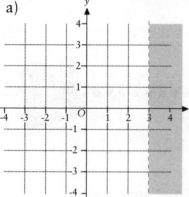

c)

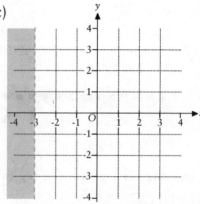

b)

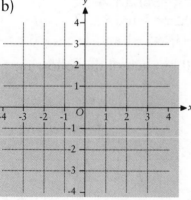

d)

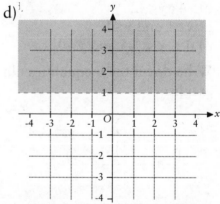

Two or more combined inequalities can be illustrated on one graph.

Example

Illustrate on a graph the region where $x\leq2$ and $y>-1$.

We draw a graph and mark on the lines $x=2$ and $y=-1$. Then we shade in the region of the graph where *both* inequalities are true.

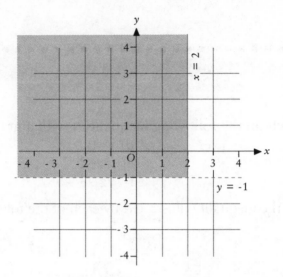

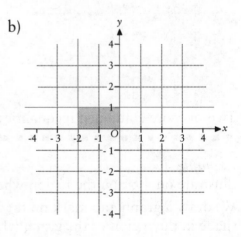

PAUSE

1 Illustrate on a graph the regions where:

a) $-1 < x \le 1$

b) $-4 < y < -2$

c) $x \ge 1$ and $y \le 1$

d) $0 \le x < 1.5$ and $y > 0$

e) $x \le -2$ and $y \le 2$

f) $x < -1$ and $y > -3$

g) $0 \le x < 3$ and $y \ge -2$

h) $-3 < x \le -1$ and $1 < y < 3$

2 Write down the inequalities illustrated by each of the following graphs:

a)

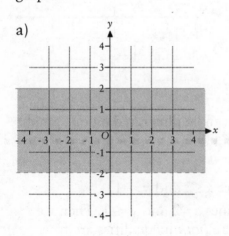

b)

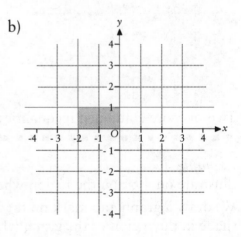

180

c)   d)

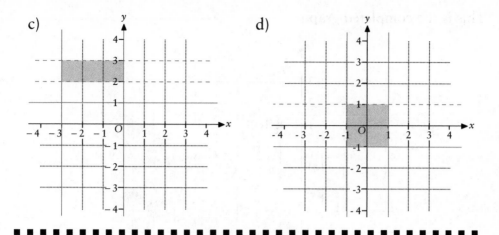

Other inequalities

Example

Illustrate on a graph the region where $2x - 3y \geq 12$.

The boundary line for the region will be $2x - 3y = 12$, so we start by constructing a table and drawing the graph.

x	0	6	3
y	−4	0	−2

The region we have been asked to illustrate will consist of all the points on one side of this line.

Our only problem is in deciding *which* side of the line.

We select any point which is not actually on the line and test to see if the inequality is true.

For example, if we select $(1,1)$ as a test point and apply the inequality at this point we have:

$$(2 \times 1) - (3 \times 1) > 12$$
$$-1 > 12$$

This is *not* true. So, we shade the side of the line which *does not* include $(1,1)$ to illustrate the region $2x - 3y > 12$.

This is the completed graph:

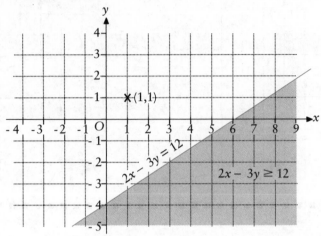

PAUSE

1 Illustrate on a graph the regions where:
 a) $y > x+1$
 b) $y > 2x-1$
 c) $x+y > 5$
 d) $5x+4y < 20$
 e) $x > 0$, $y < 0$ and $y > 2x-5$
 f) $-1 < x < -4$, $y < 4$ and $x+2y > 4$

2 Anita has £2 pocket money to spend on sweets. Her favourites are Mars bars at 16p each and Skittles at 25p a packet.
 a) Write down an expression for the cost of x Mars bars at 16p each.
 b) Write down an expression for the cost of y packets of Skittles at 25p each.
 c) Explain why Anita's spending decisions can be represented by the inequality $16x+25y \le 200$.
 d) Draw a graph to illustrate all the combinations of different numbers of Mars bars and packets of Skittles that Anita can buy.
 e) Can Anita spend the whole £2 on Mars bars and Skittles?

3 Write down the inequalities illustrated by each of the following graphs:

a)

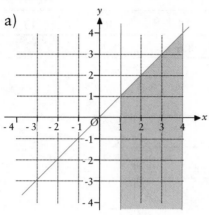

c)

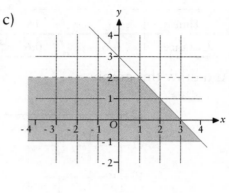

b)

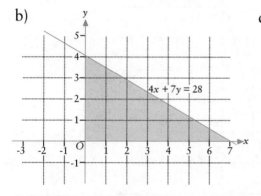

d)

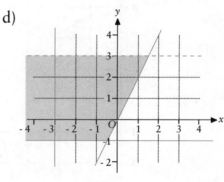

■ ■

Distance and time graphs

Here is a description of a 12 km journey from home to work lasting 40 minutes.

- The journey started with a 5 minute walk to a bus stop 600 metres from the house.
- There was a 10 minute wait for a bus.
- The bus then travelled 8 km in 10 minutes.
- The was a 5 minute wait for a second bus.
- The second bus completed the journey in the remaining 10 minutes.

From this information, we can draw up this table of times and distances.

Time	5	15	25	30	40
Distance	0.6	0.6	8.6	8.6	12

And draw this graph:

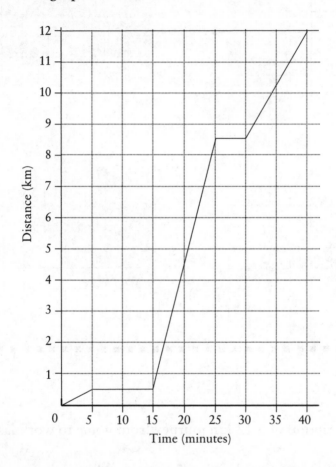

To calculate the average speed for a journey we use the formula:

$$\text{average speed} = \frac{\text{distance}}{\text{time}}$$

The journey to work was 12 km in 40 minutes.

The average speed was $\frac{12}{40} = 0.3$ km per minute.

0.3 km per minute $= 0.3 \times 60$ km per hour
$= 18$ km per hour

1 Mary Walker cycles to school each day. Leaving home she rides to the Co-op where she waits for her friend Benny Shaw. They then ride on together to the end of Pascal Drive, where they wait for their friend Lac Tran. The three then cycle together to school. This graph illustrates Mary's journey one day last week.

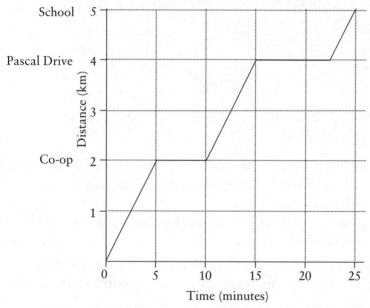

a) How far is the Co-op from Mary's house?
b) How long does Mary wait for Benny?
c) How far is the end of Pascal Drive from the Co-op?
d) How long do Mary and Benny wait for Lac Tran?
e) How far is it from Mary's house to her school?
f) What is Mary's average speed in kilometres per hour for the journey from;
   i) home to the Co-op?
   ii) the Co-op to the end of Pascal Drive?
   iii) home to school?

2 The graph shows the journeys made by a cyclist and a motorist. The cyclist left town A at 10.00 a.m. to travel to town B 40 miles

away, stopping for lunch on the way. The motorist did not leave town A until after his lunch. He had to stop on the way because of a puncture. He arrived at town B before the cyclist, in spite of the puncture.

a) At what time did the motorist leave town A?
b) How long did the cyclist take for lunch?
c) How far in front of the motorist was the cyclist at 1.15 p.m.?

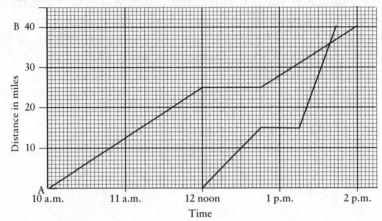

d) How far from town B was the motorist when he overtook the cyclist?
e) What was the motorist's average speed until he had to stop because of the puncture?
f) What was the cyclist's average speed for the whole journey between town A and town B? (NICCEA)

5 Jennifer walks from Corfe Castle to Wareham Forest and then returns to Corfe Castle. The travel graph of her journey is shown.

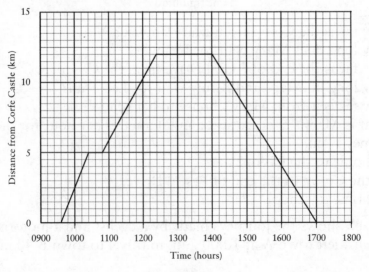

a) At what time did Jennifer leave Corfe Castle?
b) How far from Wareham Forest did Jennifer make her first stop?
c) Jennifer had lunch at Wareham Forest. How many minutes did she stop for lunch?
d) At what average speed did Jennifer walk back from Wareham Forest to Corfe Castle? (SEG)

6 Peter went on a cycle ride from home to his aunt's house and back. He had a rest on his way there. The travel graph shows part of his journey.

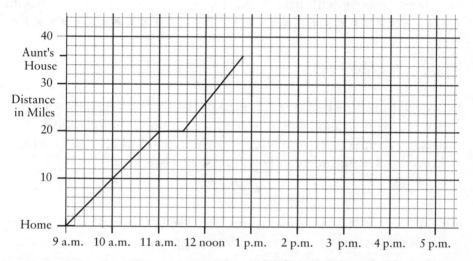

a) How far is his aunt's house from Peter's home?
b) How long did it take Peter to get from home to his aunt's house? Peter stayed for an hour at his aunt's house. Then he rode home without stopping at a speed of 12 miles per hour.
c) Use this information to copy and complete the travel graph for Peter's journey.
d) Use the graph to find at what time Peter arrived home. (ULEAC)

7 Jeremy lives at one end of a path which is 7.5 km long. He left home at 2.00 pm and ran to the other end of the path at a constant speed of 9 km/h.
a) Show that Jeremy took 50 minutes to run the full length of the path. When Jeremy reached the other end, he turned immediately and set off for home at a constant speed of 8 km/h. Thirty minutes after turning he stopped, sat down on a bench, and waited for his father to arrive.

187

b) On graph paper draw the travel-graph of Jeremy's journey so far. Jeremy's father left home at 2.30 pm and walked along the path at a steady 3 km/h until he reached Jeremy sitting on the bench. They sat chatting for 8 minutes before walking home together at a steady speed, arriving there at 4.30 pm.
c) Add to the graph the travel-graph of Jeremy's father's complete journey.
d) For how long did Jeremy sit on the bench?
e) For what fraction of the total time that Jeremy was away from home was he with his father? (NICCEA)

■■■■■■■■■■■■■■■■■■■■■■■■■■■■■■■■■■■■■

Conversion graphs

Many relationships can be expressed in the form $y = mx + c$ and illustrated with straight line graphs. For example, the relationship between a temperature measured in Fahrenheit and a temperature measured in Centigrade can be expressed in the form:

$$F = \frac{9C}{5} + 32$$

If we are interested in temperatures between 0°C and 100°C, we can construct this table of values and draw a graph.

°C	0	50	100
°F	32	122	212

The graph can be used as a *conversion graph*.
For example, we can use the graph to convert:

1. 75°C into 167°F
2. 100°F into 38°C

The lines drawn on the graph show how these conversions were made.

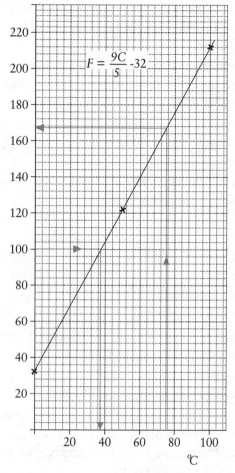

$$F = \frac{9C}{5} - 32$$

1 The Rick Dastardly Detective Agency charges £40 a day to hire one of their detectives. This can be expressed as the equation:

$$C = 40d$$

where C is the total cost and d is the number of days the detective is hired for.

a) The Purple Panther Detective Agency charges a fixed fee of £80 plus a charge of £20 a day to hire one of their detectives. Express this as an equation.

b) Prepare a graph on 2 mm graph paper, with a horizontal axis from 0 to 7 days and a vertical axis from 0 to £280. Use 2 cm to represent 1 day on the horizontal axis and 2 cm to represent £40 on the vertical axis.

c) Plot lines to represent the cost equations for both companies. What advice would you offer to somebody who intended to hire a detective from one of the two companies?

2 The cost of building a stretch of motorway is estimated at £5 million per kilometre, plus a fixed cost of £15 million. Express this as an equation.

a) Draw a graph to illustrate the cost equation, with a horizontal axis from 0 to 15 km and a vertical axis from 0 to £120 million. Use a scale of 1 cm to represent 1 km on the horizontal axis and 1 cm to represent £5 million on the vertical axis.

b) Use your graph to calculate the cost per kilometre of building a 12 km stretch of motorway.

c) Use your graph to calculate the cost per kilometre of building a 4 km stretch of motorway.

3 Tulip bulbs are offered for sale on the following terms:
  ■ 10p each for orders of up to 20 bulbs
  ■ 9p each for orders of more than 20 bulbs
  ■ 8p each for orders of more than 40 bulbs.

a) Prepare a graph with a horizontal axis from 0 to 80 bulbs and a vertical axis from 0 to 650 pence. Use a scale of 1 cm to represent 10 bulbs on the horizontal axis and 1 cm to represent 50 pence on the vertical axis.

b) Plot lines to represent the cost of buying between 1 and 20 bulbs, between 21 and 40 bulbs and between 41 and 80 bulbs.

c) What advice would you offer to somebody buying either 20 bulbs or 40 bulbs?

4 The Hire-A-Heap car hire company offer an executive saloon at three different daily hiring rates.
  ■ Scheme A: £20 per day plus 4 pence per mile.
  ■ Scheme B: £25 per day plus 2 pence per mile.
  ■ Scheme C: £33 per day with no mileage charges.

The daily cost of hiring a car under Scheme A can be expressed as the equation:

$$C = 20 + 0.04m$$

Where C is the daily cost and m in the number of miles covered.

a) Write down an equation to express the daily cost of hiring a car under Scheme B.
b) Prepare a graph with a horizontal axis from 0 to 500 miles and a vertical axis from 0 to £50. Use a scale of 1 cm to represent 50 miles on the horizontal axis and 1 cm to represent £5 on the vertical axis.
c) Plot lines to represent the daily hire costs under each of the three schemes.
d) Jeff Collinson wants to hire a car for a holiday during which he antici pates he will travel an average of 150 miles each day. Which hiring scheme will be the cheapest for Jeff?
e) Susan Taylor wants to hire a car for a holiday during which she anticipates she will travel an average of 300 miles each day. Which hiring scheme will be the cheapest for Susan?
f) Aruna Patel wants to hire a car for a day to make a business trip of 500 miles. Which hiring scheme will be the cheapest for Aruna?

■ ■

Graphs that model real situations

An electrician has a fixed charge of £10 to call at a house plus a charge of £12 for each hour of work he does.

We can draw this graph to illustrate the cost of up to 6 hours work.

The gradient of the graph is $\dfrac{72}{6} = \dfrac{12}{1}$.

The meaning of this gradient is that labour costs are £12 per hour.

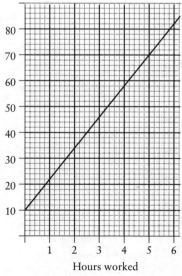

Hours worked

1 The graph shows the cost of using various amounts of electricity.
   a) What is the cost of using 500 units of electricity?
   b) Why does the graph start at £10 on the cost axis?
   c) The graph passes through the points (0,10) and (500, 40).
      i) What is the horizontal change between these points?
      ii) What is the vertical change between these points?
      iii) What is the gradient of the graph?
   d) What is the physical meaning of this gradient?

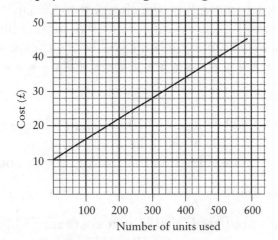

2 This graph shows the length of a spring as it is loaded with various weights.

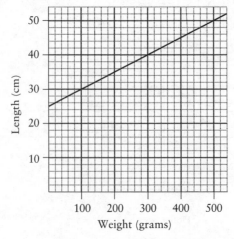

a) How long is the unstretched spring?
b) What weight is required to stretch the spring to a length of 38 cm?
c) The graph passes through the points (0,25) and (500,50).
   i) What is the horizontal change between these points?
   ii) What is the vertical change between these points?
   iii) What is the gradient of the graph?
d) What is the physical meaning of this gradient?

3 This graph shows the change in height of a candle as it burns.

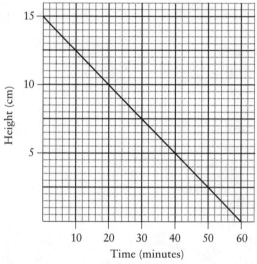

a) How high is the un-lit candle?
b) How long does it take the candle to burn down to a height of 8 cm?
c) The graph passes through the points (0,15) and (60,0).
   i) What is the horizontal change between these points?
   ii) What is the vertical change between these points?
   iii) What is the gradient of the graph?
d) What is the physical meaning of this gradient?

4

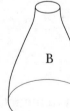

These three containers all have water poured into them at a steady rate. The graphs record how the depth of water increases with time.

Which graph belongs to which container?

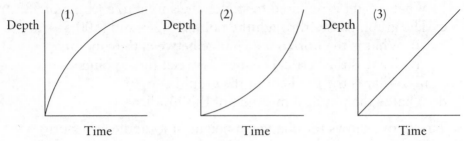

(1) Depth / Time
(2) Depth / Time
(3) Depth / Time

5 Make a copy of the axes and use them to make a sketch graph of petrol sales during the day.

The garage opens at 8am with its petrol tank $\frac{3}{4}$ full. It does a brisk trade until 9am. Trade between 9am and 12 noon is slow and steady. Between 12pm and 2pm trade picks up but it is very slow in the afternoon. No petrol is sold between 2 pm and 3 pm and very little between 3pm and 4pm. Sales are fast from 4pm until 7pm when they tail off. The garage closes at 8pm with its petrol tank nearly empty.

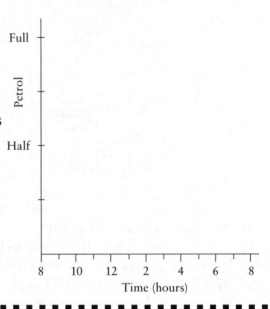

Full

Petrol

Half

8   10   12   2   4   6   8
Time (hours)

REWIND

1 Write each of the following statements in standard mathematical shorthand.

a) The number y is always 5 more than the number x.

b) 3 times the number t, minus 4 times the number s is equal to the number r.

c) To convert a temperature in degrees centigrade (C) into a temperature in degrees fahrenheit (F), multiply by 9, divide by 5 and then add 32.

d) The final velocity (v) of an object which accelerates is equal to its initial velocity (u) plus the product of its acceleration (a) and the time for which it accelerates (t).

2 If $a=6$, $b=3$ and $c=5$, find the value of:

a) $a+b$
b) $b-c$
c) $2a+3b$
d) $5c-2b$

e) ab
f) $2ac-ab$
g) ab^2
h) a^3b

i) $17+2cb^2$
j) $a(b+c)$

3 If $a=-2$, $b=-1$ and $c=5$, find the value of:

a) $a+b$
b) $b-c$
c) $2a+3b$
d) $5c-2b$

e) ab
f) $2ac-ab$
g) ab^2
h) a^3b

i) $17+2cb^2$
j) $a(b+c)$

4 When an electrical appliance of W watts power is used for h hours, in an area where eletricity costs c pence per unit, the price of the electricity used in P pence is given by:

$$P=\frac{Whc}{1000}$$

Find the cost of running a 750 watt iron for 4 hours in an area where electricity costs 12p per unit.

5 Simplify each of the following expressions:

a) $17x+y+3x+5y$
b) $3x+2y-2x+6y$
c) $x-12y+7x+9y$
d) $-x+2y+7x+5y$

e) $2m^2+8m-5m+27$
f) $2a^2-14a-2a+14$
g) $9v^3+5v^2-3v^2+2v$
h) $14s^2t+3t^2s-4st^2-7ts^2$

6 Expand the brackets and simplify:

a) $4(3r+5)$
b) $4(7r-5)$
c) $-(3m-5n)$
d) $-a(1-b)$
e) $4r(3r-2r^2+9)$

f) $5(8r+5)+3(4r-9)$
g) $(4m-3n)-(3n-4m)$
h) $a(b-1)-(ab-1)$
i) $4n(3n+2)-5(6n+1)$
j) $x(2y-x)-y(2y-x)$

7 Expand:

a) $(x+4)(x+1)$
b) $(x-3)(x+9)$
c) $(5x+3)(x-3)$

d) $(b-1)^2$
e) $(x+y)(x-y)$
f) $(x+3y)^2$

8 A boat leaves harbour with 12 000 litres of fuel oil on board. Each day the engines use 800 litres of oil. Write down a formula to calculate the oil that is left after the boat has been at sea for t days.

9 Factorise completely each of the following expressions.

a) $20x+35$

b) $b-6ab$

c) $15e^2-5ef$

d) xy^2+x^2y+xy

e) $abcd+acd+bcd$

f) $14m^2n-21mn^2+7mn$

10 Solve:

a) $6m=36$

b) $108=12t$

c) $12x=-26$

d) $t+45=104$

e) $18=r-7$

f) $15=5-r$

g) $\frac{x}{4}=12.5$

h) $\frac{t}{5}=-13$

i) $-21=\frac{7v}{2}$

j) $2x+7=11$

k) $13+7m=48$

l) $9-8m=49$

m) $\frac{x}{4}-5=1$

n) $3(2x-1)=51$

o) $4(1-2x)=8$

p) $2(2x-1)-3(2x+1)=7$

11 Solve:

a) $9p+4=5p-12$

b) $x-25=5-5x$

c) $5m+13-m=3m-23$

d) $\frac{x}{5}+4=x-6$

e) $9+\frac{a}{2}=a-7$

f) $7(x-2)+2=x-6$

g) $5(1-2p)=5(p+2)-5$

h) $4+9(2x+8)=3(8+2x)-x$

12 Solve, giving answers correct to 2 decimal places.

a) $2.3x+11.9=23.4$

b) $3(y-4.6)+5.2=7.6$

13 Solve these inequalities and illustrate your solution on a number line.

a) $x+5<11$

b) $y-8>-23$

c) $9e<-72$

d) $\frac{y}{8}>3.25$

e) $5x-7>53$

f) $8.5+\frac{w}{4}<4.75$

g) $\frac{5e}{3}+9>-2$

h) $7(2x-7)<0$

14 Rearrange each of these formulae to make the letter given in brackets the subject.

a) $y=x+5$ ⠀⠀(x)

b) $p=t-7$ ⠀⠀(t)

c) $C=2\pi r$ (r) g) $s=u+vt$ (u)

d) $d=\dfrac{C}{\pi}$ (C) h) $y=\dfrac{7x}{8}$ (x)

e) $y=4x-1$ (x) i) $A=\dfrac{h\,(a+b)}{2}$ (a)

f) $m=tx-tr$ (r) j) $s=\dfrac{t(u+v)}{2}$ (t)

15 Solve the following simultaneous equations.

a) $x+y=9$
$x-y=1$

b) $2x+y=10$
$x+y=7$

c) $p=7+q$
$2p=44-q$

d) $5a+3b=7$
$5a+8b=2$

e) $3p-2q=0$
$3p+2q=30$

16 Solve the following simultaneous equations.

a) $4a-3b=6$
$4a-2b=8$

b) $3e-2f=10$
$2e-2f=8$

c) $6m-2n=-4$
$8m-2n=6$

d) $-3=q-p$
$27=5q-p$

17 Solve the following simultaneous equations.

a) $m+5n=9$
$3m+n=13$

b) $m+4n=10$
$6m-3n=-21$

c) $4x-3y=15$
$5x+2y=13$

d) $4x+4y=30$
$3x+6y=30$

e) $5t-7s=25$
$7t-5s=11$

18 Find a solution (correct to one decimal place) to the following quadratic equations in the given range of values for x.

a) $x^2-2x=1$ between 2 and 3

b) $x^2-6x+7=0$ between 1 and 2

c) $2x^2+3x=6$ between 1 and 2

d) $3x^2-9x+5=0$ between 2 and 3

19 a) The 11th term in the sequence of even numbers is .....

b) The 2000th term in the sequence of even numbers is .....

c) The 2000th term in the sequence of odd numbers is .....

d) The 10th term in the sequence of prime numbers is .....

20 Write down the first five terms and the tenth term of the sequences produced by the following formulae:

a) $T_n = 3n - 1$

c) $T_n = \dfrac{360}{n}$

b) $T_n = 4n + 5$

d) $T_n = n^3$

21 Find a formula for the nth term of the following sequences:
   a) 1, 5, 9, 13, 17, ...
   c) 9, 18, 36, 72, 144, ...
   b) 6, 17, 28, 39, 50, ...

22 Draw diagrams to show each of the following mappings applied to the set of numbers −3, −2, −1, 0, 1, 2, 3
   a) $x \rightarrow x + 5$
   d) $x \rightarrow x(x+1)$
   b) $x \rightarrow 3x$
   e) $x \rightarrow x^2 - 3x + 4$
   c) $x \rightarrow \dfrac{x}{2} + 3$
   f) $x \rightarrow \dfrac{6}{x}$

23 Draw a graph and mark on it the lines with equation;
   a) $x = 3$ b) $y = -2$ c) $x = 0$ d) $y = 0$ e) $x = -3.5$ f) $y = 2.5$

24 Write down the equation of the line which passes through each of the following pairs of points:
   a) (2, −3) and (5, −3)
   b) (5, −2) and (5, 5)

25 Draw a graph and find the gradient of the line which connects each of the following pairs of points:
   a) (0,0) and (4,8)
   d) (−3,1) and (1,−3)
   b) (0,0) and (8,4)
   e) (−6,−2) and (−1,−1)
   c) (0,0) and (5,5)
   f) (−5,2) and (−4,−3)

26 Draw a graph to illustrate each of the following equations using x values from −4 to 4. Write down the gradient of the graph and the co-ordinates of the point where the graph cuts the y axis.
   a) $y = x + 5$
   e) $y = 2x - 7$
   b) $y = x - 6$
   f) $y = \dfrac{x}{2} + 1$
   c) $y = 2.5x + 1.5$
   g) $y = 6 - 2x$
   d) $y = \dfrac{x}{6}$
   h) $y = \dfrac{2(x-1)}{4}$

27 Write down the equations of the lines which have the following gradients and cutting points on the y axis.
   a) gradient 3, cutting point (0,2)
   b) gradient $\dfrac{1}{2}$, cutting point (0,−5)
   c) gradient −2, cutting point (0,0)
   d) gradient $\dfrac{-1}{4}$, cutting point (0,5)

198

28 Blank computer discs are offered for sale on the following terms:

- £1.40 each for orders of up to 9 discs.
- £1.20 each for orders of 10 or more discs.
- £1.10 each for orders of 20 or more discs.
- A fixed handling charge of £2 is added to all orders.

a) Prepare a graph with a horizontal axis from 0 to 30 discs and a vertical axis from 0 to £40. Use a scale of 2 cm to represent 10 discs on the horizontal axis and 2 cm to represent £10 on the vertical axis.

b) Plot lines to represent the cost of buying between 1 and 9 discs, between 10 and 19 discs and between 20 and 30 discs.

c) What advice would you offer to somebody buying either 9 discs or 19 discs?

29 Draw a graph to illustrate each of the following equations:

a) $5y=x+1$ d) $3x-8y=24$
b) $6y=3x-18$ e) $x+y=0$
c) $9x+2y=18$ f) $x-y=0$

30 Solve each of the following pairs of simultaneous equations by drawing a graph.

a) $y=2x+2$ b) $4x+5y=20$
  $y=3x+4$ $6x+5y=30$

31 a) Copy and complete this table of values for $y=x^2+4x-5$

x	−5	−4	−3	−2	−1	0	1	2
y	0		−8			−5		

b) Use the table to draw the graph of $y=x^2+4x-5$ as x takes values from −5 to 2.

32 If a rectangle is formed from a loop of string 18 cm long, it can take many different shapes. It could for example be a 6 cm high by 3 cm wide rectangle or a 4 cm high by 5 cm wide rectangle.

a) Copy and complete this table showing the dimensions and areas of rectangles formed from an 18 cm loop of string.

Size of rectangle (height×width)	9×0	8×1	7×2	6×3	5×4	4×5	3×6	2×7	1×8	0×9
Area (square centimetres)	0	8	14							

b) Draw a graph to illustrate the change in area as the width of the rectangle varies from 0 cm to 9 cm.

c) What is the maximum area that can be enclosed by a rectangle formed with an 18 cm loop of string?

33 a) Copy and complete this table of values for $y=8+2x-x^2$

x	−3	−2	−1	0	1	2	3	4	5
y	−7		5		9			0	

b) Use the table to draw the graph of $y=8+2x-x^2$ as x takes values from −3 to 5.

c) Use your graph to solve the equations:
   i) $8+2x-x^2=0$ iii) $8+2x-x^2=x$
   ii) $8+2x-x^2=4$

34 Draw a graph to illustrate the function $y=\dfrac{8}{x}$ as x takes values from −8 to 8.

35 A group of students want to travel from Birmingham to London to visit the Science Museum. A 40-seater coach will cost £120 to hire for the return journey.

a) Copy and complete this table to show the cost per student for the trip for different numbers of students sharing the coach hire.

Number of students on coach	5	10	15	20	25	30	35	40
Cost per student (£)	24	12			4.80			

b) What is the minimum number of students who must go on the trip to keep the cost under £5 a head?

36 Illustrate each of the following inequalities with a graph.
   a) $x<3$ c) $x>0$
   b) $y>-2$ d) $y<0$

37 Illustrate on a graph the regions where:
   a) $x>-2$ and $x<1$ d) $x>4$ and $y>4$
   b) $y>-2$ and $y<-4$ e) $x<4$ and $y<-3$
   c) $x>0$ and $y<0$ f) $x<-3$ and $y>-1$

38 Illustrate on a graph the regions where:
   a) $y>x-3$ d) $x>0$, $y>0$ and $y>x+3$
   b) $y<2x+1$ e) $-3<x<-3$, $y<3$ and $x+3y>1$
   c) $2x+y>8$

39 Illustrate on a graph the regions where:

a) $1<x<4$, $y>x-3$ and $y<x+1$

b) $4x+6y<24$, $y>x-6$ and $y>0$

40 The Hire-a-Heap car rental company hire two types of car. The Astra has 4 seats and costs £30 a day and the Carlton has 5 seats and costs £40 a day.

a) How many seats are provided in:

   i) 6 Astras?

   ii) x Astras?

   iii) 8 Carltons?

   iv) y Carltons?

   v) x Astras and y Carltons?

b) A party of 40 people want to hire cars to transport them to a wedding. If they hire a mixture of x Astras and y Carltons, explain why:

$$4x+5y \geq 40$$

c) The party does not wish to spend more than £360 on their transport. Explain why:

$$30x+40y \leq 360$$

d) The Hire-a-Heap company only has 7 Carltons and 9 Astras. Explain why:

$$y \leq 7 \text{ and } x \leq 9$$

e) Draw a graph to illustrate all these inequations.

f) How many Astras and Carltons should the party hire if they wish to obtain their transport at the minimum cost?

41 At 11.30 a.m., a lorry travelling at 60 mph passes under a motorway bridge on which a stationary police car is standing. Five minutes later, the police car drives down on to the motorway and starts to travel in the same direction as the lorry at 70 m.p.h.

Draw a distance/time graph and estimate the time at which the police car overtakes the lorry and the distance from the bridge when this happens.

Use a scale of 2 cm to represent 5 minutes on the time axis and a scale of 2 cm to represent 10 miles on the distance axis. Your time axis should extend from 9.30 a.m. to 10.30 a.m. and your distance axis should extend from 0 to 80 miles.

42 This label gives cooking instructions for a bacon joint. Draw a graph to illustrate the cooking times required for joints from 1 to 4 pounds in weight.

PREMIUM TENDERSWEET
LOIN
ROAST
ENGLISH SMOKED BONELESS

Cooking instructions: Remove joint from bag and roast
in a pre-heated oven 180° C, 350° F, Gas Mark 4
for 35 minutes per lb, plus 35 minutes.

F A S T F O R W A R D

1 a) Write, in symbols, the rule:

'to find y, double x and add 1'.

b) Use your rule from part a) to calculate the value of x when $y=9$.

(ULEAC)

2 Theatre tickets cost £7 each.

a) Write an equation to find the cost of n theatre tickets.

b) The total cost is reduced by £10 when more than ten tickets are purchased.

Write an equation to find the cost of n theatre tickets when n is more than ten.

(SEG)

3 Use the values $x=9$ and $y=5$ to work out:

a) $x+y$

b) $2y-3$

c) y^3

d) $\sqrt{x}$

(MEG)

4 a) Expand and simplify $(2x+3)(x-4)$.

b) Factorise completely $10x^2-5x$.

(ULEAC)

5 a) Factorise completely $12p^2q-15pq^2$.

b) Expand and simplify $(2x-3)(x+5)$.

c) The cost, C pence, of printing n party invitations is given by
$$C=120+40n.$$
Find a formula for n in terms of C. (MEG)

6 a) $S=180n-360$.
Find the value of S when $n=10$.
  b) $T=\dfrac{a+m}{2}$.
Find the value of T when $a=6$ and $m=81$. (MEG)

7 The power, P watts, consumed by an electric light bulb of resistance R ohms when a current of I amps is passed through it is given by:
$$P=I^2R$$
a) Find the value of P when $R=50$ and $I=4$.
b) Express I in terms of P and R. (MEG)

8 A formula connecting T, f and g is $T=4f-5g$.
Work out the value of T when $f=6.4$ and $g=3.9$. (MEG)

9 Lilian asked her uncle how old he was. 'In 13 years, I'll be twice as old as I was 7 years ago,' he replied.
a) Taking his age now to be x years, write down:
    i) his age in 13 years, in terms of x.
    ii) an equation in x.
b) Solve your equation and find Lilian's uncle's age. (MEG)

10 The cost, C pence, of a newspaper advertisement of n words is given by the formula
$$C=12n+32$$
Find the cost of an advertisement of 16 words. (MEG)

11 The price £P, charged by 'Motif Shirts' for making sweat shirts of your own design is given by the formula $P=3N+20$, where P is the price in pounds and N is the number of shirts ordered.
a) Work out the price of 40 shirts.
b) Work out the price per shirt when 40 shirts are ordered. (MEG)

12 Fran, Jo and Tom share some money.
Jo gets 50p more than Tom.
Fran gets twice as much as Jo.
Let x be the number of pence Tom gets.
Write expressions in terms of x for the number of pence given to:
a) Jo,

b) Fran,
c) all three together.
The money shared is £6.70.
d) Write down an equation in x and solve it to find x.

(NICCEA)

13 The diagram shows a square and a rectangle. The square has sides of length $2y$ metres. The rectangle has length $3y$ metres and breadth 3 metres.

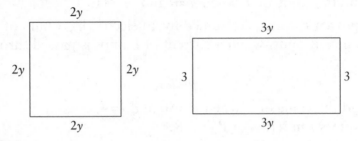

a) i) Find, in terms of y, the perimeter of the square.
   ii) Find, in terms of y, the perimeter of the rectangle.
b) The perimeter of the square is equal to the perimeter of the rectangle.
   i) Form an equation in y.
   ii) Work out the value of y. (SEG)

14 a) Solve the equation:
$$6+u=20$$
b) Using the formula:
$$v=\frac{1}{a}+\frac{1}{b}$$
calculate v when $a=8$ and $b=5$
c) Factorise completely:
$$3ax^2-6axy$$
d) Rearrange the formula:
$$s=\tfrac{1}{2}at^2$$
so that t is the subject. (ULEAC)

15 The sides of a triangle are a cm, $(a-2)$ cm and $(a+3)$ cm, as shown.

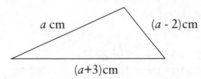

a) What is the perimeter of the triangle in terms of a?

b) The triangle has a perimeter of 19 cm. Calculate the value of a.

(MEG)

16 Solve:

$$3x-4=11 \qquad \text{(ULEAC)}$$

17 Solve these equations:
  a) $3x+2=18-5x$
  b) $2(x+3)=18-6x$

(SEG)

18 Solve the equation $4(x-3)=22$ (MEG)

19 Solve the equation

$$11x+5=x+25 \qquad \text{(ULEAC)}$$

20 Ruchi is $\frac{1}{3}$ the age of her Father.
  Ruchi is x years old.

  a) Write down, in terms of x,
    i) Father's age,
    ii) Ruchi's age 9 years ago,
    iii) Father's age 9 years ago.

  Nine years ago, Father was 6 times as old as Ruchi.

  b) i) Using some of your answers to a), write down an equation
       in x which represents the above statement.
     ii) Solve your equation to find the value of x. (ULEAC)

21 a) Solve the equation $9x+6=36-x$
  b) Solve the simultaneous equations:

$$4p-q=15$$
$$2p-q=9 \qquad \text{(ULEAC)}$$

22 In a Video Hire shop, when n videos are hired, the profit is
  £$(pn-q)$, where p and q are constants.
  On Monday, when 70 videos were hired, the profit was £60.
  On Tuesday, when 56 videos were hired, the profit was £39.

  a) i) Use this information to write down two equations in p and q.
     ii) Solve these equations to find the values of p and q.

  On Wednesday, the shop made a loss of £12.

  b) Calculate the number of videos hired on Wednesday.

(ULEAC)

23

$$5x+3y=84$$
$$3x+5y=68$$

a) Add the equations to find the value of $(x+y)$.

b) Subtract the equations to find the value of $(x-y)$.

c) Hence, or otherwise, solve the equations for x and y.   (NEAB)

24 Solve the simultaneous equations:
$$2x+3y=23$$
$$x-\ y=\ 4$$
(ULEAC)

25 Alex is using 'trial and improvement' to solve the equation:
$$x^2-3x=1$$
Continue the working to find the solution correct to one decimal place.

Try $x=3$ $3^2-3\times3=0$ too small

Try $x=4$ $4^2-3\times4=4$ too large

Try $x=3.5$

Try $x=$ (SEG)

26 Judy is using 'trial and improvement' to solve the equation:
$$x^2+x=11$$
Complete her working and find a solution correct to one decimal place.

Try $x=3.5$ $3.5^2+3.5=15.75$ too large

Try $x=2.5$ $2.5^2+2.5=8.75$ too small

Try $x=3.0$

Try $x=$ (SEG)

27 The equation $x^2+2x=13$ has a solution for x between 2.7 and 2.8. Use trial and improvement methods to find this solution correct to two decimal places.

Show all the trials clearly.   (ULEAC)

28 a) Use the formula $S=180n-360$ to find the value of S when $n=10$.

b) Use the formula $n=\dfrac{S+360}{180}$ to find the value of n when $S=720$
(ULEAC)

29 Here is a formula for working out the perimeter of a rectangle.
$$P=2(l+w)$$
Use the formula to work out the value of P when $l=6$ and $w=4$.
(ULEAC)

30 Here is a formula:
$$v=u+10t$$
Find the value of v when:

a) $u=6$, $t=8$

b) $u=5$, $t=0$ (ULEAC)

31 Rearrange the formula:
$$V=u+ctu$$
to give c in terms of V, u and t. (ULEAC)

32 a) Re-arrange the formula $v^2=u^2+2as$ to make a the subject.
   b) Calculate a when $v=15$, $u=10$ and $s=125$. (NEAB)

33 List all the possible integer values of n such that
$$-3\leq n<2$$
(SEG)

34 a) Solve the inequality
$$4n+3<18$$
   b) Given that n is an integer,
   write down the greatest value of n for which
$$4n+3<18$$
(ULEAC)

35 Solve these inequalities.
   a) $2x+1\leq 5$
   b) $x^2>25$
   c) $7x+3>13x+15$ (ULEAC)

36 a) List all the integers which satisfy
$$-2<n\leq 3$$
   b) Ajaz said 'I thought of an integer, multiplied it by 3 then
   subtracted 2. The answer was between 47 and 62'.
   List the integers that Ajaz could have used. (MEG)

37 Find the next two numbers in each of these simple number
patterns.
   a) 10, 35, 60, 85, ..., ...
   b) 29, 22, 15, 8, ..., ... (ULEAC)

38 Here are the first four lines of a number pattern.
   Line 1   5 → 11
   Line 2   10 → 16
   Line 3   15 → 21
   Line 4   20 → ☐

   a) What is the missing number in the box?
   b) Explain how you found the missing number.
   c) Write down Line 20 of the pattern. (ULEAC)

39 Here are the first four numbers of a number pattern
$$7, 14, 21, 28, ..., ...$$

207

a) Write down the next two numbers in the pattern.

b) Describe, in words, the rule for finding the next number in the pattern. (ULEAC)

40 Here are the first four terms of a number sequence,

$$7, 11, 15, 19.$$

Write down the nth term of the sequence: (ULEAC)

41 The rule for a sequence is:

> Multiply the previous number by 3, then add 2.

The first four numbers of the sequence are:

$$3, 11, 35, 107$$

a) Work out the next two numbers in the sequence.

  x represents a number in this sequence.

b) Write down, in terms of x, an expression for the next number after x in this sequence. (ULEAC)

42 a) Write down the next term in the series

$$x, x^3, x^5, x^7,$$

b) What is the value of this term when $x=1$? (MEG)

43 a) On a grid, plot and label the points

$$P(4,1) \qquad Q(3,-2) \qquad R(-3,-2) \qquad S(-2,1).$$

b) Give the geometrical name of quadrilateral $PQRS$. (ULEAC)

44 Copy and complete the following table for the rule.

a) 'To find y, double x and add 1'.

x	y
0	
1	
3	

b) Plot the values from the table onto a coordinate grid. Join your points with a straight line.

c) Write, in symbols, the rule

'To find y, double x and add 1'.

d) Use your rule from part c) to calculate the value of x when $y=9$. (ULEAC)

45 a) Draw a graph and plot the points (1,3) and (4,6). Join the points with a straight line.

b) The point $P(a, 5)$ lies on the line.
   What is the value of a?
c) The line is extended.
   Complete the following mapping for points on the line.

$$1 \rightarrow 3$$
$$5 \rightarrow 7$$
$$10 \rightarrow \dots\dots$$
$$x \rightarrow \dots\dots$$

(SEG)

46

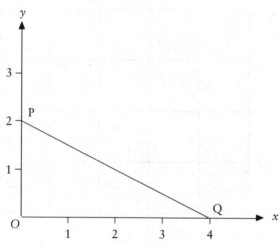

Calculate the gradient of the line PQ.

(ULEAC)

47 The diagram shows the point A with coordinates $(2, 3)$.

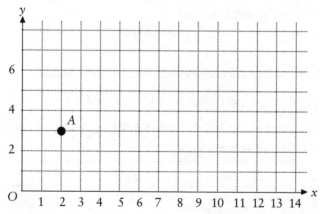

a) Copy the diagram and draw in the straight line passing through
   A with gradient $-\dfrac{1}{2}$.
b) Write down the coordinates of the point where your line crosses
   the y-axis.

(MEG)

48 Huw observes a bird flying directly away from a bird box. He starts his watch and finds out how far the bird is from the box at different times.

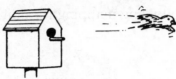

This graph is drawn from his results.

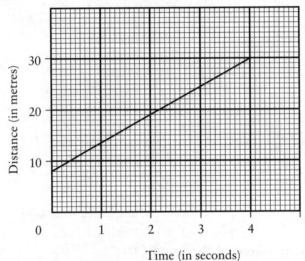

a) How far is the bird from the box when Huw starts his watch?
b) How fast is the bird flying?
c) Write down a formula for the distance, *d*, the bird is from the box in terms of time, *t*.   (WJEC)

49 The number of kilometres (*y*) approximately equivalent to a number of miles (*x*) is given by the formula $y=\dfrac{8x}{5}$.

a) Copy and complete the following table, using the formula to calculate the values of *y*.

x	0	20	40	60
y	0		64	

b) i) Draw a pair of axes using a scale of 2 cm to represent 10 units on each axis. Mark on your graph the points from your table.

ii) Draw the *straight* line which is the graph of $y=\dfrac{8x}{5}$.

c) Use your graph to find
   i) the number of kilometres equivalent to 52 miles,
   ii) the number of miles equivalent to 43 kilometres.
d) Work out the number of kilometres equivalent to 200 miles.

(MEG)

50 The equations of five straight lines are:

$$y= x-2,$$
$$y=2x+3,$$
$$y=3x+2,$$
$$y=5x+2,$$
$$y=3x-3.$$

Two of the lines go through the point (0,2).
a) Write down the equations of these two lines.
Two of the lines are parallel.
b) Write down the equations of these two lines.   (ULEAC)

51 A longlife battery and a standard life battery were both tested for their length of life.

The longlife battery lasted for x hours.

The standard life battery lasted for y hours.
a) The combined length of life of the two batteries was 14 hours.
   Explain why $x+y=14$.
b) Draw the graph of $x+y=14$
c) The longlife battery lasted 3 hours longer than the standard life battery.
   Write down another equation connecting x and y.
d) Copy and complete the table of values for your equation in c) and use it to add this equation to your graph.

x	3	7	10
y			

e) Use your graphs to find the length of life of each type of battery.
(SEG)

52 a) Draw the graphs of the equations
$$y=2x+1 \text{ and } x +y=7.$$
b) Use your graph to solve the simultaneous equations
$$y=2x+1,$$
$$x+y=7.$$
(ULEAC)

53 Describe, using inequalities, the region illustrated in the diagram below.

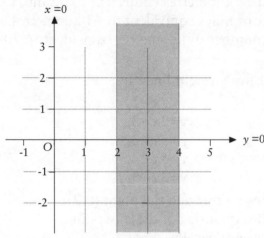

(SEG)

54 a) Copy and complete this table of values for $y=2x-1$.

x	-2	-1	0	1	2	3
y		-3				

b) Draw the graph of $y=2x-1$
c) Use your graph to find the value of x when $y=\frac{1}{2}$.
d) Shade the region of your graph which represents $y\geq 2x-1$

(ULEAC)

55 Corks released from Champagne bottles can reach great heights. For a cork leaving a bottle with speed V (in metres per second) the maximum height (in centimetres) is given by

$$h=5\times V^2$$

a) Complete this table of values.

V	0	1	2	3	4	5
h	0	5			80	

b) On a graph plot the points and join them all in a smooth curve.
c) Use your graph to find:
    i) the height reached by a cork which leaves the bottle at 3.5 m/s,
    ii) the speed a cork leaves the bottle if it reaches 30 cm. (MEG)

56 a) Given that $y=x^2$, copy and complete the following table.

x	-3	-2.5	-2	-1.5	-1	-0.5	0	0.5	1	1.5	2	2.5	3
y	9	6.25		1		0		1				6.25	9

b) Using a scale of 2 cm to represent 1 unit on each axis, draw the graph of $y=x^2$ for values of x from -3 to 3.

c) i) Using the same scales and axes as in part b), draw the straight line $y=3-x$.

   ii) Write down the gradient of the straight line.

d) Using your graphs of $y=x^2$ and $y=3-x$ and making your method clear, find the solutions of the equation $x^2=3-x$. (MEG)

57 A pebble is thrown upwards from the edge of a sea-side cliff and eventually falls into the sea. The height of the pebble above the sea after t seconds is h metres, where h is given by the formula

$$h=24+8t-2t^2$$

a) Copy and complete the table for the values of h.

t	0	1	2	3	4	5	6
h							

b) Using a scale of 2 cm for 5 m on the h-axis and 2 cm for 1 second on the t-axis, draw a graph of h against t for $0\leq t\leq 6$.

c) Find:
   i) the height of the cliff,
   ii) how high the pebble rises above the level of the cliff-top,
   iii) after how many seconds the pebble lands in the sea,
   iv) by drawing a suitable line, an estimate for the speed of the pebble after 5 seconds. (ULEAC)

58

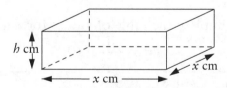

A rectangular block has a square base of side x cm and a height of h cm. The total surface area of the block is 72 cm^2.

a) Express h in terms of x.

b) Show that the volume, V cm^3, of the block is given by

$$V=18x-\tfrac{1}{2}x^3$$

c) Copy and complete the following table to show corresponding values of x and V.

x	0	1	2	3	4	5	6
V	0			40.5	40		0

213

d) Using a scale of 2 cm to represent 1 unit on the x-axis and 2 cm to represent 10 units on the V-axis, draw the graph of $V=18x-\frac{1}{2}x^3$ for values of x from 0 to 6 inclusive.

e) A block of this type has a volume of 30 cm³. Given that $h>x$, find the dimensions of the block. (MEG)

59 The stopping distance s feet for a car travelling at v mph is given by the formula $s=v+\frac{v^2}{20}$.

a) Copy and complete the table.

v	10	30	40	60	80
s		75		240	400

b) Draw the graph of s against v.

c) What is the maximum speed at which a car can travel in order to be able to stop safely within 200 feet?

Two cars, travelling in the same direction, are approaching a set of traffic lights. Car B is directly behind car A.

Car A is travelling at 35 mph.

Car B is travelling at 45 mph.

The traffic lights change and both cars brake at exactly the same time.

d) What is the minimum safe distance between the cars when the lights change? (NEAB)

60 a) Copy and complete the table of values for y given that:
$$y=3x+\frac{12}{x}-10.$$

x	1	1.5	2	3	4	5	6
y	5	2.5			5		10

b) Draw the graph of $y=3x+\frac{12}{x}-10$, for values of x from 1 to 6. Use a scale of 2 cm to represent 1 unit on each axis.

c) i) Using the same axes and scales, draw the graph of $y=x+3$.

ii) Use your graphs to find the x coordinates of the points of intersection of the graphs of
$$y=3x+\frac{12}{x}-10 \text{ and } y=x+3.$$

d) From the given equations, obtain and simplify a quadratic equation which has these x coordinates for its roots. (Do not attempt to solve this equation). (ULEAC)

61 a) Copy and complete the table of values for the function:
$$y=(x+1)(x-4).$$

x	-2	-1	0	1	2	3	4	5
$x+1$	-1		1	2			5	6
$x-4$	-6		-4	-3			0	1
y	6		-4	-6			0	6

b) Draw the graph of:
$$y=(x+1)(x-4) \text{ for } -2\leq x\leq 5.$$

c) Use your graph to estimate the value of x for which y has its minimum value. (NEAB)

62 The numbers 4, 6 and 9 are all factors of 36.
a) Explain why 5 is not a factor of 36.
b) i) Complete these pairs of whole numbers which multiply to give 36.
(2, ...), (3, ...), (4, 9), (6, 6), (9, 4), (12, ...)
ii) Copy the grid below. Plot these six pairs of whole numbers as points on the grid and join them with a smooth curve.

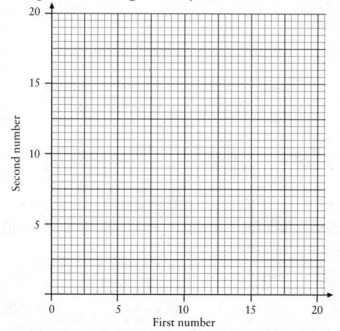

215

iii) By drawing lines on your graph show how to find the
number which multiplied by 5 gives 36. (SEG)

63

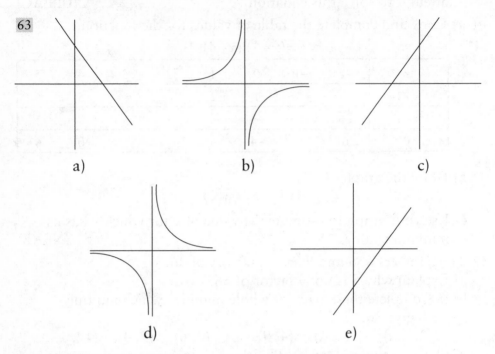

a)   b)   c)

d)   e)

Which of the above graphs could represent the functions:

    i) $y = 2x + 3$,

    ii) $y = \dfrac{6}{x}$.   (NICCEA)

64

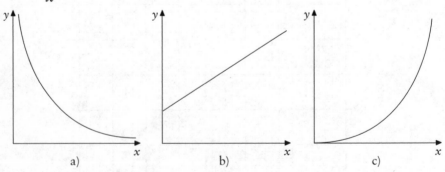

a)   b)   c)

For each graph, select the one equation which the graph could
represent (for positive x).

    i) $y = x + 3$ iv) $y = 3x^2$

    ii) $y = x - 3$ v) $y = -3x^2$

    iii) $y = 3x$ vi) $y = \dfrac{3}{x}$ (NICCEA)

65 This graph can be used to change between £'s and French Francs.

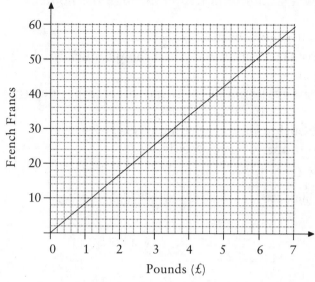

Use the graph to change:

a) i) £6 to French Francs,
    ii) £2.90 to French Francs,

b) i) 55 French Francs to £'s,
    ii) 40 French Francs to £'s.   (ULEAC)

66 5 gallons are equivalent to 22.7 litres.

a) Draw a conversion graph.

b) Use your graph, showing its use clearly, to find:
   i) how many gallons are equivalent to 12 litres,
   ii) how many litres are equivalent to $4\frac{1}{2}$ gallons.   (NICCEA)

67 A plumber earns a basic weekly wage plus extra money for working overtime. The table shows the total weekly wage when the plumber works different amounts of overtime.

Hours of overtime worked	Total weekly wage
5	£280
10	£320
20	£400

a) i) Draw a graph using a scale of 2 cm to 5 hours on the Hours of overtime axis and 2 cm to £100 on the Total weekly wage axis.

217

ii) Plot the values given in the table.
iii) Draw a straight line graph through the points you have plotted.
b) From your graph, find:
  i) the total weekly wage when the plumber works 7 hours of overtime,
  ii) how many hours of overtime the plumber needs to work to earn £340 in a week,
  iii) the hourly rate of overtime pay,
  iv) the plumber's basic weekly wage.   (ULEAC)

68 Fiona's ride from Luton to Hitchin:

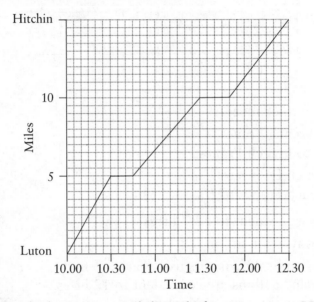

This graph shows Fiona's bike ride from Luton to Hitchin. She set out in the morning and stopped for 2 rests.
a) What time did she set out?
b) How far is it from Luton to Hitchin?
c) How long was her second rest?
d) During which part of her journey was she cycling fastest?
e) How fast was she cycling for the first half-hour?
f) At what time did she reach Hitchin?
g) What was her average speed for the whole journey?

Gary left Hitchin at 10.00 a.m. and raced to Luton, arriving at 11.20 a.m.

h) Show this journey on the graph.

i) At what time did he pass Fiona?

j) How far from Luton did they pass each other?   (MEG)

69
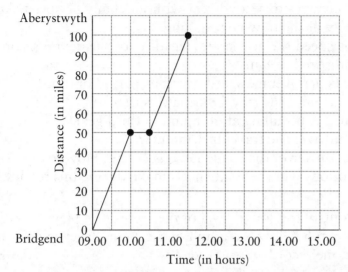

The above graph represents part of the journey of a motorist travelling from Bridgend to Aberystwyth and back to Bridgend.

a) Describe fully the section of the journey shown in the graph.

b) The motorist stays in Aberystwyth for $1\frac{1}{2}$ hours. He then drives back to Bridgend without stopping, arriving at 15.00.

Complete the graph to show this section of his journey.   (WJEC)

70 The graph shows the relationship between the rate of turning of the pedals of Anne's bicycle and the road-speed, for three gears.

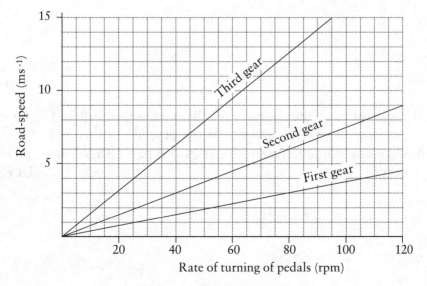

a) Anne is in first gear. She is pedalling at 35 r.p.m. What is her speed?
b) She now changes to second gear without changing speed. At what rate will she now be pedalling?
c) From this speed she accelerates steadily for the next 6 seconds to reach a speed of 8 ms⁻¹.
   i) What is her average speed over this period of time?
   ii) How far does she travel during this time?
d) Later, Anne is pedalling at 60 r.p.m. in third gear.
   i) How far will she travel in 1 minute?
   ii) The diameter of the roadwheels is 66 cm. How many revolutions will the roadwheels make during this minute?

(MEG)

71 A DJ can control the sound level of the records he plays at a disco. The sketch graph below is a graph of the sound level against the time whilst one record was played.   (ULEAC)

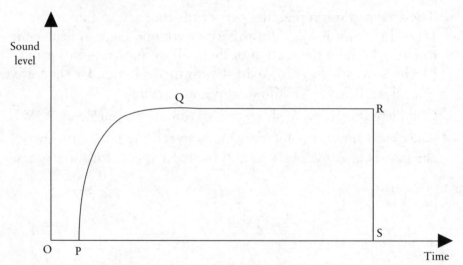

a) Describe how the sound level changed between P and Q on the graph whilst the record was being played.
b) Give one possible reason for the third part, RS, of the sketch graph.   (ULEAC)

72

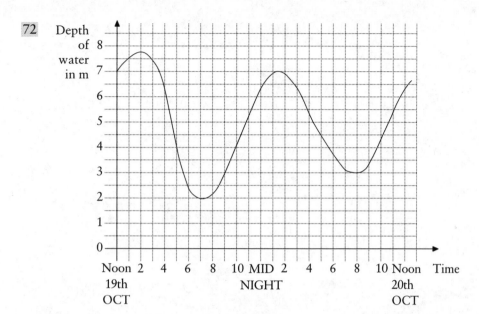

The graph shows the depth of water in a harbour from noon on October 19th to noon on October 20th 1992.

a) From the graph, write down,
  i) the depth of the water in the harbour at 5 p.m. on October 19th,
  ii) the time of high water on October 19th,
  iii) how much time there is from one high water to the next high water.

A ship can come into the harbour only if the depth of water is 4 metres or more. The ship was on its way to the harbour at noon on October 19th.

b) How much time was there before the water level would be too low for the ship to come into the harbour?   (ULEAC)

Shape and Space

3

As civilisations developed, more and more practical mathematical knowledge was needed. The construction of buildings required accurate measurement of length and angle.

The ownership of land required surveying techniques and the measurement of area. The spread of trade required the development of arithmetic and the measurement of weights and volumes.

The discoveries of many different early mathematicians were drawn together by the Greek mathematician Euclid. Euclid described the mathematics of shape and space as *Geometry*, from the Greek words 'geo' meaning earth and 'metry' meaning to measure.

Bisecting a line or an angle

These diagrams show how a ruler and a pair of compasses can be used to cut a line or an angle in half. This is called **bisecting** the line or the angle.

To bisect a straight line.

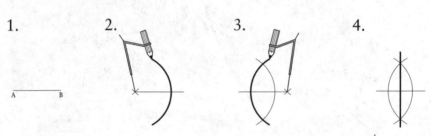

To bisect an angle.

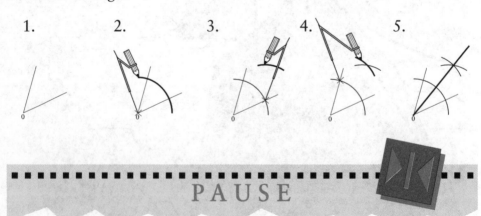

PAUSE

1 Draw lines with these lengths and then bisect them. Measure each half of the line to check the accuracy of your construction.

   a) 10 cm   c) 11 cm
   b) 12 cm   d) 8.6 cm

2 Draw the angles and then bisect them. Measure each half of the angle to check the accuracy of your construction.

a) 60°   c) 120°
b) 40°   d) 136°

3 This diagram shows a 3 cm radius circle with three chords drawn in it (a chord is any line crossing the circle).

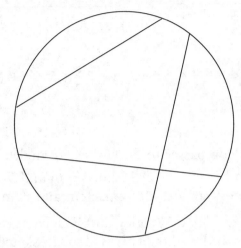

a) Copy the diagram.
b) Contruct the bisector of each chord.
c) What do you notice?

4 Draw any circle and mark in a diameter AB.

a) Contruct several angles by joining A to a point on the circle and then to B.
b) What do you notice?

■ ■

Constructing triangles

This diagram shows the stages in drawing a triangle with sides of 5 cm, 3 cm and 4 cm.

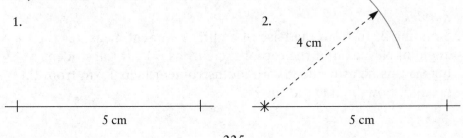

1.

2.

4 cm

5 cm

5 cm

3.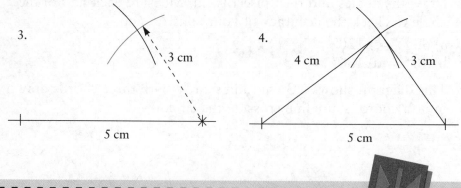

4.

PAUSE

1 Use a ruler and compasses to draw these triangles.

   a) 4 cm, 4 cm and 5 cm   c) 12 cm, 13 cm and 5 cm
   b) 8 cm, 6 cm and 9 cm   d) 9 cm, 9 cm and 9 cm

2 The villages of Smidgley, Widgley and Ridgley form a triangle with sides 4 km, 7 km and 9.5 km. This is a rough sketch map of the villages.

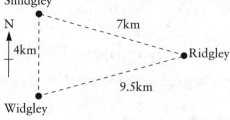

Using a scale of 1 cm = 1 km, draw an accurate map showing the three villages.

Scale drawings

Examples

1. Standing 25 m from the base of a cliff, a student measures the angle of elevation of the top of the cliff as 60°. If the student makes this measurement with an instrument held 1.5 m from the ground, how high is the cliff?

Use a scale of 1 cm = 5 m.
We can draw this diagram, using the scale given.

The height of the triangle is 8.7 cm.
8.7 cm represents $5 \times 8.7 = 43.5$ m.

Remembering to add on the
1.5 m for the height of the
instrument, this gives a height
for the cliff of 45 m.

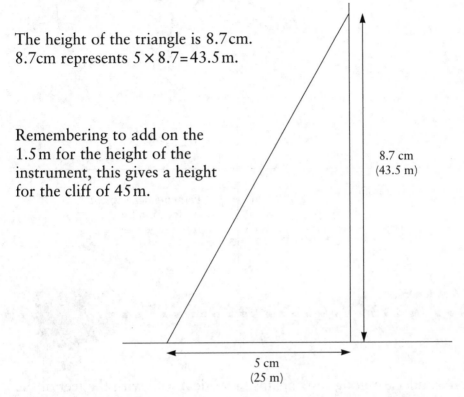

8.7 cm
(43.5 m)

5 cm
(25 m)

2. Three towns form a triangle with sides of 12 km, 15 km and
20 km. A supermarket chain wishes to build a superstore which is
the same distance from each of the three towns. Show the relative
positions of the three towns on a diagram and the position for the
superstore.

Using a scale of 1 cm = 2 km, we start by drawing the triangle
connecting the towns.

We bisect two sides of the triangle. Where these bisectors cross
must be a point which is the same distance from all three towns.

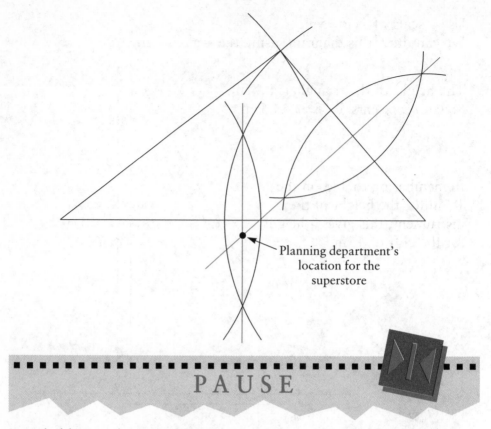

Planning department's
location for the
superstore

1 A ladder 6 m long rests against a vertical wall with the feet of the ladder 1.5 m from the base of the wall.

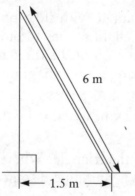

6 m

1.5 m

Make an accurate drawing using a scale of 2 cm = 1 m. How far does the ladder reach up the wall?

2 Amberbridge is 25 km from Beaverbrooke and 50 km from Smidgley. Smidgley is 40 km from Beaverbrooke. A plane flies over Amberbridge on a flight path which bisects the angle formed by Smidgley, Amberbridge and Beaverbrooke.

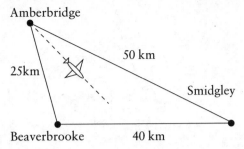

Draw an accurate map showing the three towns using a scale of 1 cm = 5 km. As the plane continues along this flight path, what is the closest that it comes to Smidgley?

3 A camping site is in the form of a triangle with straight sides 100 m, 150 m and 180 m long. Busy roads run along all three sides of the site.

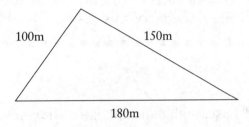

Draw an accurate scale diagram of the site using a scale of 1 cm = 10 m. Show on your diagram the best position to pitch a tent so that it is as far away as possible from all three roads.

4 The triangular framework shown is *not* drawn to scale.

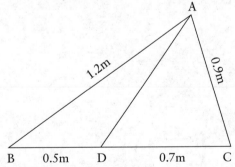

a) Make a scale drawing of the framework, taking a scale of 1 cm to 0.1 m.

Do not remove any of your construction lines.

b) Use your scale drawing to find:
   i) the size of angle BDA,
   ii) the length of AD in metres. (NICCEA)

5

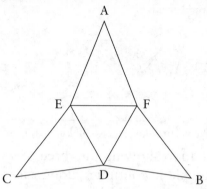

A design for a badge consists of three congruent isosceles triangles AEF, BFD, CDE placed as shown.
a) What type of triangle is DEF?
b) Given that AE = AF = 6.2 cm and EF = 4 cm, use instruments to make a full size drawing of this design. (NICCEA)

■ ■

Map scales

This is a map of East Anglia and the East Midlands drawn to a scale of 1 cm = 20 km.

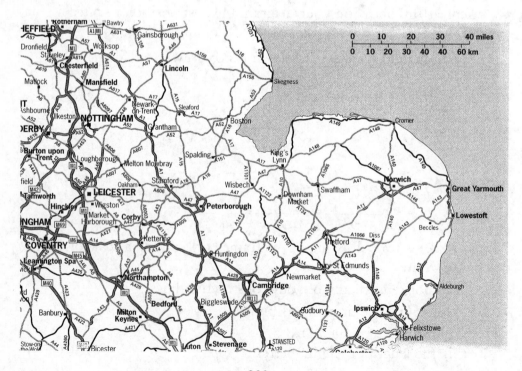

Example

1. Estimate the direct distance between Norwich and Peterborough.

   On the map the distance is 5.5 cm.

   So, the distance between Norwich and Peterborough
   = 5.5 × 20 = 110 km.

2. If two towns are 136 km apart, what will be the distance between
   them on the map?

   $$Distance = 136 \div 20 = 6.8 \, cm.$$

PAUSE

Estimate the direct distance between:

1 Kings Lynn and Swaffham,

2 Swaffham and Lowestoft,

3 Lowestoft and Felixstowe,

4 Felixstowe and Lincoln,

5 Lincoln and Leamington Spa,

6 Leamington Spa and Sleaford,

7 Sleaford and Boston,

8 Boston and Cambridge,

9 Cambridge and Worksop,

10 Worksop and Skegness.

What will be the distance on the map between towns which are:

11 178 km apart.

12 286 km apart,

13 34 km apart,

14 58 km apart,

15 242 km apart?

This is a map of part of London drawn to a scale of 1 cm = 100 m.

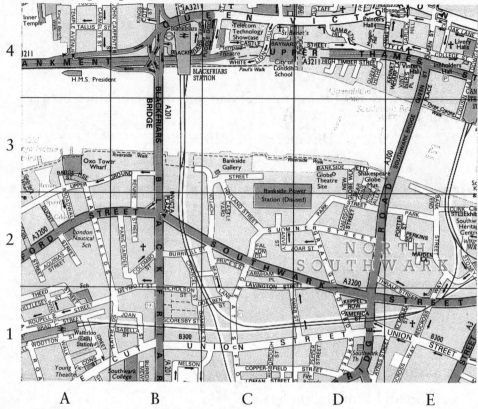

Example

1. Estimate the length of Southwark Bridge.

   Southwork Bridge is 2 cm long on the map.

   The length of Southwark Bridge = 2 × 100 = 200 m.

2. If a street is 480 metres long, how long will it be on the map?

   Length = 480 ÷ 100 = 4.8 cm.

PAUSE

Estimate the length of:

1 Scoresby St., B1

2 Sumner St., C2/D2

3 Holland St., C2/C3

4 Blackfriars Bridge, B3/B4

5 Union St., B1/C1/D1/E1

6 Great Guildford St., D1/D2

7 Hatfields, A2/A3/B1/B2

8 Park St., D2/E2

9 Temple Avenue, A4

How long will a street be on the map if its real length is:

10 597m,

11 147m,

12 831m,

13 1km,

14 1.3km?

■ ■

Bearings

Bearings are used to measure the direction in which one location lies from another. For example, this sketch map shows the locations of Cambridge and Norwich.

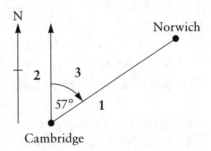

Three steps are needed to measure the bearing of Norwich from Cambridge.

1 Draw in a line connecting Cambridge to Norwich.
2 Draw in a line through Cambridge pointing due North.
3 Measure the clockwise angle between the North line and the line connecting the two places.

Bearings are always written with three figures. The bearing of Norwich from Cambridge is 057°.

The bearing of *Cambridge from Norwich* is *not* the same as the bearing of *Norwich from Cambridge*. This map shows the necessary construction to measure the bearing of Cambridge from Norwich.

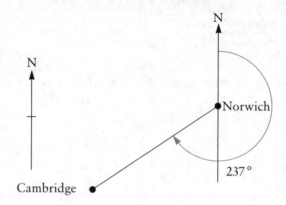

The bearing of Cambridge from Norwich is 237°.

PAUSE

1 Newcastle-on-Tyne is 70 miles to the north and 50 miles to the east of Kendal.
   a) What is the bearing of Newcastle from Kendal?
   b) What is the bearing of Kendal from Newcastle?
2 Manchester is 40 miles to the north and 40 miles to the west of Nottingham.
   a) What is the bearing of Manchester from Nottingham?
   b) What is the bearing of Nottingham from Manchester?
3 Bradford is 150 miles to the north and 40 miles to the west of Chelmsford.
   a) What is the bearing of Bradford from Chelmsford?
   b) What is the bearing of Chelmsford from Bradford?

4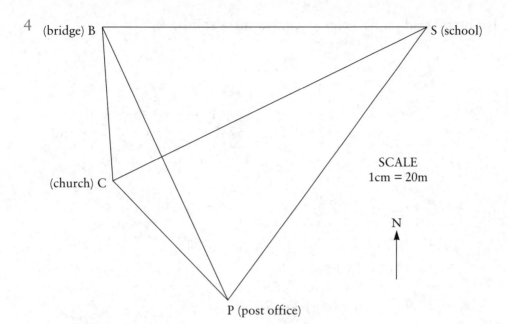

(bridge) B

S (school)

(church) C

SCALE
1cm = 20m

N

P (post office)

The diagram is an accurate scale drawing showing some features of a village. The bridge is due north of the church.

a) i) Measure the size of angle BCS.
   ii) What is the bearing of the school from the church?
b) i) Measure the length of PS. Give your answer in centimetres.
   ii) How far is the school from the post-office?
c) Taking measurements from the drawing, find:
   i) the distance, BP, of the post-office from the bridge,
   ii) the bearing of the post-office from the bridge.   (NICCEA)

5 In the village of Little Marsh, the church is 100 m due north of the post office.
   a) Use a scale of 1cm to 20m to draw the positions of the church and post office.
   b) The village Hall, H, is south-east of the church and 130 m from the post-office. On your drawing find by construction the position of H.

■ ■

Solids

These are some common solids, with some words we use to describe them. Hidden edges are often shown by dotted lines to make the diagrams clearer.

Isometric grid paper helps to draws these solids.

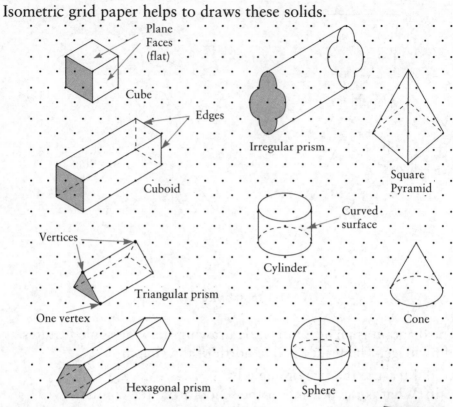

Plane
Faces
(flat)

Cube

Edges

Cuboid

Irregular prism

Square
Pyramid

Vertices

Curved
surface

Triangular prism

Cylinder

One vertex

Cone

Hexagonal prism

Sphere

PAUSE

1 Make a copy of each
diagram on isometric
grid paper.

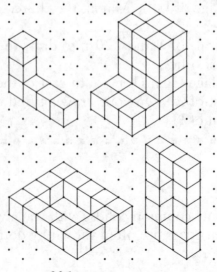

2 List the names of each basic solid which have been used to build up the diagram.
Use isometric grid paper to copy the diagram.

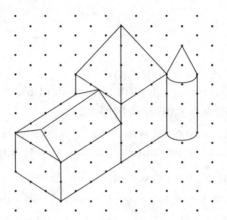

3 Draw an imaginary building on isometric grid paper.

■ ■

Nets for solids

A net when cut out and folded forms the outside of a hollow solid.
For example, this is one possible net for a cube.

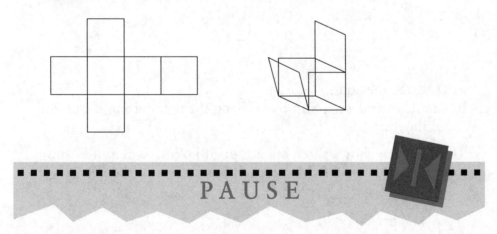

■ ■
PAUSE

1 a) Here are two more patterns of six squares:

237

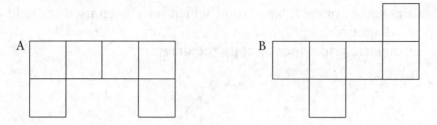

Which one is the net of a cube?

b) Draw on squared paper five more nets for a cube. Check each one by cutting it out and folding it.

2 Draw a net for a cuboid which is 3 cm wide, 4 cm high and 5 cm long. Check your net by cutting it out and folding it.

3 Draw a net for a cuboid which is 3 cm wide, 3 cm high and 4 cm long. Check your net by cutting it out and folding it.

4 Draw a net for a cuboid which is 2 cm wide, 2.5 cm high and 3 cm long. Check your net by cutting it out and folding it.

5 This is the net for a solid drawn on a grid.

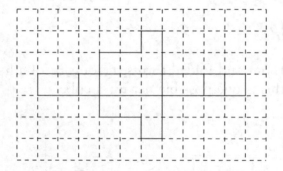

a) Describe the solid.

b) Check your description by copying the net, cutting it out and folding it.

6 This diagram shows three open topped boxes, A, B and C and three nets, D, E and F. Match each box to one of the nets.

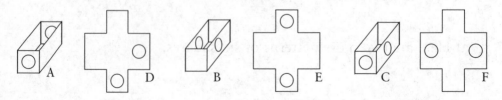

7 Describe the solids that you think each of the following nets will make.

a)

c)

b)

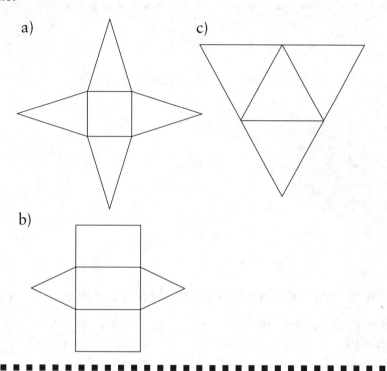

■ ■

Plans and elevations

This diagram shows three different views of the same car.

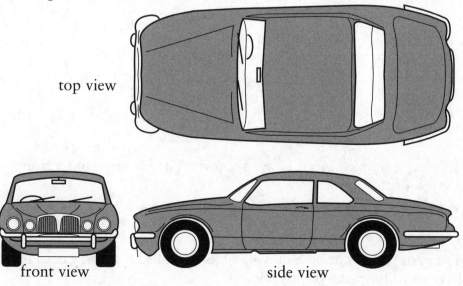

top view

front view

side view

P A U S E

1 A, B and C are three views of the triangular prism.

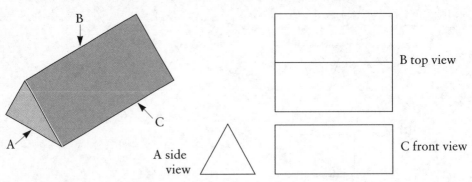

A side view

B top view

C front view

Sketch the required view in each of these questions.

a) Show a side view of this cuboid.

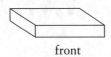

front

c) Show a top view of this hexagonal prism.

b) Show a top view of this cube.

front

d) Show a side view of this cylinder.

2 Select a simple object. Draw a top view, a side view and a front view of the object.

Loci: *paths of moving points*

The path traced out by a moving object is called a *locus*. Locus is the latin word for place or position.

To draw a locus, build up possible positions with a set of dots. Join the dots to get the locus.

Example
Point P moves so that it is always 2 cm from point O.

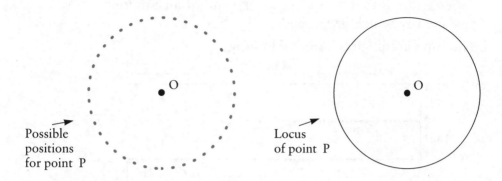

Possible positions for point P

Locus of point P

The locus is the circumference of a circle with centre O and radius 2 cm.

PAUSE

1 The line AB is 8 cm long. Point P moves so that it is always 3 cm from point A. Copy the diagram and mark on it the locus of P.

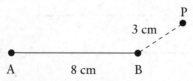

P

3 cm

A   8 cm   B

2 The line AB is 8 cm long. Point P moves so that it is always 3 cm from point B. Copy the diagram and mark on it the locus of P.

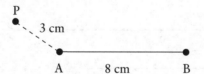

P

3 cm

A   8 cm   B

3 The line AB is 8 cm long. Point P moves so that it is always 3 cm from the line AB. Copy the diagram and mark on it the locus of P.

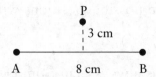

P

3 cm

A   8 cm   B

4 A fierce dog is tied to a stake at one end of an 8 m long
   passageway which is 2 m wide.

   The rope holding the dog is 5 m long.

8 m

1 m

5 m

1 m

   a) Copy the diagram which shows the passageway seen from
      above (use a scale of 1 cm = 1 m).
   b) Mark on the diagram the locus of the dog's movements as it
      strains against the rope.

5 Copy these diagrams and mark on them the locus of the dog's
   movements as it strains against the rope.

   a)

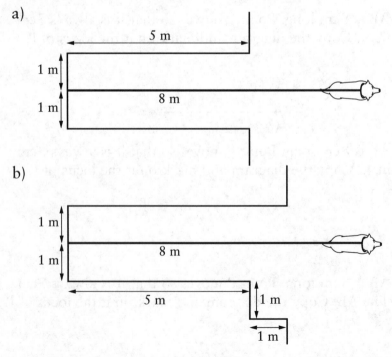

5 m

1 m

8 m

1 m

   b)

1 m

8 m

1 m

5 m

1 m

1 m

242

6 This diagram shows the square side of a packing case resting on a level floor. The packing case is rolled along the floor, tipping it over each corner in turn. The diagram shows the locus of the point A as the case is tipped once.

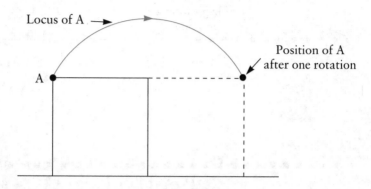

Copy the diagram and show the locus of corner A as the case is tipped two more times.

■ ■

Loci which are the bisectors of lines or angles

Examples

1. The line AB is 6 cm long. Point P moves so that its distance from point A is always equal to its distance from point B.

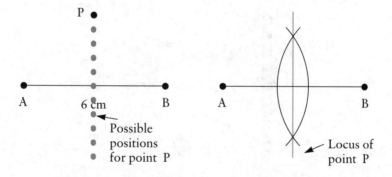

2. Point P moves so that its distance from the line AC is always equal to its distance from the line AB.

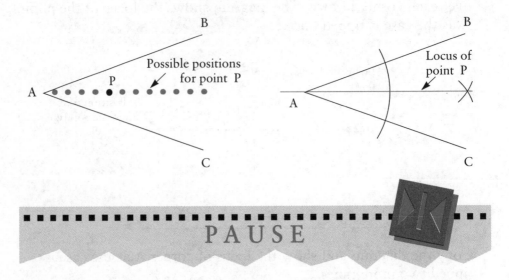

1 The line AB is 10 cm long. Point P moves so that its distance from point A is always equal to its distance from point B. Draw a diagram and mark on it the locus of P.

2 Angle BAC is 60°. Point P moves so that its distance from the line AC is always equal to its distance from the line AB. Draw a diagram and mark on it the locus of P.

3 Susan is about to release her catapult. This is the plan view.

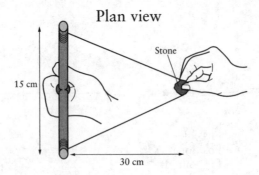

Plan view

a) Copy the Plan view (use a scale of 1 cm = 5 cm).
b) Mark on your diagram the locus of the stone after Susan releases it.

244

4 A hedgehog is at the corner of Angle Close.

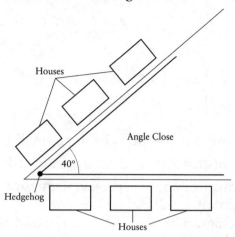

The hedgehog moves off along Angle Close keeping as far away
from all the houses as possible.

Copy the diagram and mark on it the locus of the hedgehog's
movement.

5 The diagram shows a field ABCD.

The line AB is 80 m long. The line BC is 50 m long. Draw ABCD
using a scale of 1 cm to 10 m.

Treasure is hidden in the field.

a) The treasure is at equal distance from the sides AB and AD.
   Construct the locus of points for which this is true.
b) The treasure is also 60 m from the corner C.
   Construct the locus of points for which this is true.
c) Mark with an X the position of the treasure.

(SEG)

6

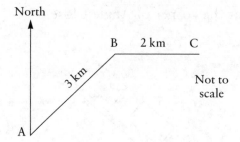

The diagram shows the positions of three small islands A, B, C.

B is 3 km north east of A. C is 2 km due east of B.

a) i) Using a scale of 2 cm to represent 1 km, draw the triangle ABC accurately.

   ii) Use your diagram to find the distance and bearing of island C from island A.

b) A yacht Y sails so that it is always 1.5 km from C.
   In your diagram, draw and label the path of Y.

c) A speedboat S moves so that it is always the same distance from A as it is from B.
   In your diagram, draw and label the path of S.

d) i) Mark clearly in your diagram the point Y_1 on the path of Y and the point S_1 on the path of S such that Y_1S_1 is the shortest distance between the two paths.

   ii) Hence find, in km, the shortest distance between the two paths. (MEG)

■ ■

Loci and regions

In some cases, a locus may be a region rather than a line.

For example, this diagram shows a goat tethered by a 5 m long rope to a post 3 m from the corner of a building. A grass lawn surrounds the building.

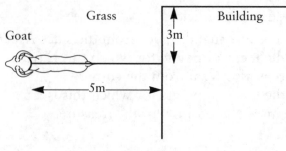

The goat can move about and eat the grass at any point in the shaded region shown below. This region is the *locus* of the goat's movements.

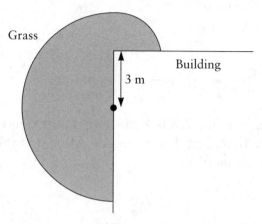

The edge of the region is shown with a solid line. This is because points on this edge (or boundary) are included in the region.

If points on the boundary are *not* included in the region, the boundary is drawn with a dotted line.

Example
Point P moves so that it is always less than 2 cm from point O.

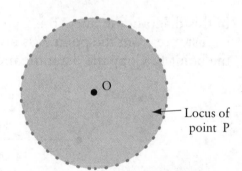

Locus of point P

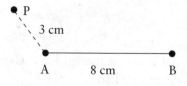

PAUSE

1 In this diagram the line AB is 8 cm long. Point P moves so that is is always less than 3 cm from point A. Copy the diagram and mark on it the locus of P.

```
● P
  ＼ 3 cm
    ●————————————●
    A   8 cm     B
```

2 In this diagram the line AB is 8 cm long. Point P moves so that it is always less than 3 cm from the point B. Copy the diagram and mark on it the locus of P.

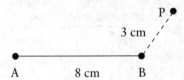

3 In this diagram the line AB is 8 cm long. Point P moves in so that it is always more than 3 cm from the line AB. Copy the diagram and mark on it the locus of P.

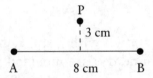

4 In this diagram the line AB is 8 cm long. The point P moves so that its distance from the point A is always less than its distance from the point B. Copy the diagram and mark on it the locus of P.

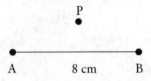

5 In this diagram, point P moves so that its distance from the line AC is always less than its distance from the line AB. Copy the diagram and mark on it the locus of P.

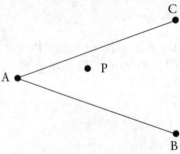

6 The map shows part of a coastline and a coastguard station. 1 cm on the map represents 2 km.

A ship is 12 km from the coastguard station on a bearing of 160°.
a) Copy the map and plot the position of the ship from the
   coastguard station, using a scale of 1 cm to represent 2 km.
   It is not safe for ships to come within 6 km of the coastguard
   station.
b) Shade the area on the map which is less than 6 km from the
   coastguard station.   (ULEAC)

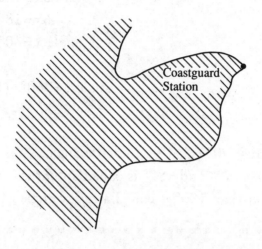

7

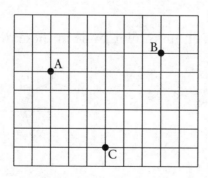

Scale: 1 cm represents 100 km.
A, B and C represent three radio masts on a one centimetre grid.
Copy the plan.
Signals from mast A can be received 300 km away, from mast B
350 km away and from mast C 200 km away.
Show by shading, the region in which signals can be received from
all 3 masts.
                                                         (ULEAC)

Lines and angles

Examples

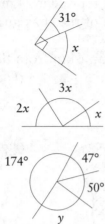

angle $x = 90° - 31° = 59°$

$x + 2x + 3x = 180°$
$6x = 180°$
angle $x = 30°$

angle $y = 360° - (174° + 47° + 50°)$
angle $y = 360° - 271° = 89°$

An angle less than 90° is called an *acute* angle.

An angle between 90° and 180° is called an *obtuse* angle.

An angle greater than 180° is called a *reflex* angle.

■■■■■■■■■■■■■■■■■■■■■■■■■■■■■■■
PAUSE

1 Find the size of each unknown angle in these diagrams:

a)

d)

g)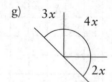

j)

b)

e)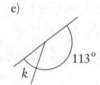

h)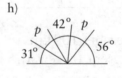

k)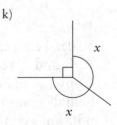

c)

f)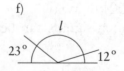

i)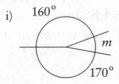

When a pair of lines intersect, then opposite angles are equal.

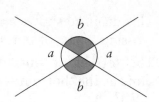

Example
Find the angles marked with letters.

$a = 180 - 50 = 130°$ (*a* and 50° are on a straight line)
$b = 50°$ (*b* and 50° are opposite angles)
$c = a = 130°$ (*c* and *a* are opposite angles)
$a = 130°$, $b = 50°$ $c = 130°$

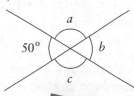

■ ■

PAUSE

Find the angles marked with letters.

1
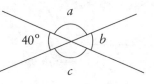

4
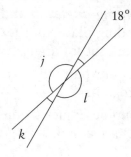

2
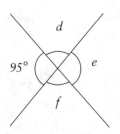

5
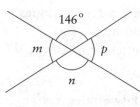

3
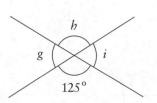

6
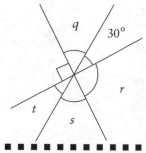

This diagram shows a third line crossing a pair of parallel lines.

The 'top set' of four angles is the same as the 'bottom set' of four angles.

All the angles marked with the letter *a* are equal.

All the angles marked with the letter *b* are equal.

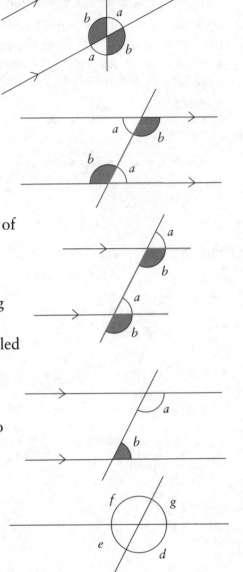

This diagram shows two pairs of **alternate angles**, *a* and *a* are on alternate sides of the crossing line.

Alternate angles (sometimes called 'Z' angles) are equal.

This diagram shows two of the pairs of **corresponding angles**, *b* and *b* are in corresponding places in the 'top set' and 'bottom set' of angles.

There are four pairs of corresponding angles altogether.

Corresponding angles (sometimes called 'F' angles) are equal.

a and *b* are **interior angles**. Interior angles add up to 180°. There are two pairs of interior angles.

Examples

1. Calculate angles *a*, *b*, *c*, *d*, *e*, *f* and *g*.

$a = 60°$
$b = 120°$
$c = 60°$
$d = 120°$
$e = 60°$
$f = 120°$
$g = 60°$

252

2. Find angles *a* and *b*.

  a = 40° (alternate angles)
  b = 180° − 110° = 70° (interior angles)

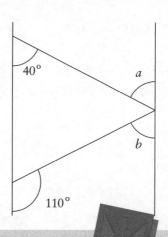

■■■■■■■■■■■■■■■■■■■■■■■■■■■■■■■■■■■■■■ ■ ■■ ■■■

PAUSE

1 Find the size of each unknown angle in these diagrams.

a)

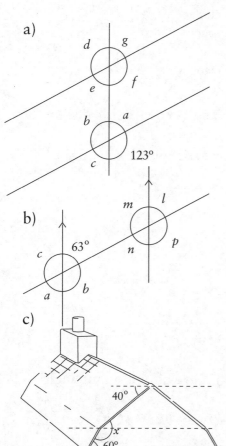

b)

c)

d)

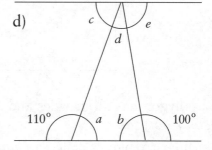

e)

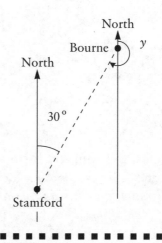

Polygons

A *polygon* is a shape made from straight lines.

These are all examples of polygons:

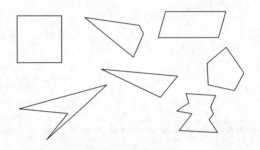

Number of sides	Name of polygon
3	triangle
4	quadrilateral
5	pentagon
6	hexagon
7	heptagon
8	octagon

If a polygon has all its sides and all its angles equal it is called a *regular polygon*.

A regular triangle is called an *equilateral* triangle. A square is a regular quadrilateral.

This is a regular octagon.

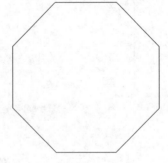

Triangles

Triangles, the simplest form of polygon, have an important angle property.

The three angles inside a triangle always add up to 180°.

Examples

1. Find angle x.

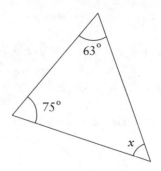

    angle $x = 180° - (75° + 63°)$
    angle $x = 180° - 138°$
    angle $x = 42°$

2. Find the size of each angle in this triangle.

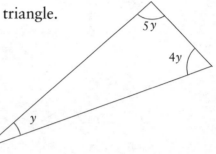

    $4y + 5y + y = 180°$
        $10y = 180°$
          $y = 18°$,
the angles are 18°, 90° and 72°.

An *isosceles* triangle has two equal sides. The equal sides are usually marked on a diagram by dashes.

The angles formed between the equal sides and the third side are always equal. In this isosceles triangle, AB and AC are the equal sides and angles ABC and ACB are the equal angles.

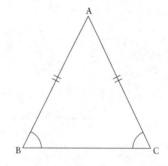

A *scalene* triangle has no equal sides or angles.

Example

Find angle x and angle y.
As the triangle is isosceles:

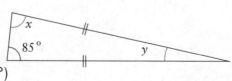

        angle $x = 85°$
          $y = 180° - (85° + 85°)$
           $= 10°$

255

1 Find the size of each unknown angle in these diagrams:

a)

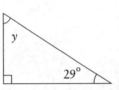

b)

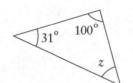

c)

d)

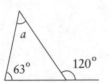

e)

f)

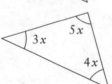

g)

h)

i)

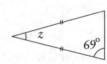

j)

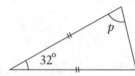

k)

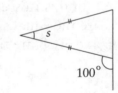

l)

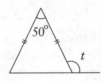

2

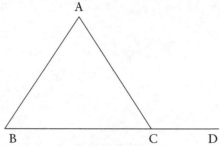

Triangle ABC is drawn on a straight line BCD.
The sides of the triangle are all the same length.

a) What special name is given
   to triangle ABC?
b) Find the size of
    i) angle BAC,
    ii) angle ACD.
c) Name an obtuse angle in
   the diagram above.

(ULEAC)

3

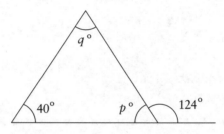

For the triangle above work out the value of:
a) p,
b) q.

(ULEAC)

4

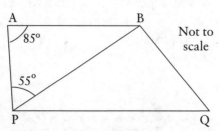

a) Work out the size of angle ABP.
b) AB is parallel to PQ.

   Work out the size of angle BPQ.

(SEG)

5

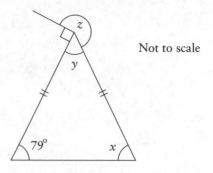

Not to scale

Work out the sizes of angle x, y and z.   (MEG)

6

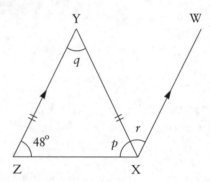

In the diagram XY = ZY and ZY is parallel to XW.

a) Write down the size of angle p.
b) Calculate the size of angle q.
c) Write down the size of angle r.   (SEG)

■ ■

Quadrilaterals

Any four-sided polygon is called a *quadrilateral*.
There are several different types of quadrilateral, each with its own
name and special properties. You are asked to discover some of these
properties in the next exercise.

■ ■

PAUSE

1 A *trapezium* is a quadrilateral with one pair of parallel sides.

258

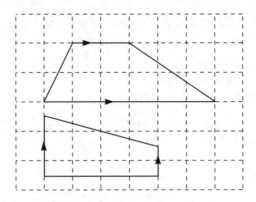

Draw three different trapeziums on squared paper.

2 A *parallelogram* is a quadrilateral with two pairs of parallel sides.

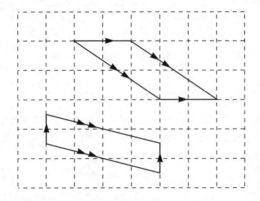

   a) Draw three different parallelograms on squared paper.
   b) By making measurements on your drawings, what can you
      discover about:
   • the angles of a parallelogram,
   • the diagonals of a parallelogram?

3 A *rhombus* is a quadrilateral with four equal sides.

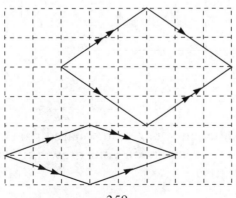

a) Draw three different rhombuses on squared paper.
b) By making measurements on your drawings, what can you discover about:
  • the angles of a rhombus,
  • the diagonals of a rhombus?

4 A *rectangle* is a quadrilateral with two pairs of parallel sides and interior angles of 90°.

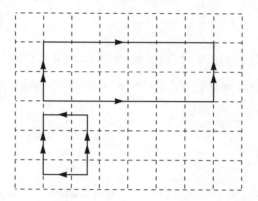

a) Draw three different rectangles on squared paper.
b) By making measurements on your drawings, what can you discover about:

  • the angles of a rectangle,
  • the diagonals of a rectangle?

5 A *square* is a quadrilateral with four equal sides and interior angles of 90°.

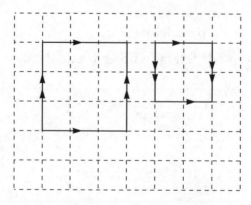

a) Draw three different squares on squared paper.
b) By making measurements on your drawings, what can you discover about:

- the angles of a square,
- the diagonals of a square?

6 A **kite** is a quadrilateral with two pairs of equal sides arranged in adjacent pairs.

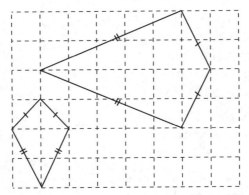

a) Draw three different kites on squared paper.
b) By making measurements on your drawings, what can you discover about:
- the angles of a kite,
- the diagonals of a kite?

■ ■

The angles in a polygon

The angles formed inside a polygon are called *interior* angles.

If each side of a polygon is extended, the angles formed are called the *exterior* angles of the polygon.

We know that the three interior angles of any triangle always add up to 180°. The interior angles of any other type of polygon also always add up to a fixed total.

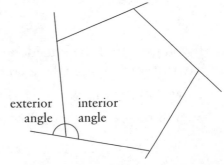

exterior angle   interior angle

Examples

1. Find the sum of the interior angles of the pentagon below.

   Starting at any corner point, divide the pentagon into triangles.

   The interior angles of the pentagon are equal to the angles of three triangles. The interior angles of the pentagon must add up to $180° \times 3$ or $540°$.

   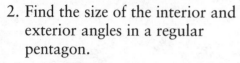

2. Find the size of the interior and exterior angles in a regular pentagon.

   From the last example, we know that the five interior angles in any pentagon total $540°$.

   In a regular pentagon, all the interior angles are equal.

   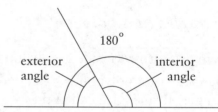

   Each interior angle must be equal to $540° \div 5 = 108°$.

   Each pair of interior and exterior form an angle of $180°$.

   So, each exterior angle must be equal to $180° - 108° = 72°$.

   180°

   exterior angle   interior angle

   If you go round the outside of any polygon, turning through each exterior angle you make a complete turn of $360°$.

   So we have the rule:

   Sum of the exterior angles of *any* polygon $= 360°$.

PAUSE

1 Find:

   a) the total of the interior angles in a quadrilateral,

b) the size of each interior angle in a regular quadrilateral,
c) the size of the exterior angles in a regular quadrilateral,
d) calculate angle x.

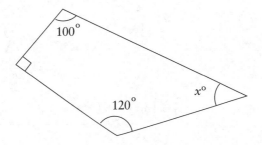

2 Find:

a) the total of the interior angles in a decagon (ten sides),
b) the size of each interior angle in a regular decagon,
c) the size of the exterior angles in a regular decagon.

3 Find:

a) the total of the interior angles in a hexagon,
b) the size of each interior angle in a regular hexagon,
c) the size of the exterior angles in a regular hexagon,
d) calculate angle x.

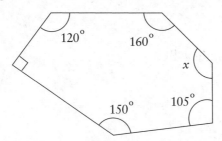

4 Find:

a) the total of the interior angles in an octagon (eight sides)
b) the size of each interior angle in a regular octagon,
c) the size of the exterior angles in a regular octagon.

5 Find the number of sides of each of these regular polygons, given exterior angles of:

a) 15°   b) 30°   c) 12°   d) 10°

6 If a regular polygon has interior angles of 162°, how many sides does it have?

7 a) Calculate the value of:
   i) p,
   ii) q.

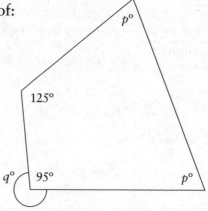

b) Calculate the value of:
   i) w,
   ii) x,
   iii) y.

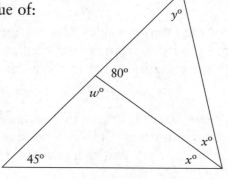

(NICCEA)

8 a)
ABCD is a kite.
Find the size of angle ABC.

b)

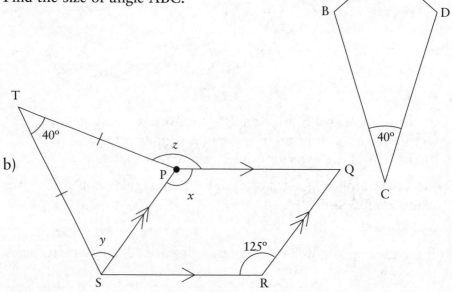

In the figure TS = TP, SP is parallel to RQ and SR is parallel to PQ.

Find the value of:

i) *x*,   ii) *y*,   iii) *z*.   (NICCEA)

9 a)

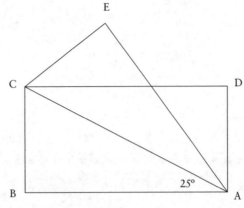

ABCD is a rectangle and ABCE is a kite.
Calculate the size of:

　i) angle ACB,　　　　　iii) angle DCE,
　ii) angle ACD,　　　　　iv) angle EAD.

b)

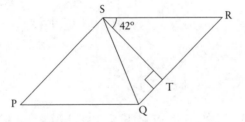

PQRS is a rhombus.
Calculate the size of:

i) angle SRT,　　ii) angle SPQ,　　iii) angle PSQ.   (NICCEA)

10

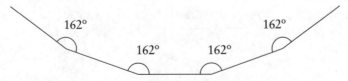

The diagram shows part of a regular polygon.
How many sides has the polygon?   (NEAB)

Tessellations

A *tessellation* is a pattern of shapes which covers a flat surface leaving no gaps or overlaps. For example, it is possible to fit regular hexagons together in this way to form a tessellation.

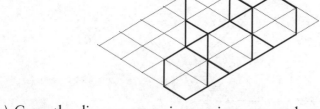

PAUSE

1 This is the start of a tessellation of equilateral triangles and hexagons.

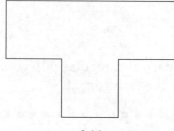

   a) Copy the diagram onto isometric paper and extend the tessellation.
   b) Draw a different tessellation of equilateral triangles and hexagons.

2 Use squared paper to draw a tessellation of squares.
   Can you discover more than one possible tessellation of squares?

3 On squared paper, draw a tessellation of this shape.

4 Regular pentagons do not tessellate. Calculate the size of the gap.

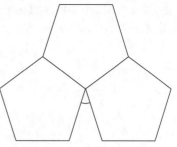

5 Regular octagons will not tessellate on their own. What other regular polygon will tessellate with regular octagons?

■■■

SECTION 2 Transformation Geometry

Transformation geometry describes the ways in which the position of an object can be changed.

All changes in position can be described in terms of three basic movements, *translations, rotations* and *reflections*.

Translations

A *translation* changes the position of an object by moving every point on the object through the same distance in the same direction.

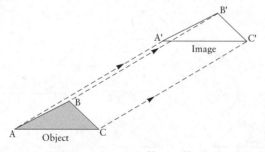

The shape in the new position is usually called the *image*.

If the object is labelled ABC the image is labelled A′B′C′.

Translations are usually applied to objects drawn on a square grid or a graph.

A translation is described using the *horizontal movement* and the *vertical movement*.

These movements are written in a bracket underneath each other so they do not look like coordinates.

This diagram shows how four translations applied to a shaded quadrilateral are described.

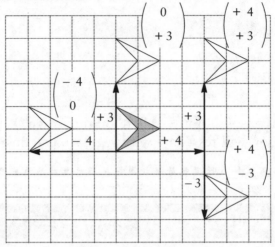

1 This diagram shows a shaded triangle moved to eight new positions.

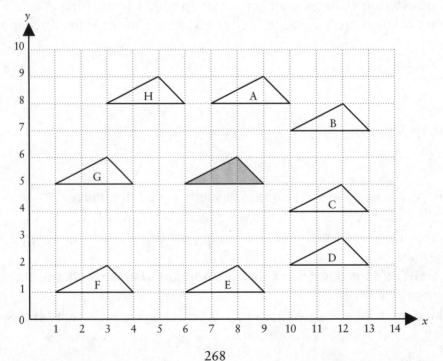

a) Copy the diagram.

b) Beside each image write the translation which moves the shaded triangle to that position.

2 Draw a graph with x- and y-axes from -6 to $+6$. Plot the triangle with corner points at $(0,1)$, $(-3,1)$ and $(-3,2)$. Show on your graph the new position of this triangle after each of the following translations.

a) $\begin{pmatrix} +5 \\ +4 \end{pmatrix}$ c) $\begin{pmatrix} +5 \\ +0 \end{pmatrix}$ e) $\begin{pmatrix} -4 \\ -0 \end{pmatrix}$ g) $\begin{pmatrix} -3 \\ -7 \end{pmatrix}$

b) $\begin{pmatrix} +3 \\ -2 \end{pmatrix}$ d) $\begin{pmatrix} 0 \\ -4 \end{pmatrix}$ f) $\begin{pmatrix} -3 \\ +3 \end{pmatrix}$ h) $\begin{pmatrix} +4 \\ -6 \end{pmatrix}$

■ ■

Rotations

A **rotation** changes the position of an object by turning it about a fixed point called the **centre of rotation**.

We describe a rotation by stating three facts.

■ The angle that the object has been rotated through.
■ The direction of rotation (anticlockwise or clockwise).
■ The position of the centre of rotation.

This rotation

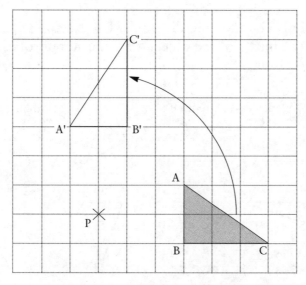

is described as a rotation of 90° anticlockwise about the point P.
This diagram shows rotations of 90°, 180° and 270° completed on
squared paper.

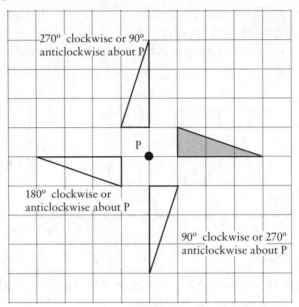

PAUSE

1 Copy each of the following diagrams. Use tracing paper to help
   you draw the image of the object after the stated rotation about
   the marked point.

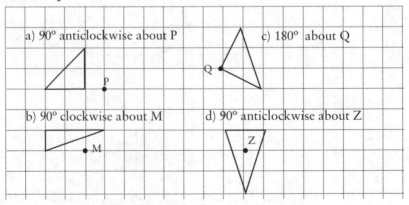

a) 90° anticlockwise about P

b) 90° clockwise about M

c) 180° about Q

d) 90° anticlockwise about Z

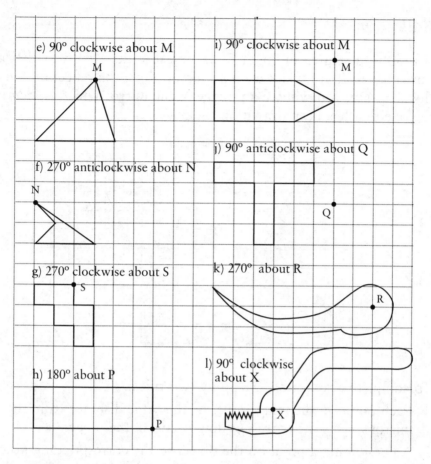

e) 90° clockwise about M

f) 270° anticlockwise about N

g) 270° clockwise about S

h) 180° about P

i) 90° clockwise about M

j) 90° anticlockwise about Q

k) 270° about R

l) 90° clockwise about X

2 Draw a graph with x- and y-axes from −6 to +6. Plot the triangle with corner points at (0,1), (−3,1) and (−3,2). Show on your graph the new position of this triangle after each of the following rotations.

a) 90° clockwise about (0,0)
b) 180° about (0,0)
c) 90° anticlockwise about (0,0)
d) 270° anticlockwise about (−3,2)
e) 270° clockwise about (0,6)
f) 180° about (−1,2)

Reflections

A reflection creates an image of an object in the same way that a mirror does. If you place a small mirror along the reflection line in this diagram, you will see an image in the mirror which is identical to the one that has been drawn on the page.

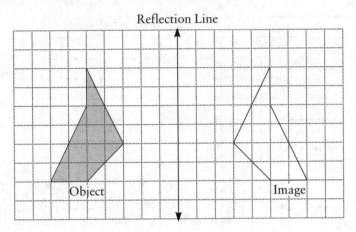

To complete a reflection, we apply two rules:

1. Each point on the image must be the same distance from the mirror line as the corresponding point on the object.

2. A line joining a point on the object to the corresponding point on the image must cross the mirror line at right angles.

Example

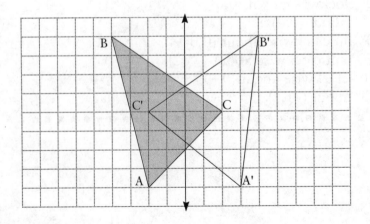

PAUSE

Copy each diagram and draw the image of the object after reflection in the mirror line.

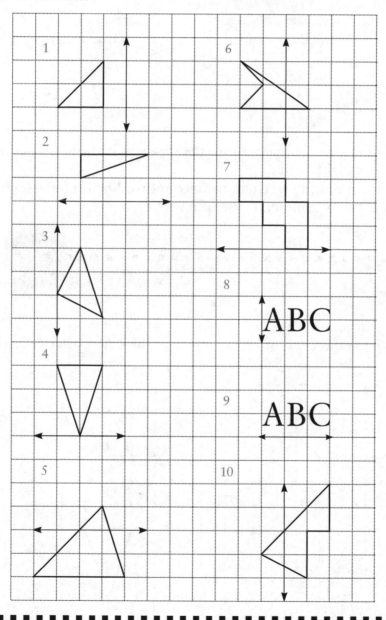

This example demonstrates an important point.

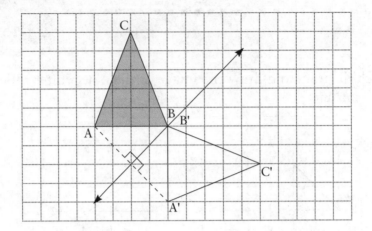

The line joining the object point to the image point *must* cross the mirror line at right angles.

This is often forgotten when an inclined mirror line is used to reflect an object and is a common cause of mistakes.

PAUSE

Copy each diagram and draw the image of the object after reflection in the mirror line.

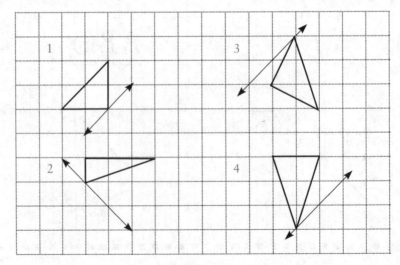

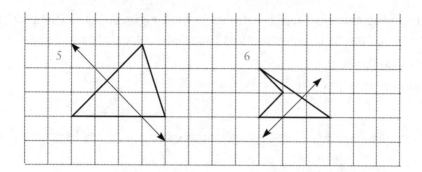

7 Draw a graph with x- and y-axes from -6 to $+6$. Plot the triangle with vertices at $(1,1),(1,2)$ and $(3,1)$. Show on your graph the new position of this triangle after reflection in each of the following mirror lines.

a) the x-axis,
b) the y-axis,
c) the line with equation $y=2$,
d) the line with equation $x=-1$,
e) the line with equation $y=x$.

8 Draw a graph with x- and y-axes from -6 to $+6$. Plot the rectangle with corner points at $(2,-1),(2-2),(5,-2)$ and $(5,-1)$. Show on your graph the new position of this triangle after reflection in each of the following mirror lines:

a) the x-axis,
b) the y-axis,
c) the line with equation $y=-2$,
d) the line with equation $x=1$,
e) the line with equation $y=-x$.

■■■■■■■■■■■■■■■■■■■■■■■■■■■■■■■■■■

Enlargements

An **enlargement** changes an object into one with the same *shape* but a different *size*.

To complete an enlargement, we draw lines outwards from the centre of enlargement through the main points on the object.

Along each of these lines, we first measure the distance of the object point from the centre of enlargement.

Multiplying this distance by the scale factor gives us the distance on that line of the image point from the centre of enlargement.

Examples

1. Enlargement, centre P, scale factor 2.

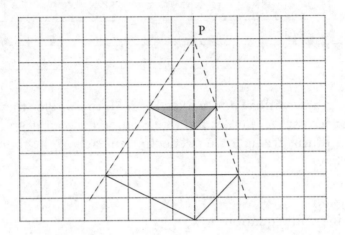

2. Enlargement, centre Q, scale factor 3.

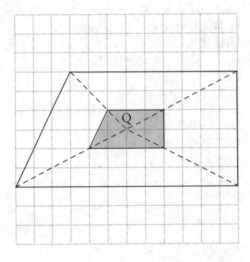

PAUSE

1 Copy each of the following diagrams and draw in the image of the object after the stated enlargement.

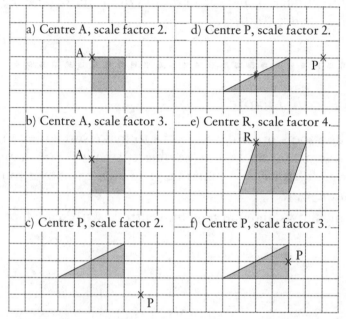

a) Centre A, scale factor 2.   d) Centre P, scale factor 2.

b) Centre A, scale factor 3.   e) Centre R, scale factor 4.

c) Centre P, scale factor 2.   f) Centre P, scale factor 3.

2 Draw a graph with x-and y-axes from 0 to 9. Plot the triangle with vertices at (1,1), (1,2) and (2,1). Show on your graph the image of this triangle after an enlargement with centre (0,0) and scale factor:

a) 2   b) 3   c) 4

3 Draw a graph with x-and y-axes from −6 to +6. Plot the object with vertices at (0,0), (0,1) and (2,0). Show on your graph the image of this object after an enlargement scale factor 2 and centre of enlargement:

a) (0,0)   c) (4,−1)
b) (2,2)   d) (0,5)

4 Draw a graph with x and y values from −6 to 12.
   a) Plot the points A (−1,1), B(4,2), C(4,−1), D(−1,−2).
   b) Join the points A, B, C, D in order. What type of quadrilateral is ABCD?

c) Draw the enlargement of ABCD with centre O and scale factor 2.

(NICCEA)

5 a) Draw a graph with x and y values from 0 to 14.
 b) Mark on the grid points R(7,4), Q(6,2) and P(2,2).
    A point S is to be marked on the grid so that PQRS is a
    parallelogram.
 c) Mark S on the grid and join up parallelogram PQRS.
 d) Draw the image of parallelogram PQRS after enlargement, scale
    factor 2, centre (0,0).

(ULEAC)

■ ■

Fractional scale factors

In normal use, the word enlarge means to make something bigger. In
maths, enlargements can also make objects smaller! This happens
when the scale factor is less than one. The image is closer to the
centre than the object.

This diagram shows an enlargement, centre P, with scale factor $\frac{1}{2}$.

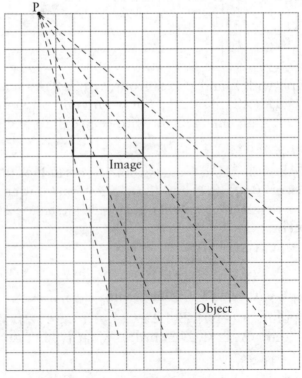

Multiplying all the distances by $\frac{1}{2}$ is the same as dividing them by 2. So all the distances are halved and the image is closer to the centre of enlargement and half the size of the object.

PAUSE

1 Copy each of the following diagrams and draw in the image of the object after the stated enlargement.

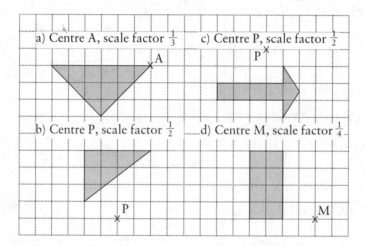

a) Centre A, scale factor $\frac{1}{3}$

b) Centre P, scale factor $\frac{1}{2}$

c) Centre P, scale factor $\frac{1}{2}$

d) Centre M, scale factor $\frac{1}{4}$

2 Draw a graph with x- and y-axes from 0 to 12. Plot the triangle with corner points at (3,6), (6,3) and (3,3). Show on your graph the image of this triangle after an enlargement with centre (0,0) and scale factor:

a) $\frac{1}{2}$　　　b) $\frac{1}{3}$

3 Draw a graph with x-and y-axes from 0 to 20. Plot the object with corner points at (3,6), (6,6) and (6,3). Show on your graph the image of this object after an enlargement with centre (0,0) and scale factor:

a) 2　　c) $\frac{1}{2}$
b) 3　　d) $\frac{1}{3}$

4 Four straight lines and a quarter circle form the shape below.
Draw an enlargement of the shape scale factor $1\frac{1}{4}$, centre A.

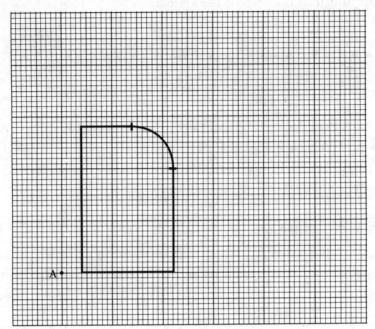

(ULEAC)

5 Draw a graph with x and y values from 0 to 14.

The parallelogram ABCD has vertices (6,3), (9,3), (12,9) and (9,9)
respectively.

Show the parallelogram on your graph.

An enlargement scale factor $\frac{1}{3}$ and centre (0, 0) transforms
parallelogram ABCD on to parallelogram A'B'C'D'.

a) i) Draw the parallelogram A'B'C'D' on your graph.
    ii) Calculate the area of parallelogram A'B'C'D'.
b) The side AB has length 3 cm. The original shape ABCD is now
   enlarged with a scale factor of $\frac{2}{3}$ to give A"B"C"D".
   Calculate the length of the side A"B".

(SEG)

■■■■■■■■■■■■■■■■■■■■■■■■■■■■■■■■■■■■■■■

Congruent and similar shapes

If two objects have the same shape and size they are **congruent**.
These are pairs of congruent objects.

If two objects have the same shape but different sizes they are *similar*. One is an enlargement of the other.

These are pairs of similar objects.

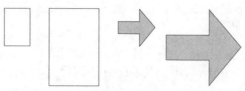

If two objects are similar there is a fixed ratio or scale factor between each pair of sides.

Example
Triangles ABC and MNO are similar with AC similar to MO. Find the length of MN.

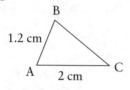
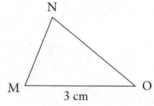

AC:MO = 2:3
so, AB:MN = 2:3 = 1:1.5
the scale factor is 1.5
MN = AB × 1.5 = 1.8 cm

PAUSE

1 Triangles ABC and PQR are similar.

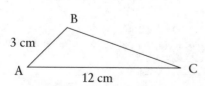
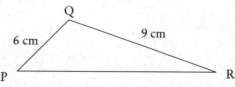

a) Find the ratio and hence the scale factor using sides AB and PQ.
b) Use the scale factor to find BC and PR.

2 Quadrilaterals CDEF and MNOP are similar.

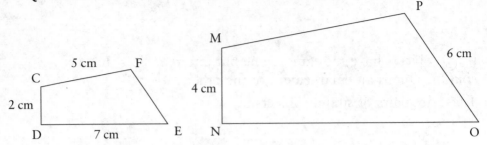

a) Use side CD and side MN to find the ratio and the scale factor.
b) Find the lengths of MP, FE and NO.

3 Pentagons JKLMN and ABCDE are similar.

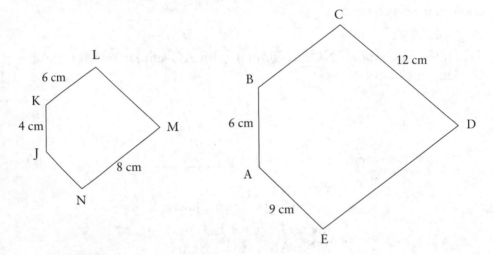

Find the lengths of sides LM, NJ, BC, and DE.

4 A manufacturer of garden furniture makes table tops in four
similar sizes.

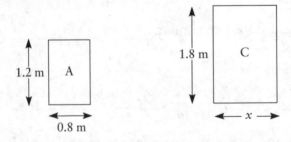

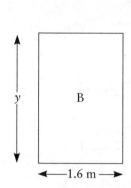

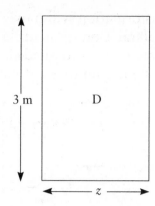

a) Find the missing lengths x, y and z.
b) Calculate the area of each table top.
c) The lengths of table top A and D are in the ratio $2:5$. What is the ratio of the areas of table tops A and D?

Mixed transformations

Many examination questions will require a mixture of translation, rotation, reflection and enlargement. The following exercise contains questions like this.

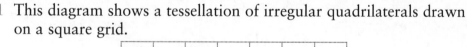

PAUSE

1 This diagram shows a tessellation of irregular quadrilaterals drawn on a square grid.

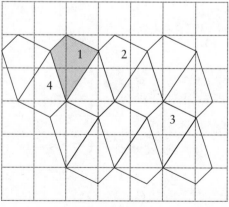

What transformation will move:
a) quadrilateral 1 on to quadrilateral 2,
b) quadrilateral 1 on to quadrilateral 3,
c) quadrilateral 1 on to quadrilateral 4?

2

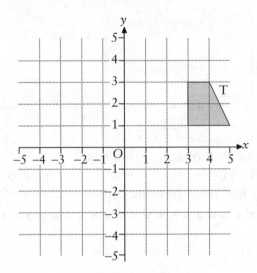

The diagram shows a shaded trapezium T.

a) Copy the diagram and show the new position of the trapezium T when it is rotated $\frac{3}{4}$ of a turn anticlockwise about O. Label this A.

b) Trapezium T is also reflected in the line $x=0$. Show this position and label it B.

c) Write down a single transformation which would map trapezium B on to trapezium A.

(NEAB)

3

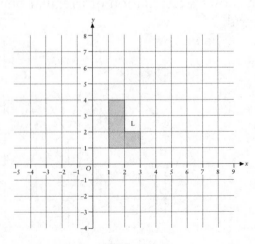

Copy the grid and show the results of the following transformations of the shape L.

a) A reflection of L in the x-axis. Label your answer A.
b) A half turn of L centre (0,0). Label your answer B.
c) An enlargement of L scale factor 2, centre (0,0). Label your answer C.

(ULEAC)

4

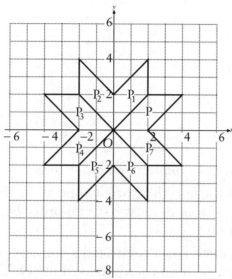

The star shape is formed by transformations of the parallelogram marked P on to the parallelograms marked P_1, P_2, P_3...P_7.

a) Describe fully a single transformation which would map
   i) P on to P_3,
   ii) P on to P_1,
   iii) P on to P_6.
b) Describe fully two *different* transformations which would map P onto P_4.
c) Describe fully a pair of *successive* transformations which would map P on to P_5.
d) i) Copy the grid and draw the enlargement of the star shape, scale factor $\frac{3}{2}$, centre (0,2).
   ii) State the ratio of the area of the enlarged star shape to the area of the original star shape.

(NICCEA)

■ ■

Rotational symmetry

Shapes which have rotational symmetry fit back into their original position as they are rotated.

The shape below fits back into its original position three times as it is rotated through 360°.

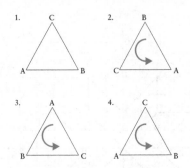

The shape has *rotational symmetry of order three.*

All shapes must fit back into their original position after a full rotation of 360°, so all shapes must have an order of rotational symmetry of at least one.

Sometimes a shape which has a rotational symmetry of order one may be called a shape with no rotational symmetry. This shape has rotational symmetry of order one (no rotational symmetry).

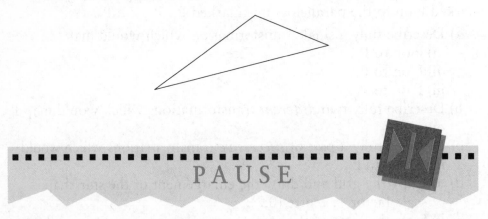

PAUSE

1 What is the order of rotational symmetry of each of the following shapes?

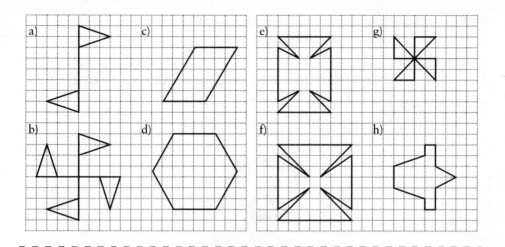

Reflectional symmetry

Shapes which have reflectional symmetry fit back into their original position if they are reflected in a mirror line across the shape.
For example this shape has reflectional symmetry because it fits back into its original position after a reflection in the dotted mirror line.

The mirror line is called *an axis of symmetry* or *a line of symmetry*. Some shapes have several different lines of symmetry. For example, this shape has four lines of symmetry.

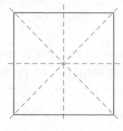

Reflectional symmetry in solid objects

It may be possible to imagine a 'slice' through a solid object
leaving two parts which are exact mirror reflections of each other.
In this case the solid object has reflective symmetry.
The mirror 'slice' is called a *plane of symmetry*.

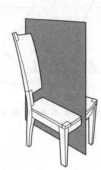

Example
Draw diagrams to show all the planes of symmetry for this cuboid.

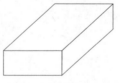

These are the required diagrams.

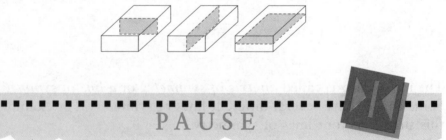

PAUSE

1 Copy each of the diagrams in the previous exercise. Mark on your
  diagrams all the lines of symmetry of each shape.
2 State how many planes of symmetry each of the following objects
  has.

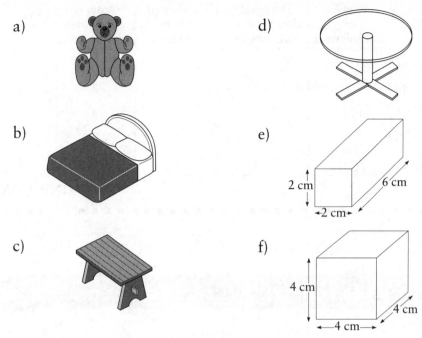

a)

b)

c)

d)

e)

2 cm

6 cm

2 cm

f)

4 cm

4 cm

4 cm

3 a) Write down the number of lines of symmetry of a
      regular hexagon.
   b) Copy this diagram and shade two triangles of the regular
      hexagon so that the resulting figure has only one line of
      symmetry.

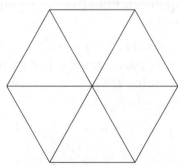

   c) Make a new copy of the diagram and shade two triangles of the
      regular hexagon so that the resulting figure has rotational
      symmetry of order two.   (MEG)

4 Copy and complete this table. The names of the quadrilaterals may
   be chosen from the following list: square, parallelogram, rhombus,
   kite and trapezium.

	Name of quadrilateral	Diagonal always cut at right angles	Number of axes of symmetry	Order of rotational symmetry
a)	rectangle			
b)		yes		2
c)			0	2
d)		yes	1	
e)				4

(NICCEA)

▪▪▪▪▪▪▪▪▪▪▪▪▪▪▪▪▪▪▪▪▪▪▪▪▪▪▪▪▪▪▪▪▪▪▪▪

SECTION 3 Perimeter, Area and Volume

Measuring area

The *area* of a shape is a measurement of the size of its surface.

In the metric system of measurements, the standard squares used to measure area are the square millimetre, the square centimetre, the square metre and the hectare.

Other shapes cannot be divided into standard squares, but we can find their area by counting part squares. For example, this is a picture of a shape drawn on a centimetre grid.

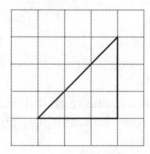

It has an area of 3 whole squares plus 3 half squares. The area of the shape is $4.5\,\text{cm}^2$.

Find the area of each shape. This is a picture of a shape drawn on a centimetre grid.

Area and perimeter of a rectangle

The perimeter of a shape is the distance all the way round the outside of a shape.

The area of a rectangle can be calculated by multiplying the base length by the height.

Examples

1. Find the perimeter and area of this rectangle.

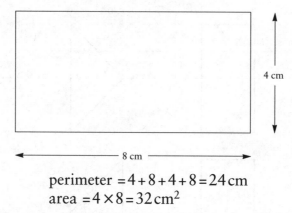

$$\text{perimeter} = 4+8+4+8 = 24\,\text{cm}$$
$$\text{area} = 4 \times 8 = 32\,\text{cm}^2$$

2. Find the perimeter and area of this rectangle.

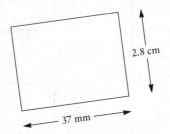

There is a mixture of units in the measurements given. We must therefore first convert all the measurements into either centimetres or millimetres.

Using millimetres:

$$\text{perimeter} = 37+28+37+28 = 130\,\text{mm}$$
$$\text{area} = 37 \times 28 = 1036\,\text{mm}^2$$

3. Find the perimeter and area of this shape.

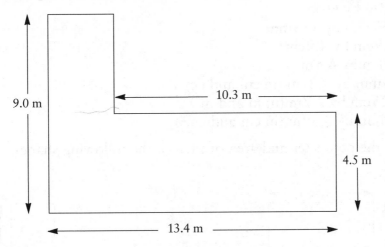

This diagram shows the extra measurements which have been filled in and how it has been divided into rectangles.

$$perimeter = 3.1 + 4.5 + 10.3 + 4.5 + 13.4 + 9.0 = 44.8\,m$$
$$area = (3.1 \times 9.0) + (10.3 \times 4.5) = 27.9 + 46.35 = 74.25\,m^2$$

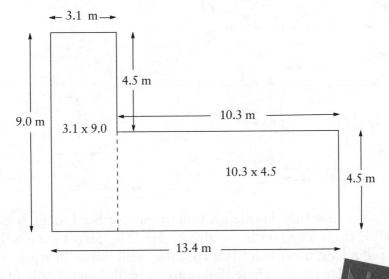

■■■■■■■■■■■■■■■■■■■■■■■■■■■■

PAUSE

1 Find the perimeter and area of rectangles with the following measurements.

a) 2 cm by 16 cm
b) 23 m by 60 m
c) 165 mm by 113 mm
d) 8.9 cm by 4.0 cm
e) 2.1 m by 4.5 m
f) 34 mm by 7.3 cm (in cm and cm^2)
g) 125 cm by 2.7 m (in m and m^2)
h) 24 cm by 0.89 m (in cm and cm^2)

2 Find the perimeter and area of each of the following shapes.

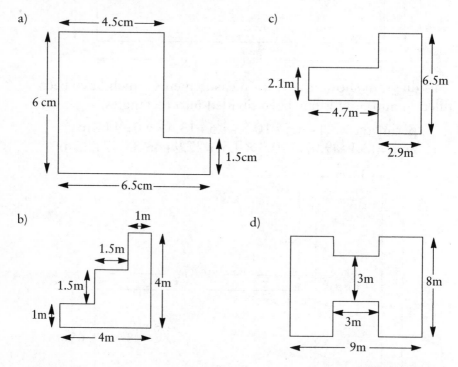

a)

4.5cm

6 cm

1.5cm

6.5cm

c)

2.1m

4.7m

6.5m

2.9m

b)

1m

1.5m

1.5m

4m

1m

4m

d)

3m

3m

9m

8m

3 The carpet that Julie Hasdell wants to buy for her lounge is only available from a roll with a width of 4 m. The carpet shop will only sell a length cut from this roll, they will not sell carpet cut to any particular shape. These diagrams show the carpet roll and the measurements of Julie's lounge. Julie does not want any joins in the carpet.

4 m

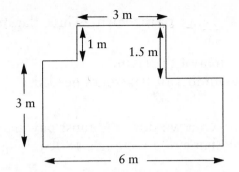

a) What length of carpet does Julie need to buy?
b) What is the area of the piece of carpet Julie buys?
c) If the carpet costs £9.99 per square metre, how much will Julie pay?
d) What area of the carpet will be wasted after it has been cut to fit Julie's lounge?
e) What is the cost of this wasted carpet?

4

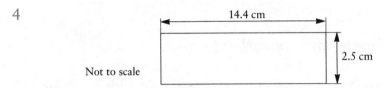

Not to scale

a) Calculate the area of the rectangle.
b) i) What is the length of each side of the square which has the same area as the rectangle above?
   ii) Draw this square accurately. (MEG)

5

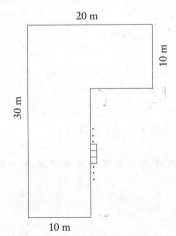

The diagram shows a lawn. All of its corners are right angles.
Using flagstones which are 1 metre square, a path 1 metre wide is

laid all around the edge of the lawn. (Three flagstones are drawn in place.)

a) Calculate the area of the lawn.

b) Calculate the number of flagstones needed.

(NICCEA)

6 Abrahim has broken a window and must buy glass to mend it. The window is in the shape of a rectangle 180cm long and 120cm wide.

a) Change the measurements to metres.

b) Work out the area of the window. Give your answer in square metres.

Glass costs £14.50 per square metre.

c) Work out the cost of the glass that Abrahim must buy.

(ULEAC)

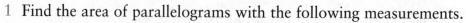

Area and perimeter of a parallelogram

Any parallelogram has the same area as a rectangle with the same base length and height.

Example
Find the perimeter and area of this parallelogram.

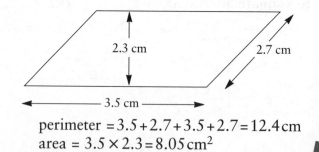

perimeter = 3.5 + 2.7 + 3.5 + 2.7 = 12.4 cm
area = 3.5 × 2.3 = 8.05 cm²

PAUSE

1 Find the area of parallelograms with the following measurements.

a) base = 5 cm height = 8 cm

b) base = 5.4 m height = 9.4 m

c) base = 3.7 mm height = 8.9 mm
d) base = 5.4 m height = 94 cm (in m²)
e) base = 456 mm height = 0.89 m (in cm²)

2 Find the area of each of the following shapes.

a)

b)

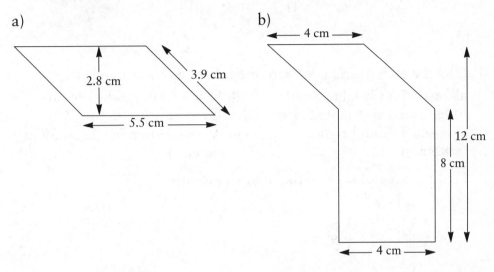

■ ■

Area and perimeter of a triangle

Any triangle with a base b and a height h is half of a parallelogram
with a base b and a height h.
Because the area of the parallelogram is $b \times h$, the area of the
triangle is $\dfrac{b \times h}{2}$, or $\dfrac{bh}{2}$.

Example
Find the perimeter and
area of this triangle.

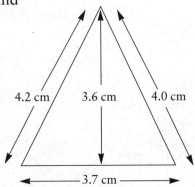

297

$$\text{perimeter} = 4.2 + 4.0 + 3.7 = 11.9\,\text{cm}$$

$$\text{area} = \frac{3.6 \times 3.7}{2} = \frac{13.32}{2} = 6.66\,\text{cm}^2$$

PAUSE

1 Find the area of triangles with the following measurements.
   a) base = 5 cm height = 8 cm
   b) base = 5.4 m height = 9.4 m
   c) base = 3.7 mm height = 8.9 mm
   d) base = 5.4 m height = 94 cm (in m²)
   e) base = 456 mm height = 0.89 m (in cm²)

2 Find the area of each of the following shapes.

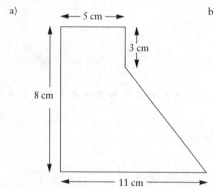

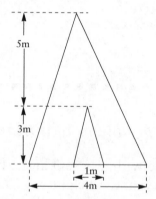

3 An arrow for a traffic diversion sign is made using an isosceles triangle and a rectangle.

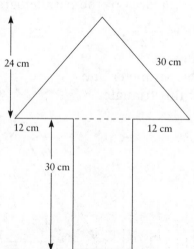

Calculate:

a) the perimeter of the arrow,
b) the area of the triangle,
c) the total area of the arrow.
                    (NICCEA)

Area and perimeter of a trapezium

A trapezium with parallel
sides a and b and height h:

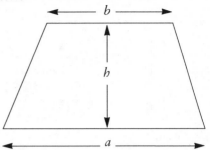

This is the area formula for *any* trapezium:

$$\text{area} = \frac{h(a+b)}{2}$$

Example
Find the area and perimeter of this trapezium.

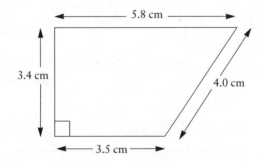

perimeter $= 3.4 + 5.8 + 4.0 + 3.5 = 16.7\,\text{cm}$

area $= \dfrac{3.4(3.5 + 5.8)}{2} = \dfrac{3.4 \times 9.3}{2} = \dfrac{31.62}{2} = 15.81\,\text{cm}^2$

PAUSE

1 Find the area of trapeziums with the following measurements.

a) $h = 5\,\text{cm}$, $a = 3\,\text{cm}$, $b = 2\,\text{cm}$

b) $h = 11.3\,\text{m}$, $a = 7.9\,\text{m}$, $b = 2\,\text{m}$

c) $h = 4\,\text{mm}$, $a = 3.9\,\text{mm}$,
   $b = 8.7\,\text{mm}$

d) $h = 5\,\text{cm}$, $a = 33\,\text{mm}$, $b = 24\,\text{mm}$
   (in cm^2)

e) $h = 15.4\,\text{m}$, $a = 398\,\text{cm}$,
   $b = 897\,\text{cm}$ (in m^2)

2 Find the area of each of the following shapes.

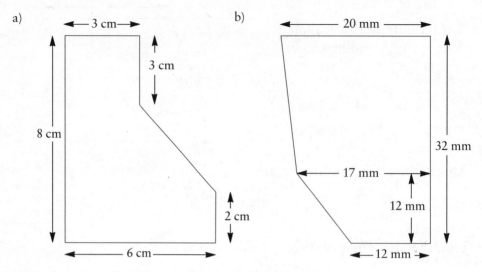

a)

b)

3 This diagram represents a farmer's field.

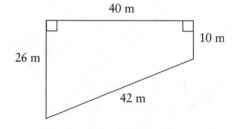

a) Work out the perimeter of the field.
   The farmer decides to put up a new fence around the perimeter
   of the field.
   It costs £1.50 to put up one metre of fence.
b) How much would it cost to fence the whole field?
c) Work out the area of the field.

The farmer wants to plant seed in the field.
Seed is sold in sacks.
Each sack contains 20 kg of seed.
One kg of seed covers 14 m².

d) How many kg of seed does it take to cover the whole field?
e) How many sacks does the farmer need to buy to plant this field
   with seed?

(ULEAC)

Area of a shape with cut out sections

Example

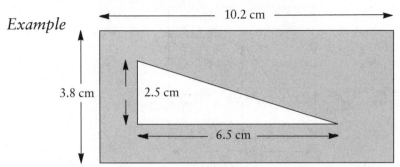

This shape is a rectangle with a triangle cut out.

area of rectangle $= 3.8 \times 10.2 = 38.76\,\text{cm}^2$

area of triangle $= \dfrac{6.5 \times 2.5}{2} = 8.125\,\text{cm}^2$

area of shape $= 38.76 - 8.125 = 30.635\,\text{cm}^2$

area of shape $= 30.6\,\text{cm}^2$ (correct to 1 decimal place)

■ ■

PAUSE

Find the area of each shape.

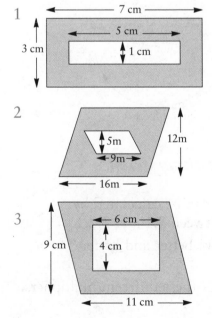

1

2

3

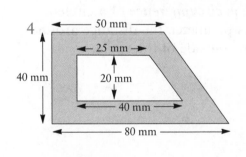

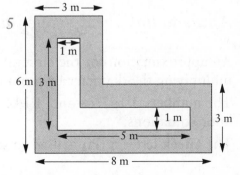

4

5

6 The diagram shows a special board for a game.

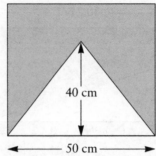

It has a triangle on a square of side 50 cm.

a) What is the area of the square?

b) What is the area of the shaded part of the board?

c) Find the ratio of the shaded area to the area of the whole board.

■■■■■■■■■■■■■■■■■■■■■■■■■■■■■■■■■■■■

The circumference of a circle

The *radius* of a circle is a line joining the centre to a point on the circle.

The *diameter* of a circle is a line which joins two points on the circle and passes through the centre.

The *circumference* of a circle is its perimeter, the distance around the outside of the circle.

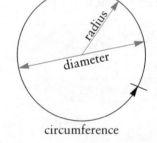

A rule to link the diameter and the circumference

An approximation for the circumference can be calculated by multiplying the diameter by a number between 3.1 and 3.2.

The numbers 3.1, 3.14 and 3.142 will give better and better approximations.

The Greek letter π (Pi) is used to stand for these different multipliers.

This allows us to write the formula:
$$C = \pi d$$
In most examination questions you will be told which value of π to use.

Example

Find the circumference of this circle using $\pi = 3.14$.

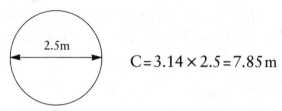

$$C = 3.14 \times 2.5 = 7.85 \, \text{m}$$

If you have a scientific calculator, it will have a special button to give a value for π.

The π button on an eight digit calculator gives the value 3.1415927.

We can use the formula 'in reverse' to find the diameter, if we are given the circumference. The reverse formula is:
$$d = \frac{C}{\pi}$$

Example

Find, correct to one decimal place, the diameter of a circle with a circumference of 36.5 cm ($\pi = 3.14$).
$$\text{diameter} = \frac{36.5}{3.14} = 11.6 \, \text{cm}$$

PAUSE

1 Find, correct to one decimal place, the circumference of the following circles.

Use $\pi = 3.14$ or use the π button on your calculator.

a) diameter = 8 cm
b) radius = 8 cm
c) diameter = 3.5 m
d) radius = 10.8 mm
e) diameter = 240 m
f) radius = 43 mm

2 Using the value 3.14 for π or the π button on your calculator, find, correct to one decimal place, the diameter and radius of the following circles.
   a) circumference = 314 cm
   b) circumference = 100 mm
   c) circumference = 59.5 cm
   d) circumference = 37.2 mm

3 A bicycle is fitted with wheels which have a diameter of 65 cm.
   a) What is the circumference of a wheel? (π = 3.14)
   b) How far would you expect the bike to travel as the wheels turn 10 times?

4

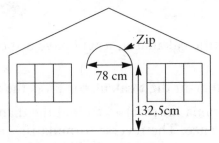

A flap door is made for a tent by using a zip as shown in the diagram. The zip is made up of a semicircle and a straight line.
What length of zip is required? (you may take π = 3.14)

(NICCEA)

5

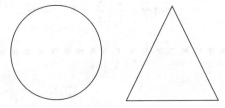

   a) Calculate the circumference of a circle of radius 18 cm.
      (you may take π as 3.14)
   b) An equilateral triangle has the same perimeter as the circle.
      Calculate the length of one side.   (NICCEA)

304

6 Andre is rolling a hoop along the ground.

The hoop has a diameter of 60 cm.

a) What is the circumference of the hoop?

Take π to be 3.14 or use the π key on your calculator.

b) What is the minimum number of times the hoop must rotate to cover a distance of 5 m?

(SEG)

7

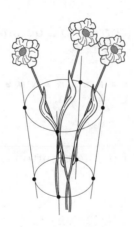

The diagram shows a wire frame to support flowers.
The top circle has diameter 20 cm.
The bottom circle has diameter 15 cm.
The four straight wires are each 90 cm long.
Calculate the total length of the wire in the wire frame.

(ULEAC)

■■■■■■■■■■■■■■■■■■■■■■■■■■■■■■■■■■■■■■■

The area of a circle

The *area* of a circle can be found with the formula:

$$A = \pi \times r \times r$$

which can be written:

$$A = \pi r^2$$

Examples

1. Find the area of this circle, correct to 2 decimal places, using the value 3.14 for π.

radius = 2.5 ÷ 2 = 1.25
area = 3.14 × 1.25 × 1.25 = 4.91 m² (correct to 2 decimal places)

2. Find, correct to 2 decimal places, the radius of a circle with an area of 49.55 m² (π = 3.14)

$$r^2 = \frac{49.55}{3.14} = 15.780255$$

$r = \sqrt{15.780255} = 3.97$ m (correct to 2 decimal places).

PAUSE

1 Using the value 3.14 for π or the π button on your calculator find the area of each circle.

a) 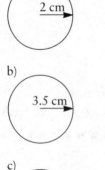 2 cm

d)  13.5 m

g) 3 cm

j) 15 mm

b) 3.5 cm

e) 55 cm

h) 9 cm

c) 5.5 cm

f) 2 cm

i) 6 m

2 Using the value 3.14 for π find, correct to 1 decimal place, the radius of each circle.

a)
A=
314 cm^2

b)
A=
15.7 m^2

c)
A=
12.6 m^2

d)
A=
87.6 mm^2

3 The instructions for a garden fertilizer state:

'Apply 100 grams per square metre'.

Estimate how many kilograms of fertilizer should be applied to a circular flower bed with a diameter of 12 m ($\pi=3$).

4 A circular plaque with a radius of 13.7 cm is to be plated with silver.

a) Find the area of the plaque ($\pi=3.14$).

b) Silver plating costs 35p per square centimetre. Find the total cost of silver plating the plaque.

5

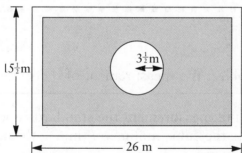

The diagram represents a rectangular garden measuring 26 m by $15\frac{1}{2}$ m.

The garden is to be sown with grass seed except for a 1 m strip around the edge and a circular rose bed of radius $3\frac{1}{2}$ m in the centre.

(You may take π to be 3.14 or use the π button on your calculator.)

a) What are the length and breadth of the inner rectangle?

b) What is the area of the rose bed?

c) What is the area to be sown with grass seed?

A box of grass seed covers 25 m^2.

d) How many boxes of grass seed are required?

(NICCEA)

307

6

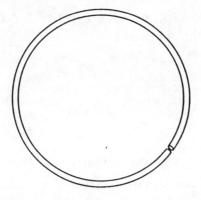

A piece of rope is 12 m long. It is laid on the ground in a circle, as shown in the diagram.

a) Using 3.14 as the value of π calculate the diameter of the circle.
b) Explain briefly how you would check the answer to part a) mentally.

The cross-section of the rope is a circle of radius 1.2 cm.

c) Calculate the area of the cross-section.

■ ■

Volume of a prism

The **volume** of an object is a measurement of the space that it occupies.

In the metric system of measurement the standard cubes used to measure volumes are the cubic millimetre (mm^3), the cubic centimetre (cm^3) and the cubic metre (m^3).

The volumes of liquids are usually measured in litres (l).

A litre is 1000 cubic centimetres. The cubic centimetre is also called a *millilitre* (ml).

A prism is a solid which has the same shape along its whole length.

Cuboids and cylinders are special types of prism.

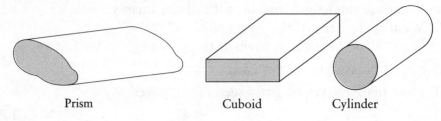

Prism　　　　　Cuboid　　　　　Cylinder

The following formula applies to any prism.

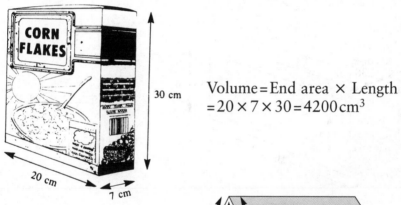

End area

Length

$$\text{Volume} = \text{End area} \times \text{Length}$$

Examples

1. Find the volume of this cornflake packet.

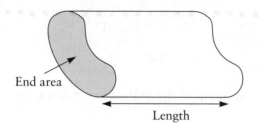

30 cm

20 cm

7 cm

$\text{Volume} = \text{End area} \times \text{Length}$
$= 20 \times 7 \times 30 = 4200 \, \text{cm}^3$

2. Find the volume of this tent.

$\text{Volume} = \text{End area} \times \text{Length}$
$= \dfrac{1.5 \times 1}{2} \times 2.1 = 1.575 \, \text{m}^3$

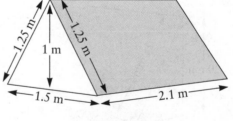

1.25 m

1 m

1.25 m

1.5 m

2.1 m

3. This diagram shows a tin of beans. Calculate the volume, correct to $1 \, \text{cm}^3$, of the can. ($\pi = 3.14$)

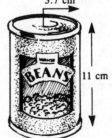

3.7 cm

11 cm

$\text{Volume} = \text{End area} \times \text{Length}$
$= 3.14 \times 3.7 \times 3.7 \times 11 = 473 \, \text{cm}^3$ (to the nearest cm^3)

1 Find the volume of each of the following prisms. ($\pi = 3.14$)

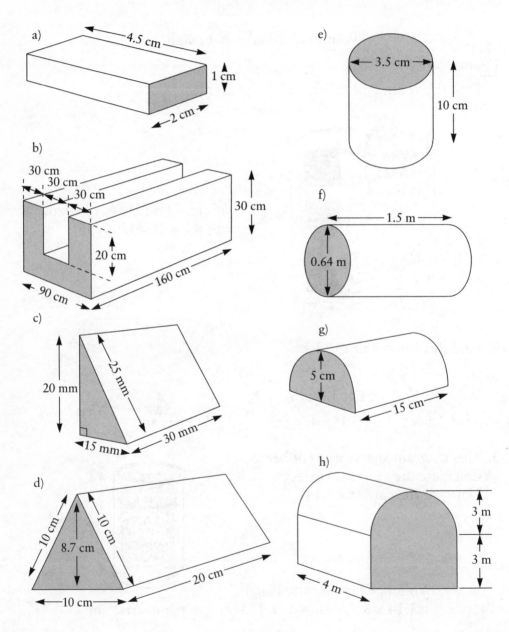

2 a)

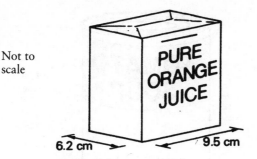

Not to scale

6.2 cm   9.5 cm

When full, a cardboard carton, in the shape of a cuboid, holds 1 litre of orange juice.
The base of the carton measures 9.5 cm by 6.2 cm.
What is its height?

b)

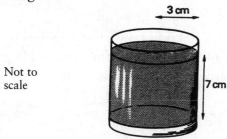

3 cm

Not to scale

7 cm

The orange juice is poured into cylindrical glasses.
The radius of each glass is 3 cm and the depth of orange juice is 7 cm.
Taking π to be 3.142, calculate the volume of orange juice in each glass.

c) How many glasses of orange juice can be filled to a height of 7 cm from the 1 litre carton?   (MEG)

3

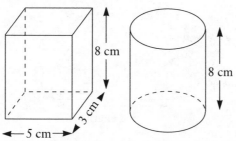

8 cm   8 cm

3 cm

←5 cm→

The cuboid and the cylinder have the same height and the same volume.
Calculate the radius of the cylinder.   (NEAB)

311

4

Tomato soup is sold in cylindrical tins.

Each tin has a base radius of 3.5 cm and a height of 12 cm.

a) Calculate the volume of soup in a full tin.
Take π to be 3.14 or use the π key on your calculator.

b) Mark has a full tin of tomato soup for dinner. He pours the soup into a cylindrical bowl of radius 7 cm.

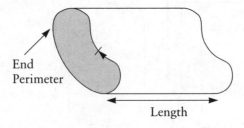

What is the depth of the soup in the bowl?   (SEG)

■■■■■■■■■■■■■■■■■■■■■■■■■■■■■■■■■■■■■

Surface area of a prism

The following formula applies *any* prism.

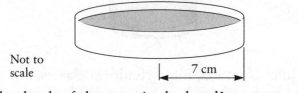

Surface area (not including ends) = End perimeter × Length

Examples

1. Find the surface area of this cornflake packet.

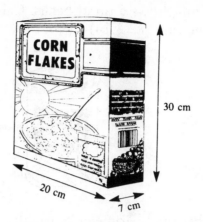

Surface area (not including top and bottom)

$$= \text{End perimeter} \times \text{Length}$$
$$=(20+7+20+7)\times 30$$
$$=1620\,\text{cm}^2$$

area of top $=20\times 7=140\,\text{cm}^2$
area of bottom $=20\times 7=140\,\text{cm}^2$

Total Surface Area $=1620+140+140=1900\,\text{cm}^2$

2. Find the area of nylon used to make this tent.

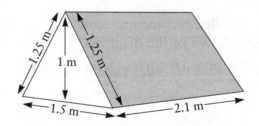

Surface area (not including ends) $= \text{End perimeter} \times \text{Length}$
$$=(1.5+1.25+1.25)\times 2.1$$
$$=8.4\,\text{m}^2$$

area of each end $=\dfrac{(1.5\times 1)}{2} =0.75\,\text{m}^2$

Total Surface Area $=8.4+0.75+0.75=9.9\,\text{m}^2$

3. This diagram shows a tin of beans. Calculate the area of the label stuck to the can. ($\pi = 3.14$)

area of label = end perimeter × length

area of label = $3.14 \times 7.4 \times 11 = 256 \, \text{cm}^2$ (nearest cm^2)

PAUSE

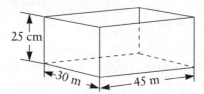

1 Find the total surface area of each of the prisms in question 1 of the previous exercise. (Use the value $\pi = 3.14$ or use the π button on your calculator.)

2 An open thin-walled rectangular tank whose base is horizontal has dimensions as shown in the diagram.

a) Calculate the internal surface area of the tank.

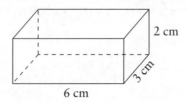

b) Calculate the capacity of the tank in litres.

3 a) Calculate the volume of the cuboid shown.

b) Make an accurate drawing of a net of this cuboid.

c) Calculate the surface area of the cuboid.

(NICCEA)

■■■■■■■■■■■■■■■■■■■■■■■■■■■■■■■■■■

Checking the dimensions of a formula

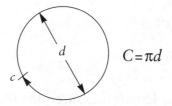

$$C = \pi d$$

This formula gives a length (the circumference of a circle). The formula contains *one* length, it is *one* dimensional. The length would be measured in units like cm or m.

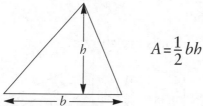

$$A = \frac{1}{2} bh$$

This formula gives an area. It contains a length times a length and is *two* dimensional.

The area would be measured in units like cm^2 or m^2.

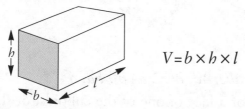

$$V = b \times h \times l$$

This formula gives a volume. It contains *three* lengths multiplied together and is *three* dimensional.

The volume would be measured in units like cm^3 or m^3.

PAUSE

1 State whether each of these is a length, an area or a volume.
   a) $8\,cm^3$,   e) $0.64\,cm^2$,
   b) $5\,cm^2$,   f) $1000\,mm^3$,
   c) $2\pi\,cm$,   g) $9\pi\,cm^2$,
   d) $15\,mm^3$,   h) $500\,m$.

2 If a, b, c each represent a length measured in cm, state whether each formula gives a length, an area or a volume.
   a) $T=abc$ e) $P=2a+2b$
   b) $N=3a^2$ f) $S=4\pi c^2$
   c) $D=\pi b^2$ g) $Q=2bc$
   d) $L=\frac{4}{3}\pi a^3$ h) $M=6c^2$

3 The dimensions of four cuboids are shown.

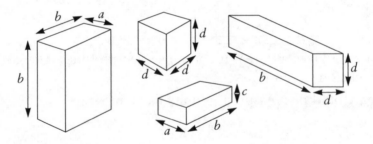

These expressions:
$$abc,\ 4d,\ d^2,\ 2(a+b),\ bd,\ d^3$$
give the perimeter of a face of one of the cuboids, or the area of a face of one of the cuboids, or the volume of one of the cuboids.

Copy and complete the statements below by writing perimeter, or area, or volume on the dotted lines.

abc gives a ....

$2(a+b)$ gives a ....

bd gives a ....

(ULEAC)

4 The letters a, b, h, l and r represent lengths.
  Consider the following formulae.

$$ab, \tfrac{2}{3}\pi r^3, \pi r, 4\pi rl, \frac{\pi r^2 h}{3}, \sqrt{a^2 + b^2}$$

a) Which of these formulae represent areas?
b) Use dimensions to explain how you can tell which formulae
   represent areas.   (SEG)

■ ■

SECTION 4 Pythagoras and Trigonometry

Pythagoras

Squares

The *square* of a number is calculated by multiplying the number by
itself.

> The square of 4 $= 4 \times 4 = 16$.
> The square of 10 $= 10 \times 10 = 100$.

There is a special way to write the square of a number:

> 4×4 can be written as 4^2.
> 10×10 can be written as 10^2.

We read 4^2 as '*4 squared*' and 10^2 as '*10 squared*'.

A calculator can be used to find squares like 8.73^2.

All calculators will allow you to find the answer in the most obvious
way by pressing:

8 . 7 3 × 8 . 7 3 =

Some calculators will allow you to save time by pressing:

8 . 7 3 × =

Scientific calculators usually have a function button marked x^2, this
will allow you to press:

8 . 7 3 x^2

Try these methods with your calculator, the answer should be
76.2129.

1 Find the squares of:
   a) 65   c) 1.3
   b) 12.9 d) 231

2 Find the area of a square with sides 53.5 cm long.

■■

Square roots

The **square root** of a number n is the number which when squared is equal to n.

The square root of 16 is 4, because 4 squared is equal to 16.

There is a special way to write the square root of a number:

The square root of 16 is written as $\sqrt{16}$, so: $\sqrt{16} = 4$.

Most calculators have a square root button marked $\sqrt{}$ or $\boxed{\sqrt{x}}$.

You enter the number you wish to square root and then press this button.

Many square roots are long decimals.

For example, $\sqrt{20} = 4.4721359$ (on an eight digit calculator).

These answers are usually rounded to a number of decimal places.

$$\sqrt{20} = 4.47 \text{ (correct to 2 decimal places)}$$

Calculate:

1 the square root of 64

2 $\sqrt{4}$

3 $\sqrt{81}$

4 $\sqrt{9}$

5 $\sqrt{1}$

6 the side length of a square with an area of $36\,cm^2$,

7 the side length of a square with an area of $49\,cm^2$,

8 the perimeter of a square with an area of $400\ cm^2$.

9 Find, correct to 2 decimal places, the square roots of:
  a) $\sqrt{60}$ c) $\sqrt{5.7}$
  b) $\sqrt{734}$ d) $\sqrt{936.5}$

10 Find, correct to 2 decimal places, the side length of a square with an area of $45\,cm^2$.

■ ■

The Theorem of Pythagoras

Pythagoras was a Greek mathematician and philosopher. He lived in Crotona, a Greek city in what is now Southern Italy, about 2500 years ago.

He was the founder of a secret society called the Pythagoreans.

Look at this right-angled triangle:

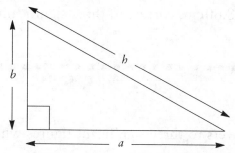

The Pythagoreans proved that in every right-angled triangle like this:
$$h^2 = a^2 + b^2$$

The longest side of a right-angled triangle is called the *hypotenuse*.

So, the Theorem of Pythagoras can be stated as:

'*The square of the hypotenuse of a right-angled triangle is equal to the sum of the squares of the other two sides.*'

319

Calculating the length of the hypotenuse

If we know the lengths of two of the sides in a right-angled triangle, we can use the Theorem of Pythagoras to calculate the length of the third side.

Example

Calculate the length of the side PQ in this triangle.

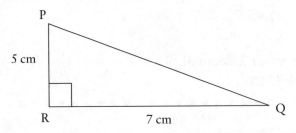

PQ is the hypotenuse of the triangle, so, using the Theorem of Pythagoras, we can write:

$$PQ^2 = PR^2 + RQ^2$$
$$PQ^2 = 7^2 + 5^2$$
$$PQ^2 = 49 + 25$$
$$PQ^2 = 74$$
$$PQ^2 = \sqrt{74}$$

$PQ^2 = 8.60\,cm$, correct to three significant figures.

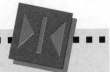

PAUSE

Round your answers to three significant figures where necessary.

1 Find the length of the side RT in this triangle.

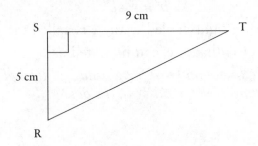

2 Find the length of the side XZ in this triangle.

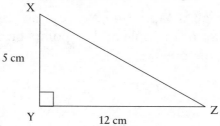

3 Find the length of the side AB in this triangle.

4 This is an isosceles triangle with a base of 6 cm and a height of 10 cm.

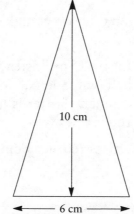

Calculate the length of the equal sides of the triangle.

5 Calculate the length of the equal sides of an isosceles triangle with a base of 10 cm and a height of 6 cm.

6 Paris is 140 km to the west and 220 km to the south of Brussels.

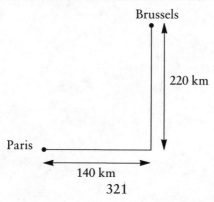

Find the shortest direct distance between Paris and Brussels.

7 A flag pole 5 m high is supported by wires attached to points 1 metre from the top of the pole and to points on the ground 2.5 m from the base of the pole.

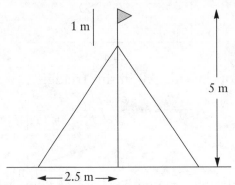

Find the length of the wires.

8 Find the length of a diagonal in a rectangle 8.4 m long and 1.3 m high.

9 A rectangular lawn is 40 m by 15 m. Instead of walking along a path down two side of the lawn, Sue Naylor takes a short cut across the diagonal. How much shorter is her direct route than the path down two sides of the lawn?

10 ABC is a triangle with vertices (corner points) at (–2,2), (1,1) and (2,–2).

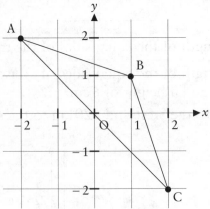

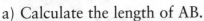
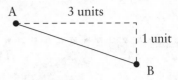

a) Calculate the length of AB.
b) Calculate the length of AC.
c) Calculate the length of BC.
d) What is the special name for a triangle like ABC?

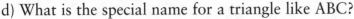

Calculating the length of the other sides

In all the calculations in the last exercise, the side to be found was the *hypotenuse*. If the hypotenuse is one of the given sides, our calculations are slightly changed.

Example

A ladder 5 m long rests against a wall with its foot 1.5 m from the base of the wall.

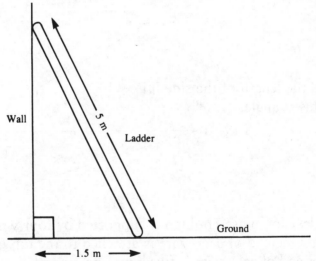

How far up the wall will the ladder reach?

If we call the distance we wish to calculate x, using the Theorem of Pythagoras we can write:

$$5^2 = x^2 + 1.5^2$$
$$25 = x^2 + 2.25$$
$$x^2 = 25 - 2.25$$
$$x^2 = 22.75$$
$$x = \sqrt{22.75}$$
$$x = 4.77 \text{ m, correct to the nearest centimetre.}$$

Round your answer to three significant figures where necessary.

1 Find the length of the side CD in this triangle.

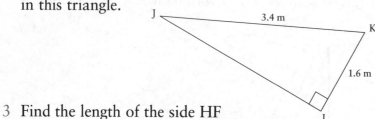

2 Find the length of the side JL in this triangle.

3 Find the length of the side HF in this triangle.

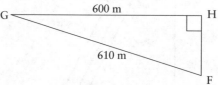

4 A television mast 7 m high is supported by four wires each 10 m long. One end of each wire is attached to the top of the pole. The other end of each wire is fixed into the ground.

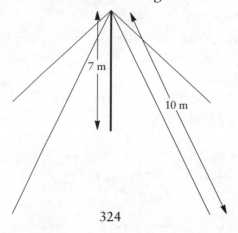

324

How far from the base of the mast are the wires fixed into the ground?

5   A tent pole 2.1 m high is supported by three 2.6 m guy ropes. The guy ropes are connected to the top of the pole and to pegs fixed into the ground. How far is each peg from the base of the tent pole?

6   A stepladder has legs 2.5 m long. The bottoms of the legs are 1.5 m apart.

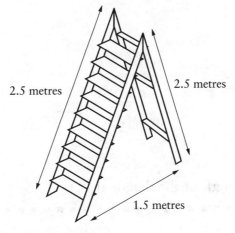

2.5 metres   2.5 metres

1.5 metres

How high is the top of the stepladder?

7   If the legs of the stepladder in question 6 are 1 m apart, now high is the top of the stepladder?

8   This isosceles triangle has a base of length 8 cm and equal sides of length 6 cm.

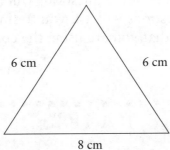

6 cm   6 cm

8 cm

a) Calculate the height of the triangle.
b) Calculate the area of the triangle.

9   Calculate the area of an isosceles triangle which has a base of length 10 cm and equal sides of length 8 cm.

10 The kite ABCD has AC = 4 cm, AB = BC = 3 cm and CD = DA = 6 cm.

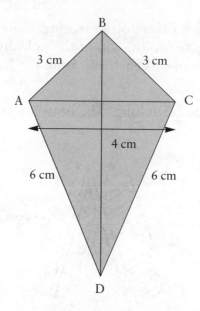

Calculate the length of the diagonal BD.

■■■■■■■■■■■■■■■■■■■■■■■■■■■■■■■■■■■

Mixed examples

We have learnt two techniques to using the Theorem of Pythagoras.
- Finding the length of the hypotenuse.
- Finding the length of one of the other sides.

In the next exercise some of the questions will ask you to find the hypotenuse. Other questions will ask you to find one of the other sides. Check carefully that you are using the correct technique for each question.

■■■■■■■■■■■■■■■■■■■■■■■■■■■■■■■■■

PAUSE

1

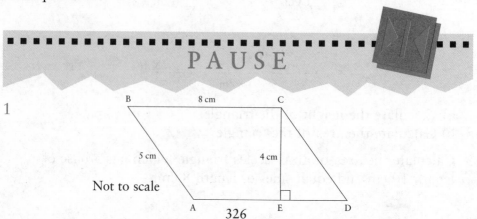

Not to scale

ABCD is a parallelogram.

The foot of the perpendicular from C to AD is E.

AB = 5 cm.

BC = 8 cm.

CE = 4 cm.

a) Calculate the perimeter of the parallelogram.

b) Calculate the area of the parallelogram.

c) Calculate the distance DE.   (ULEAC)

2   The diagram shows a symmetrical trapezium.

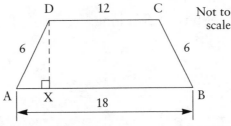

AB = 18 cm, DC = 12 cm, AD = CB = 6 cm, angle AXD = 90°.

a) Write down the length of AX.

b) Calculate the length of DX.

c) Calculate the area of the trapezium.   (SEG)

3

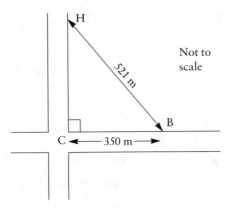

Mohamed takes a short cut from his home (H) to the bus stop (B)
along a footpath HB. How much further would it be for
Mohamed to walk to the bus stop by going from H to the corner
(C) and then from C to B?

Give your answer to the nearest metre.   (MEG)

4 Two pegs A and B are driven into a lawn. A loop of rope is placed around both pegs and pulled to form a triangle APB.

a)

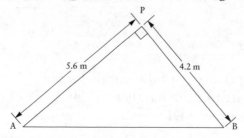

In one position the triangle is right-angled at P, and has the measurements shown.

Show that the loop is 16.8 m long.

b)

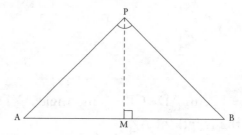

In a second position for P the triangle is isosceles with PA = PB.

Calculate:

  i) the length of PA,

 ii) the perpendicular distance PM,

iii) the area of triangle PAB.

(NICCEA)

5 ABC is an isosceles triangle with AB = AC.
The area of ABC is 48 cm².
Calculate the distance AB.

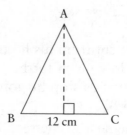

(NICCEA)

6

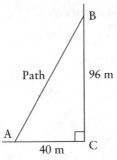

A and B are posts near to C, the corner of a park.

a) Calculate the length of the straight path connecting A and B.

The triangle ACB is one lap. Gail starts at post A and jogs round and round the triangular lap ACB in an anticlockwise direction.

b) When Gail has jogged for 1 kilometre,
    i) which lap will she be on?
    ii) where will she be at that moment?

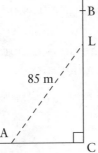

A litter bin, L, is attached to the park railings between B and C; it is 85 m from A.

c) Calculate the distance of the litter bin L from the post B.

(NICCEA)

■ ■

Trigonometry

The word ***trigonometry*** is formed from two Greek words meaning ***triangle*** and ***measure***.
We already know from the Pythagoras unit that the longest side in a right-angled triangle is called the ***hypotenuse***.

The other two sides are called the ***opposite*** and the ***adjacent*** (the word 'adjacent' means 'next to').

Which side is which depends on which angle you select.

In this triangle, where angle *A* is our selected angle:

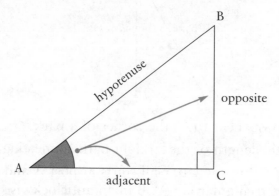

side AB is called the *hypotenuse,*

side BC is called the *opposite* because it is directly opposite the selected angle,

side AC is called the *adjacent* because it is next to the selected angle.

If we change our selected angle, the adjacent and opposite side swap over. So, in this triangle where angle *B* is our selected angle side AC is the opposite and side BC is the adjacent.

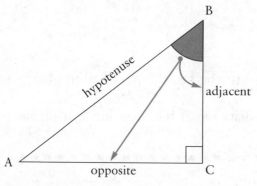

PAUSE

Copy each of these triangles and label their sides with the names hypotenuse, opposite and adjacent. The selected angle in each triangle has been coloured.

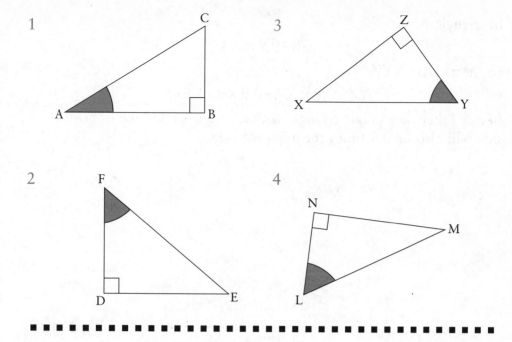

1

2

3

4

■ ■

Similar triangles

Example
These two triangles are similar:

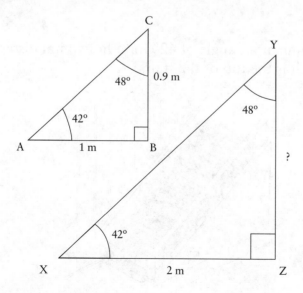

Find the length of side YZ.

331

In triangle ABC,

$$CB = 0.9 \times AB$$

So, in triangle XYZ,

$$YZ = 0.9 \times XZ = 0.9 \times 2 = 1.8 \, m.$$

In any other right-angle triangle with an angle of 42° the opposite side will also be 0.9 times the adjacent side.

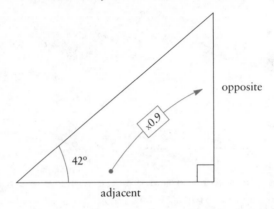

opposite

×0.9

42°

adjacent

P A U S E

1 These stairs climb at an angle of 42° for a horizontal distance of 10 m. How high is the top of the stairs?

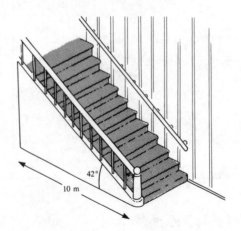

42°

10 m

2 Measured at a point 20 m from its base, the angle of elevation of the top of a fir tree is 42°. How high is the tree?

3 The diagram shows three cottages. How far is it along the path from cottage A to cottage B?

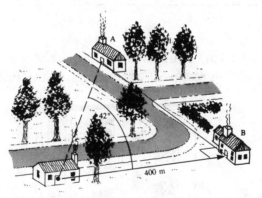

■■■■■■■■■■■■■■■■■■■■■■■■■■■■■■■■■■■■■

The tangent ratio

The last exercise demonstrated that if we know the ratio between the opposite and adjacent sides for a specific angle we can solve many different problems which involve similar triangles.

This ratio is called the *tangent ratio*.

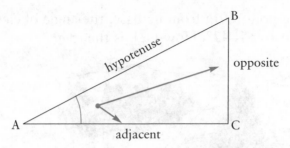

$$\text{tangent ratio of angle } A = \frac{\text{length of BC}}{\text{length of AC}} = \frac{\text{opposite}}{\text{adjacent}}$$

(To simplify things, the tangent ratio of angle A is usually written as tan A.)

This is a table of tangent ratios for various angles:

Angle	10	20	30	40	50	60	70	80
Tangent	0.176	0.364	0.577	0.839	1.192	1.732	2.747	5.671

Example
From a distance of 15 m, a surveyor measures the angle of elevation of the top of a building to be 60°. How high is the building?

The height of the building is the opposite side in a right-angled triangle with an adjacent side 15 m long.

Therefore:

$$\tan 60° = \frac{\text{height of building}}{15}$$

multiply both side by 15:

$$15 \times \tan 60° = \text{height of building}$$
$$15 \times 1.732 = \text{height of building}$$
$$\text{height of building} = 25.98 \text{ m (to 2 d.p.)}$$

PAUSE

1 Find the length of the side AB in each of these triangles:

a)

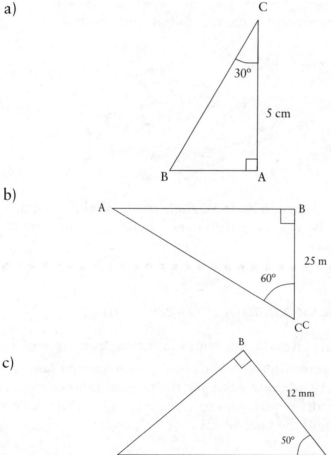

b)

c)

2 How far will this ladder reach up the wall?

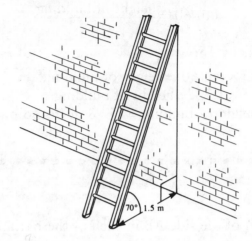

3 How far from the ground is the top end of this see-saw?

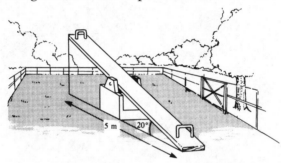

4 After take-off, a plane climbs at an angle of 40° to the ground.
   How high will it be when it passes over a building 1 km from the
   end of the runway?

■ ■

Using a scientific calculator for tangent ratios

Our simple table only gives the tangent ratio for angles in steps of 10°.
A *scientific calculator* however, will have a button marked *tan*,
which can give a very accurate value for the tangent ratio of *any*
angle. Most calculators require you to enter the angle before pressing
this button. So, to find the tangent ratio of 42°, press:

4 **2** **tan** ,

this should give an answer of 0.9004040.

Example
This triangular wooden framework requires a brace fitted half way along the base (shown with dashed lines). Calculate the length of this brace.

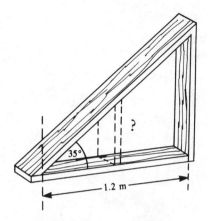

The brace is the opposite side in a right angled triangle with an adjacent side of 0.6 metres. Therefore:

$$\tan 35° = \frac{\text{length of brace}}{0.6},$$

multiply both sides by 0.6:

$$0.6 \times \tan 35° = \text{length of brace}.$$

With most scientific calculators we can complete the calculation in one step if we press:

Length of brace = 0.42 metres (to two decimal places).

■■■■■■■■■■■■■■■■■■■■■■■■■■■■■■■■■
PAUSE

1 Use your calculator to find:
  a) tan 15°   c) tan 37.5°
  b) tan 76°   d) tan 60.55°

2 Find the length of the side AB in each of these triangles:

a)

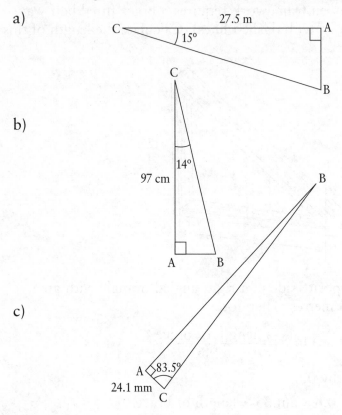

27.5 m

C   15°   A

C

b)

97 cm   14°

B

A   B

c)

A   83.5°

24.1 mm   C

B

3 The steepest normal railway in the World, in Guatemala, climbs at an angle of 5.2° to horizontal ground. What is the gain in height of a train travelling 900 m along this line?

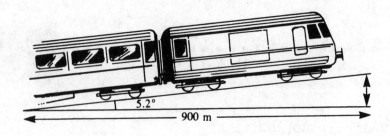

5.2°

900 m

4   On a sunny day, a flagpole casts a shadow 3.7 m long when the sun's rays are at an inclination of 56° to the ground. How high is the flagpole?

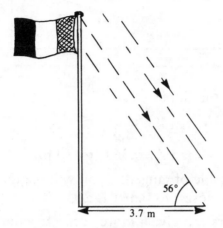

5   This diagram shows a sketch for a triangular framework made from steel wire. Calculate the total length of wire needed to make the framework.

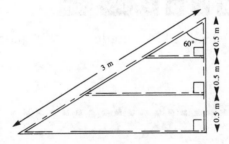

■ ■

Finding the size of an angle using the tangent ratio

In all the problems so far, we have calculated the opposite side. With a slightly different technique, we can also calculate a missing angle.

Example
The ramp leading up to the second floor of a car park climbs 3 m vertically over a horizontal distance of 8.3 m. What is the angle of the slope of the ramp?

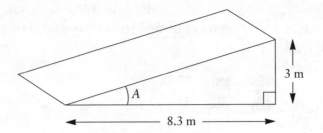

If A is the angle of the slope:

$$\tan A = \frac{3}{8.3}$$

$$\tan A = 0.361 \text{ (to 3 d.p.)}$$

If you look at the table of tangents, you will see that the angle which has a tangent ratio closest to 0.361 is 20°.

Scientific calculators can give a more accurate value for angle A.

The calculator usually has a button marked **INV** or **2nd**.

With most calculators we press:

This gives a value of 19.872176°, angle $A = 19.9°$ (correct to one decimal place).

PAUSE

1 Use a scientific calculator, to find the angles which have the following tangent ratios:

a) 0.25   d) 11.8

b) 0.745   e) 0.0174

c) 1.5

2 Calculate the angle ABC in each of these triangles.

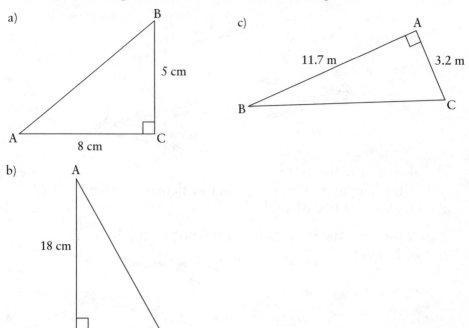

a)

B

5 cm

A
8 cm
C

c)

11.7 m
3.2 m

A

B
C

b)

A

18 cm

C
15 cm
B

3 This cone is 15 cm high and has a base radius of 8 cm. Calculate the angle between the sloping sides of the cone and the base.

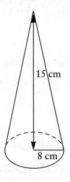

15 cm

8 cm

4 The diagonals of a rhombus always cross at an angle of 90°. Calculate the four angles of this rhombus.

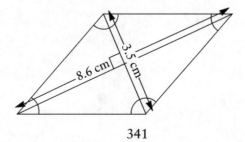

8.6 cm

3.5 cm

5 This diagram shows a framework made from steel wire.

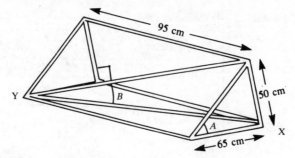

a) Calculate the size of angle *A*.
b) Use the Theorem of Pythagoras to calculate the length XY.
c) Calculate the size of angle *B*.

6 Newcastle-on-Tyne is 70 miles to the north and 50 miles to the east of Kendal.

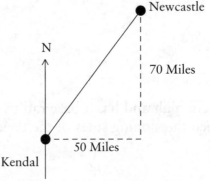

a) Calculate the bearing of Newcastle from Kendal.
b) Calculate the bearing of Kendal from Newcastle.

7 Manchester is 40 miles to the north and 40 miles to the west of Nottingham.

a) Calculate the bearing of Manchester from Nottingham.
b) Calculate the bearing of Nottingham from Manchester.

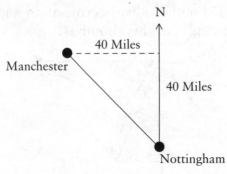

8 Bradford is 140 miles to the north and 40 miles to the west of
   Chelmsford.

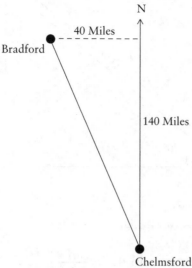

   a) Calculate the bearing of Bradford from Chelmsford.
   b) Calculate the bearing of Chelmsford from Bradford.

■ ■

The sine ratio

The *sine ratio* links the *hypotenuse* and the *opposite*.

The sine ratio is the value obtained when the length of the *opposite* is
divided by the length of the *hypotenuse*.

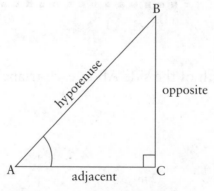

$$\text{sine ratio of angle } A = \frac{\text{length of BC}}{\text{length of AB}} = \frac{\text{opposite}}{\text{hypotenuse}}$$

(The sine ratio of angle A is usually just written as sin A).

Example

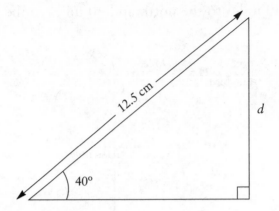

Find the length of side *d*.

Applying this sine ratio we have:

$$\sin 40° = \frac{d}{12.5}$$

Multiplying both sides by 12.5:

$$12.5 \times \sin 40° = d$$
$$12.5 \times 0.642788 = d$$
$$d = 8.04 \text{ (to 2 d.p.)}$$

We can use a scientific calculator to obtain our answer directly by pressing:

PAUSE

1 Calculate the length of the side AB in each triangle.

a)

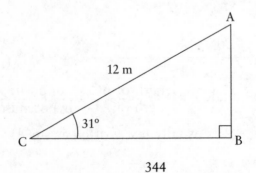

b)

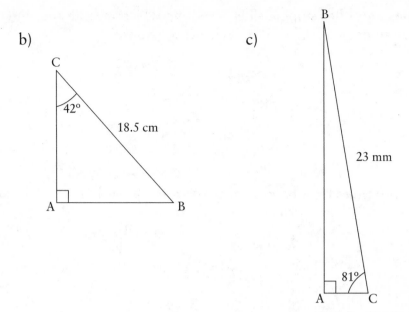

C

42°

18.5 cm

A   B

c)

B

23 mm

81°

A   C

2 A ladder 5 m long rests against a wall, making an angle of 70°
   with the ground. How far up the wall does the ladder reach?

5 m

70°

3 A plane climbs at an angle of 40° after take off. How high is the
   plane after it has flown 900 m through the air?

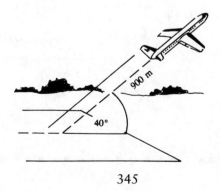

900 m

40°

4 This house roof has sloping sides 8 m long which meet in the centre at an angle of 130°. How wide is the house?

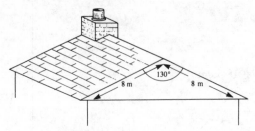

5 York is 140 miles on a bearing of 040° from Shrewsbury.

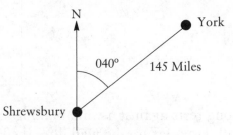

N

York

040°   145 Miles

Shrewsbury

How far is York to the east of Shrewsbury?

6 Gloucester is 140 miles on a bearing of 240° from Peterborough.

N

Peterborough   240°

140 miles

How far is Gloucester to the south of Peterborough?

Gloucester

7 Liverpool is 216 miles on a bearing 318° from London.

How far is Liverpool to the west of London?

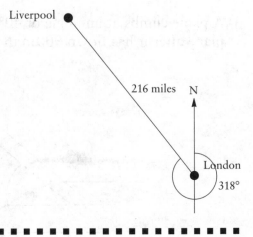

Liverpool

216 miles   N

London

318°

Finding the size of an angle using the sine ratio

In all the problems we have tackled so far, we have calculated the opposite side. With a slightly different technique, we can also calculate a missing angle.

Example
A ladder 3.1 m long rests against a wall with its top end resting 1.98 m up the wall. What angle does the foot of the ladder make with the ground?

If A is the angle required:

$$\sin A = \frac{1.98}{3.1}$$

$$\sin A = 0.639 \text{ (to 3 d.p.)}$$

Scientific calculators can give an accurate value for angle A.

With most calculators we press:

 1 **.** **9** **8** **÷** **3** **.** **1** **=** INV sin

This gives a value of 39.695671°.

Angle $A = 39.7°$ (correct to one decimal place).

PAUSE

1 Use a scientific calculator to find the angles which have the following sine ratios:
   a) 0.25   d) 0.866
   b) 0.745   e) 0.015
   c) 0.5

2 Calculate the size of the angle ABC in each triangle.

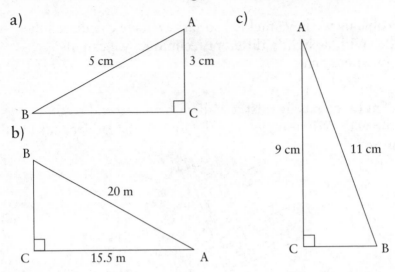

a)

5 cm 3 cm

B A C

b)

B

20 m

C 15.5 m A

c)

A

9 cm 11 cm

C B

3 The manufacturer of a type of roofing tile recommends that they should not be used on roofs with a slope of less than 21°. With a slope less that 21°, the tiles will allow water to enter the roof in severe weather. Can the tiles be used on this roof design?

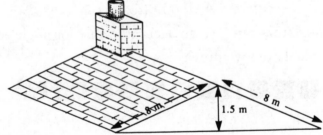

8 m

1.5 m

8 m

4 Rack and pinion railways can climb steeper slopes than normal railways. They obtain extra 'grip' from toothed wheels which engage a toothed rail. The maximum slope that can be climbed is about 7.1°. Could a rack and pinion railway be built to climb this incline?

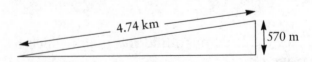

4.74 km

570 m

5 Wires 42.5 m long are used to support a radio mast. If the wires are fixed from ground anchors to a point 32.7 m up the mast, what angle do they make with the ground?

6 The sloping sides of a tent are 1.7 m long. The tent poles are 1.5 m high.

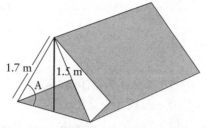

Find the size of angle *A*, the angle the sloping sides make with the ground.

■ ■

The cosine ratio

The *cosine ratio* links the *hypotenuse* and the *adjacent*.

The sine ratio is the value obtained when the length of the *opposite* is divided by the length of the *hypotenuse*.

The cosine ratio is the value obtained when the length of the *adjacent* is divided by the length of the *hypotenuse*.

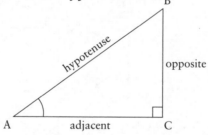

$$\text{cosine ratio of angle } A = \frac{\text{length of BC}}{\text{length of AB}} = \frac{\text{adjacent}}{\text{hypotenuse}}$$

(The cosine ratio of angle *A* is usually just written as cos *A*).

Example

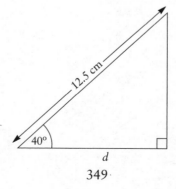

349

Find the length of side d.

$$\cos 40° = \frac{d}{12.5}$$

Multiply both sides by 12.5:

$$12.5 \times \cos 40° = d$$
$$12.5 \times 0.7660444 = d$$
$$d = 9.58 \text{ (to 2 d.p.)}$$

We can use a scientific calculator to obtain our answer directly by pressing:

PAUSE

1 Calculate the length of the side AB in each triangle.
   a)

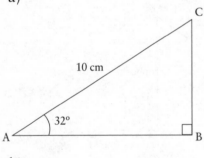

   c)

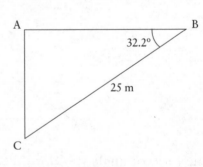

   b)

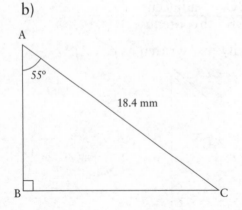

2 A ship leaves harbour and sails for 50 km on a bearing of 050°.
   How far to the East of the harbour is the ship?

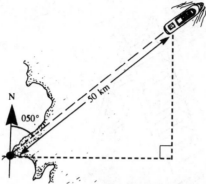

3 A ladder 3.4 m long rests against a wall, making an angle of 35°
   with the ground. How far is the foot of the ladder from the base of
   the wall?

4 Calculate the width of this roof.

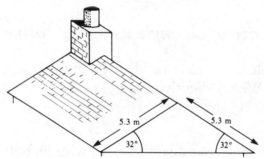

5.3 m   5.3 m
   32°   32°

5 In the kite ABCD,

   AB = BC = 3 cm, CD = DA = 6 cm,
   angle CAB = 31.2° and
   angle ADB = 15°.

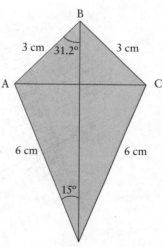

   Calculate the length of the
   diagonal BD.

6 The legs of a stepladder are 3.1 m long and make an angle of 62° with the ground.

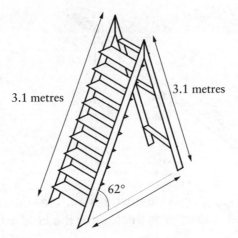

3.1 metres

3.1 metres

62°

Calculate the distance between the feet of the stepladder.

■ ■

Finding the size of an angle using the cosine ratio

In all the problems we have tackled so far, we have calculated the adjacent side. With a slightly different technique, we can also calculate a missing angle.

Example
A ladder 3.1 m long rests against a wall with its bottom end 1.98 m from the base of the wall. What angle does the foot of the ladder make with the ground?

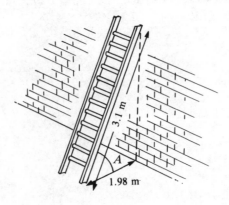

3.1 m

A

1.98 m

If A is the angle required:

$$\cos A = \frac{1.98}{3.1}$$

$$\cos A = 0.639 \text{ (to 3 d.p.)}$$

Scientific calculators can give an accurate value for angle A.
With most calculators we press:

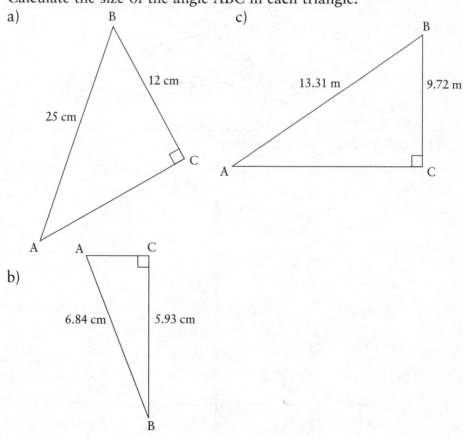

This gives a value of 50.304329°, angle $A = 50.3°$ (correct to one decimal place).

PAUSE

1 Calculate the size of the angle ABC in each triangle.

a)

12 cm

25 cm

b)

6.84 cm   5.93 cm

c)

13.31 m   9.72 m

353

2 A ship leaves harbour and sails 50 km. It has then reached a point
   that is 34.6 km east of the harbour. On what bearing did the ship
   sail?

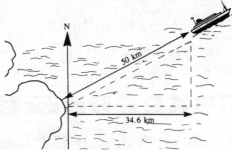

3 A ladder 3.8 m long rests against a wall. The foot of the ladder is
   1.35 m from the base of the wall. What angle does the ladder make
   with the ground?

4 Calculate angle A in the diagram of a roof.

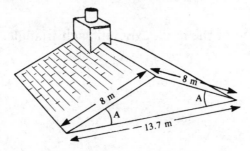

5 Preston is a direct distance of 248 miles from Southampton and
   236 miles north of Southampton.

   Calculate the bearing of Preston from Southampton.

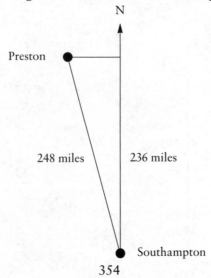

6 In this isosceles triangle, the equal sides are 25 cm long and the base is 10 cm long.

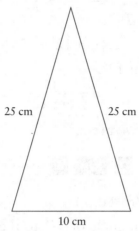

25 cm

25 cm

10 cm

Calculate the size of the equal angles in the isosceles triangle.

■ ■

Finding the side on the bottom of a ratio

In all the problems we have tackled so far, the side length we have calculated has been the side on the top of the ratio.

$$\sin A = \frac{opposite}{hypotenuse} \quad \cos A = \frac{adjacent}{hypotenuse} \quad \tan A = \frac{opposite}{adjacent}$$

A slightly different technique is required if the side length to be found is on the bottom of the ratio.

Examples

1. Ms Jones is standing at the top of a cliff which she knows is 35 m high. She sees a yacht out to sea, with an angle of depression of 12° from where she is standing. How far is the yacht from the base of the cliff?

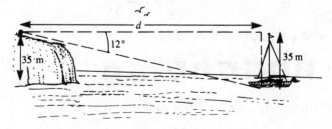

d

12°

35 m

35 m

355

If d is the distance of the yacht from the cliff:
$$\tan 12° = \frac{35}{d}$$
multiply both sides by d:
$$d \times \tan 12° = 35$$
divide both sides by $\tan 12°$:
$$d = \frac{35}{\tan 12°}$$

With most calculators we press:

| **3** | **5** | **÷** | **1** | **2** | tan | **=** |

we obtain $d = 164.7$ m (to 1 d.p.)

2. An aircraft climbs at an angle of 34° to the ground. How far will it have flown through the air before it has increased its height by 500 m?

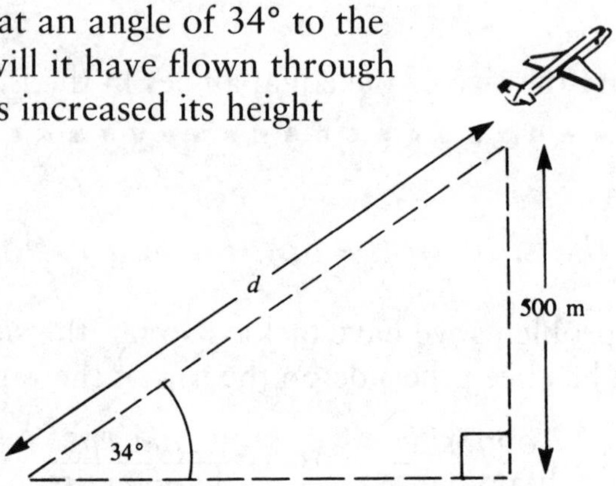

If d is the distance flown:
$$\sin 34° = \frac{500}{d}$$
multiply both sides by d:
$$d \times \sin 34° = 500$$
divide both sides of by $\sin 34°$:
$$d = \frac{500}{\sin 34°}$$

with most calculators we press:

| **5** | **0** | **0** | **÷** | **3** | **4** | sin | **=** |

$d = 894$ metres (correct to the nearest metre).

1 The wall in a loft conversion slopes at an angle of 60° to the floor. How close to the base of the wall can a wardrobe 2.1 m high be pushed?

2.1 m

60°

2 A wheelchair ramp is to be built up to a viewing platform 2.3 m above the ground. How far from the base of the platform must the ramp start if it is designed to climb at an angle of 9° to the ground?

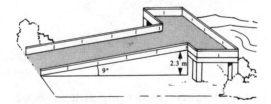

2.3 m

9°

3 The top of this step ladder is 2.1 m high when the legs make an angle of 62° with the floor.

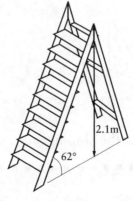

2.1 m

62°

Calculate the length of the legs of the stepladder.

357

4 An isosceles triangle has a height of 10 cm. Both equal angles are 84°.

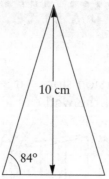

Find the length of the equal sides.

5 Calculate the length of the slope down this roof.

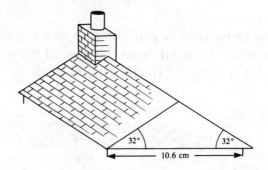

6 This tent is 1.8 m wide and the sloping sides meet the ground at an angle of 55°.

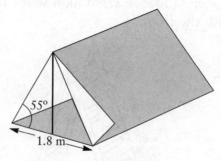

Calculate the length of the sloping sides.

1 Make an accurate drawing of these triangles.

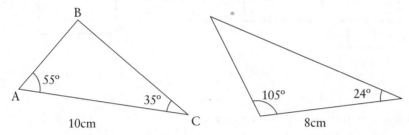

2 Chelmsford is 30 km to the west and 25 km to the south of Ipswich. Draw a map using a scale of 1 cm = 5 km. Find the bearing of Ipswich from Chelmsford and the bearing of Chelmsford from Ipswich.

3 a) Make an accurate drawing from this sketch using a scale of 1 cm = 1 m.

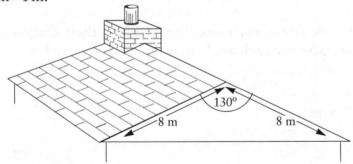

   b) Use your sketch to find the width of the roof.

4 Draw a net for a cuboid which is 6 cm wide, 4 cm high and 7 cm long. Check your net by cutting it out and folding it.

5 Describe the solids that you think each of the following nets will make. Check your answers by copying the nets, cutting them out and folding them.

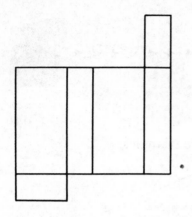

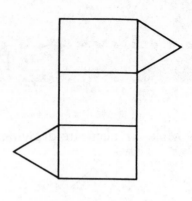

6 Draw a top view of this object and side elevations from the directions shown by the arrows.

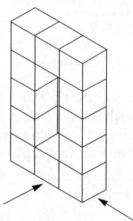

7 i) Find the size of each unknown angle in these diagrams.
ii) State whether each angle is acute, obtuse or reflex.

a)

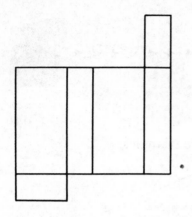
a
72°

d)
y
22°

b)
11x
7x

e)
120°
z
220°

c)
b
31.5°

8 Find the size of each unknown angle in these diagrams:

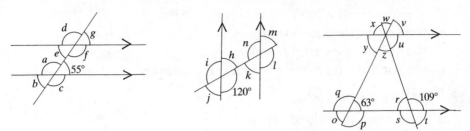

9 Find the size of each unknown angle in these diagrams:

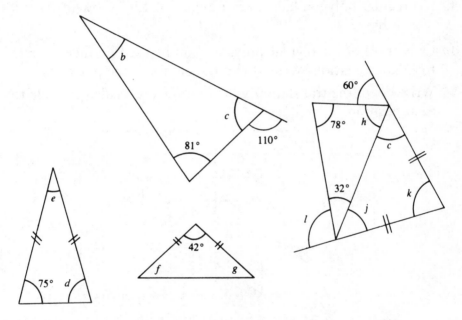

10 Write down the correct name of each of the quadrilaterals.

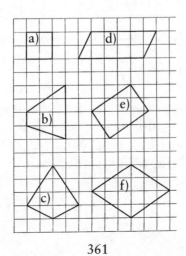

11 A heptagon is a seven sided polygon. Find:

   a) the total of the interior angles in a heptagon,
   b) the size of the interior angles in a regular heptagon,
   c) the size of the exterior angles in a regular heptagon,
   d) the total of the exterior angles in a heptagon.

12 Repeat question 11 for:

   a) a decagon (ten sided polygon),
   b) a dodecagon (twelve sided polygon).

13 If a regular polygon has interior angles of 150°, how many sides
   does it have?

14 Only three of the regular polygons can be used on their own to
   form a tessellation. Write down the names of the polygons.

15 Write down the translation which moves the shaded triangle to
   each position.

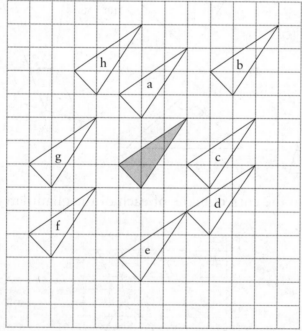

16 Draw a graph with x-and y-axes from −6 to +6. Plot the triangle
   with corner points at (0,1), (−3,1) and (−3,2). Show on your graph
   the new position of this triangle after each of the following
   translations.

   a) $\begin{pmatrix} +6 \\ 0 \end{pmatrix}$ b) $\begin{pmatrix} +6 \\ -4 \end{pmatrix}$ c) $\begin{pmatrix} -3 \\ +4 \end{pmatrix}$ d) $\begin{pmatrix} 0 \\ +4 \end{pmatrix}$

17 Copy each of the following diagrams and draw in the image of the object after the stated rotation about the marked point.

a) 90° anticlockwise about P

b) 90° clockwise about P

c) 180° about P

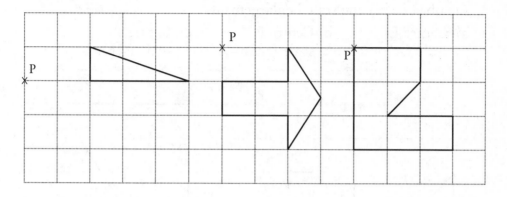

18 Draw a graph with x-and y-axes from −6 to + 6. Plot the triangle with corner points at (1,4), (3,4) and (3,1). Show on your graph the new position of this triangle after each of the following rotations.

a) 90° clockwise about (0,0)

b) 180° about (0,0)

c) 90° anticlockwise about (0,0)

d) 270° anticlockwise about (1,4)

19 Copy each of the following diagrams and draw in the image of the object after a reflection in the mirror line.

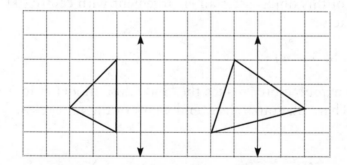

20 Draw a graph with x-and y-axes from −6 to +6. Plot the triangle with corner points at (−1,−1), (−2,−1) and (−2,−3). Show on your graph the new position of this triangle after reflection in each of the following mirror lines:

a) the x-axis,

b) the y-axis,

c) the line with equation $y=-2$,

d) the line with equation $x=1$,

e) the line with equation $y=-x$.

21 Copy each of the following diagrams and draw in the image of the object after the stated enlargement.

a) Centre P, scale factor 2.

b) Centre P, scale factor 3.

c) Centre P, scale factor $\frac{1}{2}$

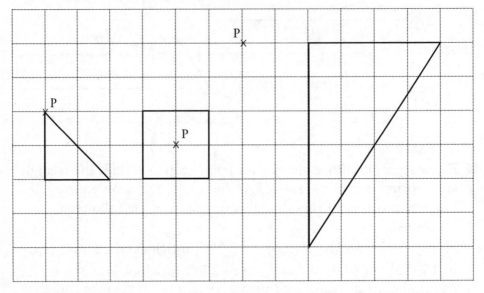

22 Draw a graph with x-and y-axes from -12 to $+12$. Plot the object with corner points at (2,2), (2,3), (4,2). Show on your graph the image of this object after an enlargement with centre (0,0) and scale factor:

a) 2   b) 3   c) $\frac{1}{2}$.

23 A tent manufacturer makes the 'Snowdon' model in four similar sizes. These diagrams show end views of the four tents.

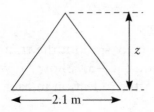

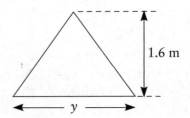

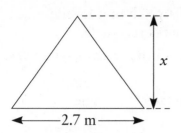

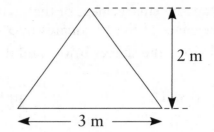

Find the lengths marked x, y and z.

24 P is the object with vertices (corner points) at (1, 3), (2, 3) and (2, 1).

> R is a rotation of +180° about (0,0)
>
> S is a reflection in the x-axis
>
> E is an enlargement with scale factor 2, centre (0,0)
>
> T is the translation $\begin{pmatrix} +2 \\ 0 \end{pmatrix}$

a) Show in one diagram, P, and its image after an enlargement with scale factor 2 centre (0,0). Call the image object E.

b) Add to your diagram the image of P after a rotation of 180° about (0,0). Call the image object R.

c) Add to your diagram the image of P after a reflection in the x-axis.

d) Add to your diagram the image of P after the translation $\begin{pmatrix} +2 \\ 0 \end{pmatrix}$.

e) Add to your diagram the image of E after the translation $\begin{pmatrix} +2 \\ 0 \end{pmatrix}$. Call this image object T.

f) Add to your diagram the image of R after a reflection in the x-axis. Call this image object S.

g) State a single transformation which moves P on to T.

h) State a single transformation which moves P on to S.

25 Tina is designing a pattern in a rectangle 4 squares by 3 squares.

She draws the small design which then has to be enlarged into a rectangle 12 squares by 9 squares.

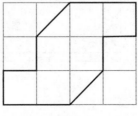

Draw an enlargement of the small shape so that its edges lie along the edges of the 12 squares by 9 squares rectangle.   (SEG)

26 a) Copy the shapes below and draw all the lines of symmetry.

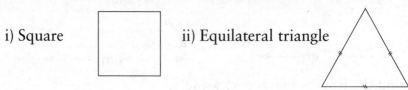

    i) Square   ii) Equilateral triangle

b) Write down the order of rotational symmetry of:
    i) the square,
    ii) the equilateral triangle.   (ULEAC)

27 The diagram shows a regular pentagon.

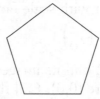

a) i) What is its order of rotational symmetry?
    ii) Copy the diagram and add to the pentagon so that the resulting diagram has only one line of symmetry.

b) Calculate the size of an interior angle of a regular pentagon.

c) Can a tessellation be formed with tiles in the shape of equally-sized regular pentagons? If so, sketch the tessellation. If not, explain why not.   (MEG)

28 Find the perimeter and area of a rectangle with a height of 5 cm and a base of 6.5 cm.

29 Find the perimeter and area of this shape:

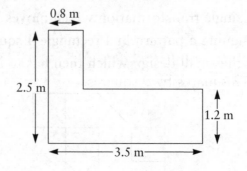

30 Find the area of a parallelogram with a height of 5.3 cm and a base of 7.5 cm.

31 Find the perimeter and area of this shape:

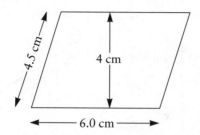

32 Find the area of a triangle with a height of 8 m and a base of 7.5 m.

33 Find the perimeter and area of this shape:

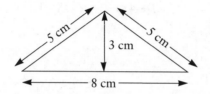

34 Find the area of a trapezium with a height of 5.3 cm and parallel sides of 7.5 cm and 8.0 cm.

35 Find the area and circumference of a circle with a radius of 5.3 cm ($\pi = 3.1$).

36 Find the area and circumference of a circle with a diameter of 4.25 m ($\pi = 3.14$).

37 Estimate the radius and diameter of a circle with an area of 507 mm^2 ($\pi = 3$).

38 Find the diameter and radius of a circle with a circumference of 37.68 cm ($\pi = 3.14$).

39 Find the volume of each of the following prisms ($\pi = 3.14$).

a)

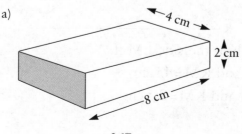

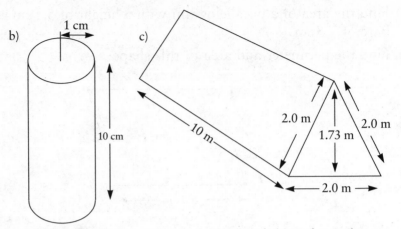

b) 1 cm, 10 cm

c) 10 m, 2.0 m, 2.0 m, 1.73 m, 2.0 m

40 If *m*, *n*, *t*, *r* and *l* represent lengths, which of these formulae represents a perimeter, which represents an area and which represents a volume:

$$R = \frac{mnt}{5} \qquad T = 4\pi r^2 \qquad S = 12l + 3m + 4n$$

41 In this diagram:

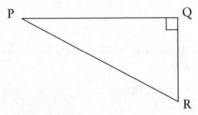

Find the length of the side PR of the triangle if:
a) PQ = 6 m and QR = 8 m,
b) PQ = 1 cm and QR = 2.4 cm.

42 In this diagram:

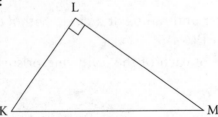

Find the length of the side LM if:
a) KL = 27 cm and KM = 45 cm,
b) KL = 2.4 cm and KM = 2.5 cm.

43 Find the length of a diagonal in a square with sides 4 cm long.

44 Find the length of the sides of a square with a diagonal 4 cm long.

45 The diagram below shows a boat moored 3 m away from a river bank. The top of the bank is 1 m above the water level and the deck of the boat is 3 m above the water level.

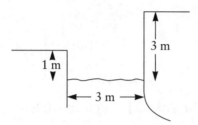

Will a 3.5 m plank of wood be long enough to make a gang-plank for the boat?

46 Chelmsford is 30 km to the west and 25 km to the south of Ipswich. Calculate the direct distance between Chelmsford and Ipswich.

47 An isosceles triangle has two sides of length 7.5 cm and one side of length 12 cm. Calculate the area of the triangle.

48 Find the length of the side d in each triangle.

a)   b)

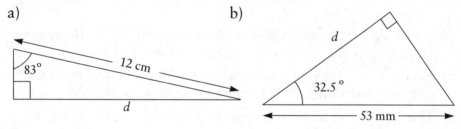

49 Find the length of the side d in each triangle.

a)   b)

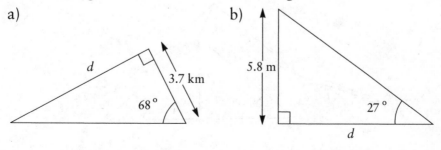

50 Find angle A in each triangle.

a)

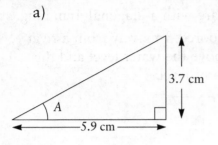

3.7 cm

A

5.9 cm

b)

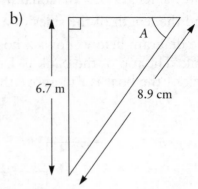

A

6.7 m

8.9 cm

51 Find the length of the side d in each triangle.

a)

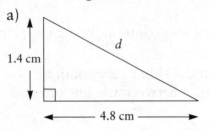

1.4 cm

d

4.8 cm

b)

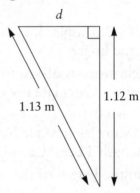

d

1.13 m

1.12 m

52 An escalator 12.5 m long rises at an angle of 37° to the horizontal. How high is the top of the escalator above the bottom?

53 Redditch is 23 km on a bearing of 060° from Worcester. Find the distance that Redditch is to the east of Worcester.

54 This ramp is to be built as part of a bike test track. It will be 3.5 m high and should have an angle of slope of 40°. What will be the total width of the ramp?

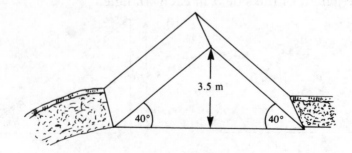

3.5 m

40°   40°

55 A plane climbs at an angle of 55°. Find its gain in height after it has covered a horizontal distance of 850 m.

56 A hiker walks 3.5 km on a bearing of 125°. How far is she to the south of her starting point?

FASTFORWARD

1 Using a scale of 1 cm to represent 2 m, mark the positions of three boys so that each boy is 12 m from the other two. (MEG)

2 Alison Taylor buys an indoor plant and is told to place it more than two metres from the fireplace in her lounge. The diagram below, drawn using a scale of 2 cm to represent 1 m, shows a plan of her lounge, including the front, AB, of the fireplace.

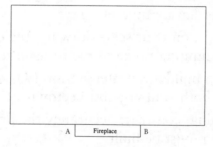

Copy the diagram and show accurately the part of the room which is more than two metres from the front, AB, of the fireplace, where Alison may stand her plant. (MEG)

3 This is a plan of baby Simon's bedroom.

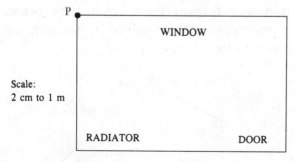

To keep warm, Simon must be within 1½ m of the wall containing the radiator. To be heard on the baby alarm at P he must be

371

within 2 m of it. Copy the diagram and shade the region in which Simon must be if he is to be warm enough and can also be heard on the alarm. (MEG)

4

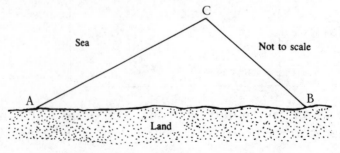

The figure shows the location of a television transmitter (T) in relation to two towns Amburg (A) and Beetown (B).

a) Using a scale of 1 cm to represent 10 km, draw an accurate scale diagram of the triangle TAB.

The transmitter has a range of 80 km.

b) Draw accurately, on your scale drawing, the curve which represents the limiting range of the transmitter.

It is planned to build a repeater station (R), which is an equal distance from both Amburg and Beetown.

c) On your drawing, construct accurately the line on which the repeater station must be built.

The repeater station is to be built at the maximum range of the transmitter.

d) i) Mark, with the letter R, the position of the repeater station on your diagram.

ii) Find the minimum transmitter range of the repeater station so that programmes can be received in Amburg. Give your answer in km, to the nearest km. (ULEAC)

5

A and B are two landmarks on a straight shore, 3.2 km apart, with A due west of B. C is the point on a bearing of 063° from A and also on a bearing of 322° from B.

a) Draw an accurate scale diagram showing A, B, C and the shore line, using a scale of 5 cm to 1 km. Shade the triangle ABC which represents a dangerous sand bank.

b) Describe the locus of points P such that angle APB equals 90°.

c) A ship P is steaming due east parallel to the shore and 600 m from it. When it is due north of A the Captain measures the angle between PA and PB and finds that it is acute. He continues on his straight course until angle APB is 90°. He then follows the locus of b) until he is again 600 m from the shore. After this he again sails due east. Draw his course on your diagram.

d) How near does he come to the sand bank?   (MEG)

6 The diagram shows the net of a cardboard carton. The carton holds a chocolate bar.

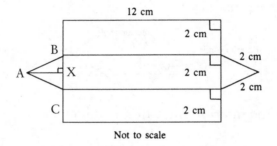

Not to scale

a) i) What type of triangle is ABC?
   ii) What is the size of angle BCA?

b) The length of AX is 1.73 cm. Calculate the area of
   i) triangle ABC,
   ii) the complete net.

c) The chocolate bars (in their cartons) are packed in layers in a box. The box is 10 cm wide, 5.4 cm high and 24 cm long. Some bars are shown in the following full-size diagram.

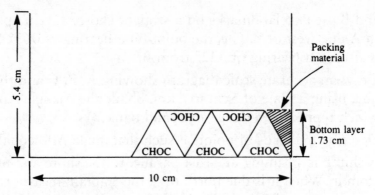

i) How many chocolate bars will there be in the complete bottom layer?

ii) How many layers will there be inside the box?

iii) How many chocolate bars will be packed inside the box?

(MEG)

7 The sum of the exterior angles of any polygon is 360°.

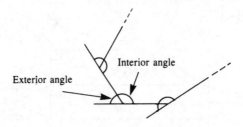

Exterior angle

Interior angle

a) Calculate the interior angle of a regular

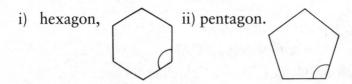

i) hexagon, ii) pentagon.

b) i) Another regular polygon has x sides. Find a formula for y, the interior angle, in terms of x.

ii) Rearrange this formula to express x in terms of y.

c) An artist is designing wallpaper patterns. She wants to cover the whole paper with a single repeated regular shape, leaving no gaps. Figure 1 shows her pattern using a regular hexagon. Figure 2 shows that she cannot use a regular pentagon.

374

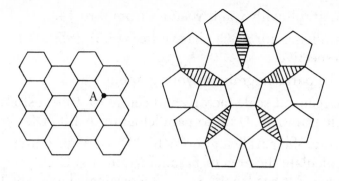

Figure 1   Figure 2

  i) How many hexagons meet at point A in Figure 1?

 ii) Use your two answers to the interior angles in part a) to
     explain why the artist can use a hexagon but not a pentagon.

iii) Can she use a regular octagon? Explain your answer.

(SEG)

8

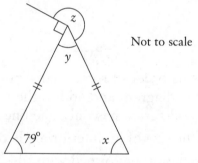

Not to scale

Work out the sizes of angle *x*, *y* and *z*.   (MEG)

9

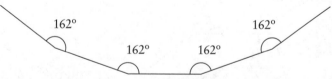

The diagram shows part of a regular polygon.
How many sides has the polygon?   (NEAB)

10 a) Draw a graph with *x* and *y* values from 0 to 14.
    b) Mark on the grid points R(7,4), Q(6,2) and P(2,2).
       A point S is to be marked on the grid so that PQRS is a
       parallelogram.
    c) Mark S on the grid and join up parallelogram PQRS.
    d) Draw the image of parallelogram PQRS after enlargement,
       scale factor 2, centre (0,0).   (ULEAC)

11 Draw a graph with x and y values from 0 to 14.

The parallelogram ABCD has vertices (6,3), (9,3), (12,9) and (9,9) respectively.

Show the parallelogram on your graph.

An enlargement scale factor $\frac{1}{3}$ and centre (0, 0) transforms parallelogram ABCD on to parallelogram A′B′C′D′.

a) i) Draw the parallelogram A′B′C′D′ on your graph.
   ii) Calculate the area of parallelogram A′B′C′D′.
b) The side AB has length 3 cm. The original shape ABCD is now enlarged with a scale factor of $\frac{2}{3}$ to give A″B″C″D″.

Calculate the length of the side A″B″. (SEG)

12 The diagram shows a regular pentagon.

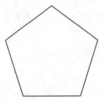

a) i) What is its order of rotational symmetry?
   ii) Copy the diagram and add to the pentagon so that the resulting diagram has only one line of symmetry.
b) Calculate the size of an interior angle of a regular pentagon.
c) Can a tessellation be formed with tiles in the shape of equally-sized regular pentagons? If so, sketch the tessellation. If not, explain why not. (MEG)

13

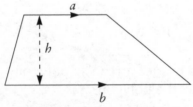

The area of this shape is found by using the formula:

$$A = \frac{h}{2}(a + b)$$

Find the area, in cm², when $a = 6$ cm, $b = 8$ cm, and $h = 5$ cm.

(NEAB)

14

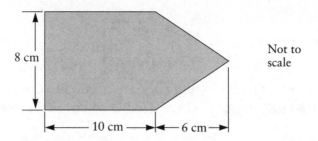

8 cm

10 cm — 6 cm

Not to scale

The diagram shows a flag.
The design on the flag consists of a rectangle and a triangle.
Calculate the area of the design. (MEG)

15 A double-glazing firm advertised the fact that its sales had doubled. It used this diagram:

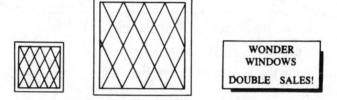

WONDER
WINDOWS
DOUBLE SALES!

The side of the smaller square is 2 cm long. The side of the larger square is double this length.

a) Calculate the area of the smaller square.

b) Calculate the area of the larger square.

c) Why is the advertisement misleading? (SEG)

16

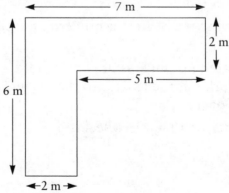

7 m

2 m

5 m

6 m

2 m

a) The diagram shows a corridor 2 m wide. Calculate the area of the floor of the corridor.

b) The floor of this corridor is covered with square carpet tiles which have a side of 0.5 m as shown.

0.5 m

0.5 m

i) How many carpet tiles are needed to cover 1 square metre?

ii) How many tiles are needed to cover the corridor?   (SEG)

17

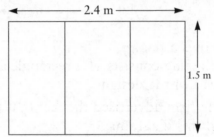

2.4 m

1.5 m

The rectangular window in the diagram (width 2.4 m and length 1.5 m) has to be glazed with three pieces of glass.

Each of the pieces is the same size.

a) Work out the width and length of one of the pieces of glass needed.

b) Work out the area of the piece of glass whose dimensions you found in a).

Glass costs £2.40 a square metre.

c) Calculate the total cost of the glass needed to glaze the window.

(ULEAC)

18 A circle has a radius of 9 cm.

a) Work out, to the nearest centimetre, the circumference of the circle.

b) An equilateral triangle is drawn with the same perimeter as the circle. Work out the length, to the nearest centimetre, of one of the sides of the triangle.   (ULEAC)

19 The diameter of a bicycle wheel is 40 cm.

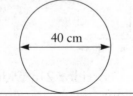

40 cm

a) Calculate the circumference of the wheel. Give your answer in centimetres.

b) Calculate how far along a road the bicycle travels while the wheel is making 50 complete turns. Give your answer in metres, to the nearest metre. (You may use $\pi = 3.14$.)   (MEG)

20

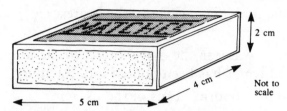

2 cm

4 cm

Not to scale

5 cm

The diagram shows a matchbox measuring 5 cm by 4 cm by 2 cm.

a) Calculate the volume of the matchbox.

12 similar matchboxes fill a carton with base dimensions of 10 cm by 8 cm.

Calculate:

b) the volume of the carton,   c) the height of the carton,

d) the total number of matches in the carton, if each box contains 48 matches.   (MEG)

21 The diagram represents a swimming pool.
The pool has vertical sides.
The pool is 8 m wide.

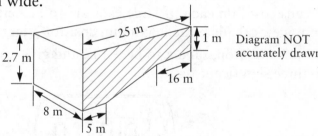

25 m

1 m   Diagram NOT accurately drawn

2.7 m

16 m

8 m

5 m

a) Calculate the area of the shaded cross section.

The swimming pool is completely filled with water.

b) Calculate the volume of water in the pool.

$64 \, m^3$ of water leaks out of the pool.

c) Calculate the distance by which the water level falls.   (ULEAC)

22 Alex buys himself a new mug.

The mug is in the form of a cylinder, with internal radius 3.5 cm and height 8 cm. Taking π to be 3.14 or by using the π key on your calculator, calculate:

a) the volume of water, correct to the nearest cm³, the mug will hold,

b) the depth of water in the mug, correct to the nearest cm, when 200 cm³ of water has been poured into the mug.　　(MEG)

23　Phiz is sold in cylindrical cans of radius 3 cm.

a) A standard can contains 350 cm³. Find its height. (Use 3.14 as the value of π or use the π button on your calculator.)

b) A large can has the same radius, but has height 22 cm. Small cans cost 20p each. If large cans are to be better value for money, calculate the maximum cost for a large can.　　(SEG)

24　(In this question, take π to be 3.142 and give each answer correct to three significant figures.)

a)

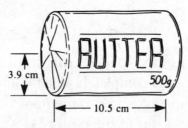

A firm sells 500 g packs of butter in the form of a circular cylinder with radius 3.9 cm and length 10.5 cm. Calculate the volume of the cylinder.

b)

The firm also sells 500g packs of butter in the form of a rectangular block with a square cross-section of side 6.5 cm. Calculate the length of the rectangular block.

c) The firm decides to sell 250g packs of butter in the form of a circular cylinder.

   i) If the radius of the (250g) cylinder is 3.9 cm, find the length of the cylinder.

   ii) If the length of the (250g) cylinder is 10.5 cm, calculate the radius of the cylinder.   (MEG)

25 A spin-dryer has a hollow cylindrical drum of radius 20 cm and height 0.45 m.

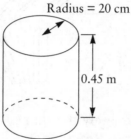

Radius = 20 cm

0.45 m

a) Calculate the length of the rim at the top of the drum. (You may take π as 3.14.)

b) Calculate the distance travelled by a point on the rim during 3600 revolutions of the drum.

c) The drum rotates at a speed of 3600 revolutions per minute. At what speed in metres per second is a point on the rim travelling?

d) Calculate the volume of the drum.   (NICCEA)

26 The diagram represents a tea packet in the shape of a cuboid.

a) Calculate the volume of the packet.

   There are 125 grams of tea in a fudl packet.

   Jason has to design a new packet that will contain 100 grams of tea when it is full.

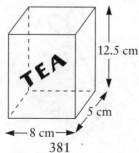

TEA

12.5 cm

5 cm

←—8 cm—→

381

b) i) Work out the volume of the new packet.

   ii) Express the weight of the new tea packet as a percentage of the weight of the packet shown.

   The new packet of tea is in the shape of a cuboid.

   The base of the new packet measures 7 cm by 6 cm.

c) i) Work out the area of the base of the new packet.

   ii) Calculate the height of the new packet.   (ULEAC)

27 This loaf of bread is cut into 30 slices, as shown.

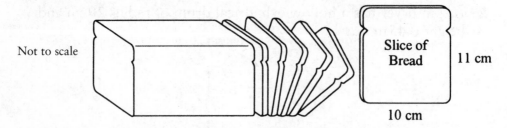

Each slice is approximately the shape of a cuboid of width 10 cm, length 11 cm and depth 8 mm.

a) Calculate the volume of the loaf in cubic centimetres.

   The same volume of bread is used to make a round loaf of length 24 cm.

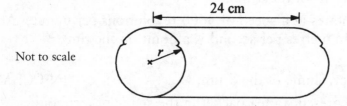

b) What is the radius of the round loaf?

   Take π to be 3.14 or use the π key on your calculator.

                                                              (SEG)

28 Sue packs sweets into a box like this:

382

This box is a square-based pyramid.

s = slant height

x = length of side of base

She uses three different formulas to calculate the packaging for this box.

Formula one:	Formula two:	Formula three:
$x^2 + 2xs$	$x^2\sqrt{s^2 - x^2}$	$4s + 2x + 20$

One formula finds the length of ribbon needed, one finds the area of card used and one finds the volume of the box. Write down which formula is used for:

a) length of ribbon   b) area of card   c) volume of box

(NEAB)

29 The expressions shown in the table below can be used to calculate lengths, areas or volumes of various shapes.

π, 2, 4 and $\frac{1}{2}$ are numbers which have no dimensions. The letters r, l, b and h represent lengths.

Copy this box and put a tick in the box underneath those expressions that can be used to calculate a volume.

$2\pi r$	$4\pi r^2$	$\pi r^2 h$	πr^2	lbh	$\frac{1}{2}bh$

(ULEAC)

30

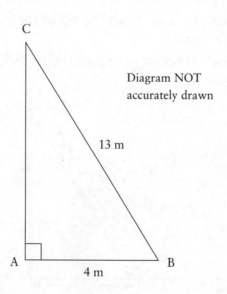

Diagram NOT accurately drawn

13 m

4 m

ABC is a right angled triangle.

AB is of length 4 m and BC is of length 13 m.

a) Calculate the length of AC.

b) Calculate the size of angle ABC.   (ULEAC)

31

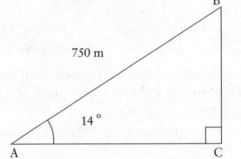

750 m

14°

A   C

B

Diagram NOT accurately drawn

Jessica's car travels a distance of 750 m in a straight line up the hill.

The angle of inclination of the hill is 14° to the horizontal AC.

a) Calculate the vertical height BC.

Give your answer to the nearest metre.

Jessica's car travelled up the hill at an average speed of 60 km/h.

b) Calculate the time the car took to travel from A to B. Give your answer in seconds.   (ULEAC)

32 The lengths of the hour hand and the minute hand of a clock are 10 cm and 15 cm respectively. At 10 o'clock, what is:

a) the shortest distance from the tip of the hour hand across to the minute hand,

b) the distance between the tip of the hour hand and the tip of the minute hand?   (NEAB)

33

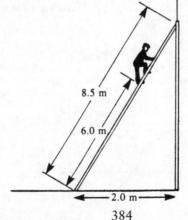

8.5 m

6.0 m

Not to scale

←—2.0 m—→

384

A window cleaner uses a ladder 8.5 m long. The ladder leans against the vertical wall of a house with the foot of the ladder 2.0 m from the wall on horizontal ground.

a) Calculate the size of the angle which the ladder makes with the ground.

b) Calculate the height of the top of the ladder above the ground.

c) The window cleaner climbs 6.0 m up the ladder (see diagram). How far is his lower foot from the wall? (MEG)

34

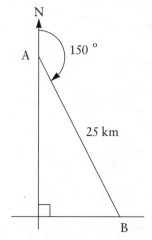

Beckthorpe House (B) is 25 km from Appletree Inn (A) on a bearing of 150°.

a) What is the bearing of Appletree Inn from Beckthorpe House?

b) Calculate how far, in km correct to three significant figures, Beckthorpe House is South of Appletree Inn. (ULEAC)

35

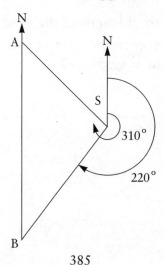

The diagram shows two navigation lights A and B with A due north of B. From a ship, S, the bearing of A is 310° and the bearing of B is 220°.

a) Calculate:

   i)   angle ASB,

   ii) angle ABS.

b) The distance AB is 2000 m; calculate the distances SA and SB.

(SEG)

36

a)

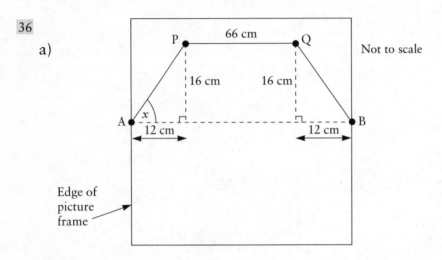

The diagram shows how Ravi intends to hang a picture on a wall. He will fasten a cord to the point A and B on the picture frame. The cord will pass over two hooks P and Q on the wall.

APQB is a trapezium. AB and PQ are horizontal.

   i)   Calculate the length of AP.

   ii)   Calculate the total length of the cord.

   iii) Calculate, correct to the nearest degree, the size of the angle marked x in the diagram.

b)

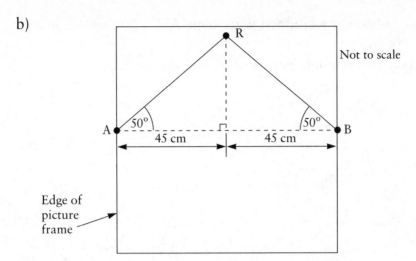

As an alternative, Ravi considers using just one hook R.

Given that angle RAB = angle RBA = 50°, calculate the length of the cord. (MEG)

37 A desk lamp of the type shown in the diagram can be positioned by varying the directions of the hinged rods AB and BC which lie in a vertical plane.

Give all answers correct to 3 significant figures.

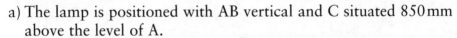

a) The lamp is positioned with AB vertical and C situated 850 mm above the level of A.

   Calculate the angle which BC makes with the vertical.

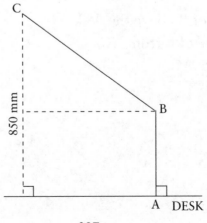

387

b) The lamp is repositioned with angle ABC=90° and C directly above A.

Calculate the angle between AB and the desk.

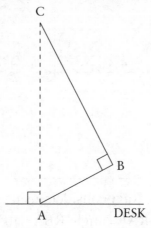

c) The lamp is repositioned with C directly above A and AB inclined at 33° to the desk.

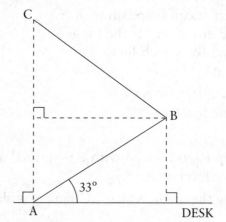

Calculate:

i) the height of B above the desk,

ii) the distance of B from AC,

iii) the distance AC.

(NICCEA)

38 *In this question, give at least 3 figures in each of your answers.*

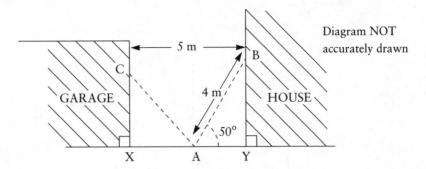

Diagram NOT accurately drawn

In the diagram, AB represents a ladder leaning against a wall of a house.

The ladder is 4 m long.

The angle between the ladder and the horizontal ground is 50°.

There is a garage near the house. It is 5 m from the house.

a) Calculate, in metres,

   i)   the height, BY, of the top of the ladder above the ground,

   ii) the distance, AY, of the foot of the ladder from the wall of the house,

   iii) the distance, AX, of the foot of the ladder from the wall of the garage.

   The top of the ladder is moved from the house wall at B to the garage wall at C.

   The bottom of the ladder, A, stays in the same place.

b) Calculate

   i)   the new height, CX, in metres, of the top of the ladder above the ground,

   ii) the angle CAB through which the ladder has been turned.

(ULEAC)

39 Two ladies were playing golf. The straight line distance from the tee (T) to the hole (H) was 160 metres.

Paula drove her ball from T to a point P, where angle HTP = 13° and angle HPT = 90°.

Sandra drove her ball from T to a point S, where SH = 14 metres and angle THS = 90°.

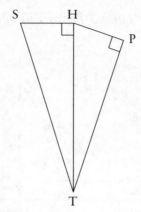

a) Calculate the distance PH.

b) Calculate the angle HTS.

c) Show that the distance TS is greater than the distance TP.

(NEAB)

40

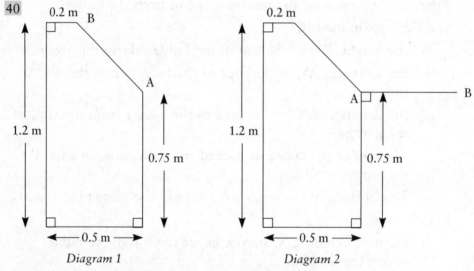

Diagram 1   Diagram 2

Diagram 1 shows the end view of a writing desk.

a) Calculate the length of AB.

The desk lid, AB, is hinged at A.

Diagram 2 shows the desk with AB opened out fully.

b) Through what angle has AB turned?   (NEAB)

41 A snail is at the point A of the closed rectangular box drawn below.

A is the mid-point of PS, and T is the mid-point of QR.

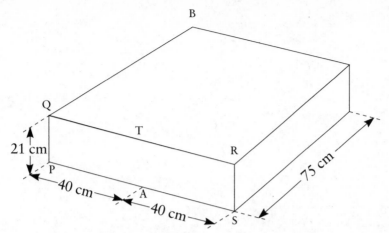

a) Calculate the distance the snail crawls to get from A to B along each of the following routes.

i) AP+PQ+QB

ii) AT+TB

b) Draw the net of the box. Calculate the shortest distance the snail would have to crawl to get from A to B. (NEAB)

Data Handling

Data Handling has been developed over the past 200 years. It is a systematic, mathematical way to answer questions about the world we live in.

To answer questions like these, we collect and organise data, present and illustrate data and make comparisons between different sets of data.

Types of variable

We collect data about *variables*. There are three types of variable.
A *qualitative variable* takes *non-numerical* values. These are some
examples of qualitative variables.

Colour of coat   Colour of paintwork
Favourite dog food   Maker's name
Breed   Type of fuel used

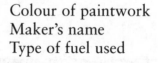

A *discrete variable* takes one of a range of *specific numerical values.* For
example, in a survey the number of children in a family may range from
0 to 6. Only whole numbers are possible, values like 1.6 or 5.78 children
are impossible. These are some more examples of discrete variables.

Number of legs   Number of wheels
Number of brothers and sisters Number of gears
Number of walks each week   Number of cylinders

A *continuous variable* can take *any value* between the start and end
of a range. For example, if the weights of the players in a team are
measured, a team member could weigh 67 kg, or 67.3 kg, or 67.37 kg
depending on the accuracy of the weighing scales. These are some
more examples of continuous variables.

Height   Top speed
Length of tail   Weight
Body temperature   Average fuel consumption

To record continuous variables we must decide on the level of accuracy required. For example, we might measure the length of a ladybird to the nearest millimetre, the length of a worm to the nearest centimetre and the height of an apple tree to the nearest metre.

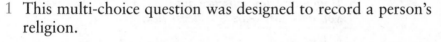

PAUSE

1 This multi-choice question was designed to record a person's religion.

Religion	Please tick:
Church of England	_____
Catholic	_____
Other Christian	_____
Muslim	_____
Hindu	_____
Sikh	_____
Jewish	_____
Other	_____

Design a multi-choice question to record:

a) A person's favourite type of pop music.
b) The method of transport a pupil uses to come to school.
c) A person's favourite breakfast cereal.

2 This multi-choice question was designed to record the number of people living in each house on an estate.

How many people live in your house?	Please tick:
1	_____
2	_____
3	_____
4	_____
5	_____
6	_____
7	_____
8	_____
More than 8	_____

Design a multi-choice question to record:

a) The number of pints of milk delivered to each house in a street.
b) The number of days a pupil was absent from school last week.
c) The shoe size of each member of your mathematics group.

3 This multi-choice question was designed to record the number of times a pupil had been absent from school in one year.

How many days absence did you have last year?	Please tick.
0 days	_____
1–5 days	_____
6–10 days	_____
11–15 days	_____
16–20 days	_____
More than 20 days	_____

Design a multi-choice question to record:

a) The number of apples that a person eats each month.
b) The number of books a person owns.
c) The number of cigarettes that a person smokes each week.

4 This multi-choice question was designed to record the length of an earthworm.

Design a multi-choice question to record:

a) The length of a ladybird (to the nearest millimetre).
b) The weight of a new-born baby (to the nearest kilogram).
c) The time taken by a Year 7 pupil to run 100 metres (to the nearest second).

Length of the worm (to the nearest centimetre)	
1	_____
2	_____
3	_____
4	_____
5	_____
6	_____
7	_____
8	_____
9	_____
10	_____
Over 10	_____

5 State whether the following are qualitative, discrete or continuous variables.

a) A person's eye colour.

b) The number of pints of milk delivered to each house in a street.

c) The number of apples that a person eats each month.

d) The shoe size of each member of your mathematics group.

e) The time taken by a Year 7 pupil to run 100 metres.

f) A person's favourite breakfast cereal.

g) The number of days holiday that an employee gets each year.

h) The height of an apple tree (to the nearest metre).

i) The number of days a pupil was absent from school last week.

j) The weight of a new-born baby.

k) The number of books a person owns.

l) The method of transport a pupil uses to come to school.

■ ■

Frequency tables

Data is often organised into *frequency tables*.

For example, some data has been collected to answer the question:

'Do the families on Housing Estate A have more children than the families on Housing Estate B?'

A door to door survey was done on each estate to discover the number of children in 40 families.

This is the 'raw' data. It is difficult to compare unless it is put into frequency tables.

Estate A
Number of children per family
0 1 2 3 2 2 2 1 1 2
1 0 0 0 2 2 3 4 3 1
1 2 1 3 0 4 2 2 2 3
0 2 2 2 4 3 2 4 0 0

Estate B
Number of children per family
4 6 5 2 4 2 2 2 1 2
3 3 1 0 4 3 2 2 2 2
1 4 2 2 3 0 2 2 2 2
5 6 6 6 2 3 5 3 1 3

A *tally table* is often used to complete a frequency table. The tally marks are done in groups of five. The fifth tally mark is used to cross out the previous four marks.

	Estate A				Estate B	
No. of children per family	Tally	Frequency		No. of children per family	Tally	Frequency
0	̶H̶H̶ III	8		0	II	2
1	̶H̶H̶ II	7		1	IIII	4
2	̶H̶H̶ ̶H̶H̶ ̶H̶H̶	15		2	̶H̶H̶ ̶H̶H̶ ̶H̶H̶ I	16
3	̶H̶H̶ I	6		3	̶H̶H̶ II	7
4	IIII	4		4	IIII	4
5		0		5	III	3
6		0		6	IIII	4
	Total	40			Total	40

Some variables have so many different values that we must arrange them into groups. For example, if we collect data on the weekly earnings of a sample of 500 people, we can expect a large number of different results. We might use a *grouped table* like this:

Earnings per week	Frequency
Less than £100	
£101 – £150	
£151 – £200	
£201 – £250	
£251 – £300	
Over £300	

When we collect continuous data we need to decide how accurate the measurements are going to be. For example, in an experiment to measure the lengths of 200 slugs, we decide to measure each slug and record the results correct to the nearest centimetre. We might use a table like this:

Length of slug (nearest cm)	Tally	Frequency
1		
2		
3		
...		

The table makes the length look like a discrete variable. We must remember that the headings are correct to the nearest centimetre.

Any slug from 1.5 cm to 2.5 cm goes in the 2 cm group. A slug which is exactly equal to 2.5 cm, goes in the 2.5 cm to 3.5 cm group. Inequality signs are sometimes used to state the exact range of a continuous variable.

For example, if the letter L represents the length of a slug, the table above could be written like this

Length of slug (L cm)	Frequency
$0.5 \leq L < 1.5$	
$1.5 \leq L < 2.5$	
$2.5 \leq L < 3.5$	

$1.5 \leq L < 2.5$ means L is between 1.5 and 2.5, but could be equal to 1.5 as $\leq$ includes an equal sign.

PAUSE

1 This table shows the membership of organisations for young people (source: Social Trends 1994).

Youth organisation	Members (in thousands)	
	1971	1994
Cub, Scouts and Venture Scouts	480	540
Brownies, Guides and Rangers	692	642
Sea, Army and Combined Cadets Force, Air Training Corps	135	132
Boys Brigade and Girls Brigade	237	173
NABC – Clubs for Young People	164	198
Youth Clubs UK	319	715
Methodist Youth Clubs	115	60
YMCA (males and females)	48	80

a) How many members did the Cub Scouts have in 1971?

b) How many members did the Cub Scouts have in 1994?

c) How many members did the Youth Clubs UK have in 1971?

d) How many members did the YMCA have in 1994?

e) Copy the table and add a third column for 'Change in Membership 1971–1994 (in thousands)'.

f) In e) most clubs changed to have more members. A few like Methodist Youth Clubs changed to have fewer members. Think of a way of marking those with fewer members in the 'Change in Membership column'.

2 These tables show details of people killed and seriously injured in road accidents (source: Annual Abstract of Statistics 1994).

Type of road user	Number
Child Pedestrians	4901
Adult Pedestrians	9119
Child Pedal Cyclists	1195
Adult Pedal Cyclists	2751
Moped Riders	807
Motor Scooter Riders	152
Motor Scooter Passengers	11
Motor Cycle Riders	5784
Motor Cycle Passengers	569
Car and Taxi Drivers	15406
Car and Taxi Passengers	9718
Users of Buses and Coaches	655
Users of Goods Vehicles	1967
Users of Other Vehicles	214

Age group	Number
0–4	1141
5–9	2588
10–14	3328
15–19	7509
20–24	8016
25–29	6133
30–39	7383
40–49	5194
50–59	3746
60 and over	7748

a) How many people aged from 20–24 were killed or seriously injured?
b) How many people aged from 30–39 were killed or seriously injured?
c) How many motor cycle riders were killed or seriously injured?
d) 569 motor cycle passengers were killed or seriously injured, compared to 9718 car or taxi passengers. Does this mean that travelling as a passenger on a motor cycle is safer than travelling as a passenger in a car or taxi?
e) 1141 people aged 0–4 were killed or seriously injured, compared to 2588 people aged 5–9. Does this mean that children aged 4 years and under are more careful road users than children aged 5 years to 9 years?
f) Are insurers right to charge more to insure drivers under 30? Use the numbers in the table to help you with your answer.

3 These are the final scores in some cup matches. Complete a table with tally marks to show the frequency with which teams scored no goals, 1 goal, 2 goals etc.

FA Cup Third Round			
Barnsley	4	Chelsea	0
Birmingham	0	Wimbledon	1
Blackpool	0	Bournemouth	1
Bradford	1	Tottenham	0
Brighton	1	Leeds	2
Cardiff	1	Hull	2
Carlisle	0	Liverpool	3
Charlton	2	Oldham	1
Crewe	2	Aston Villa	3
Derby	1	Southampton	1
Hartlepool	1	Bristol City	0
Huddersfield	0	Sheffield Utd	1
Kettering	1	Halifax	1
Manchester City	1	Leicester	0
Manchester United	0	QPR	0
Middlesbrough	1	Grimsby	2

4 Some letters of the alphabet occur more frequently than others. The letter *e*, for example, occurs far more frequently than the letter *z*.

If a large sample of English is checked, the order of frequency of each letter is usually:

ETOANIRSHDLCWUMFYGPBVKXQJZ

Carry out your own survey to check this order of frequency. Use a paragraph chosen from any book. Design a tally chart table and use it to collect your results. Comment on your results. How well do they match the predictions?

5 Two groups of 40 pupils take a Mathematics test marked out of 100. These are the raw results.

GROUP 1
75 67 34 91 23 32 45 63
82 71 63 44 87 31 92 55
45 67 87 98 37 64 63 59
56 67 89 93 63 54 45 29
65 83 66 82 67 81 66 83

GROUP 2
23 47 56 73 66 49 34 22
37 62 29 17 36 63 37 22
41 29 42 28 34 27 43 26
51 36 50 37 49 38 48 39
47 40 46 41 26 55 61 12

a) Copy and complete this tally chart table for the pupils in Group 1.

Mark	Tally	Frequency
0–10		
11–20		
21–30		
31–40		
⋮		

b) Complete a similar tally chart table for the pupils in Group 2.
c) Comment on any differences between the two groups of pupils.

6 Two different samples of 50 apples were weighed. These are the raw results.

Sample 1 (weights to the nearest gram)

83	87	92	103	121	107	88	94	103	112
90	93	108	117	123	122	89	98	100	120
129	84	95	96	99	120	115	118	117	121
116	109	108	119	118	122	127	125	130	131
93	125	94	124	95	123	125	116	112	123

Sample 2 (weights to the nearest gram)

117	119	125	143	131	144	132	145	147	151
132	115	107	99	122	125	141	134	137	138
132	139	141	145	151	98	107	105	118	119
112	124	127	128	139	130	138	137	136	152
120	134	145	146	137	138	141	133	132	137

a) Copy and complete this tally chart table for the apples in
   Sample 1. Remember that an apple weighing exactly 90g would
   go in the group $90 \leq w < 100$.

Weight (w grams)	Tally	Frequency
$80 \leq w < 90$	卌 I	6
$90 \leq w < 100$	卌 卌 I	11
$100 \leq w < 110$	卌 I	6

b) Complete a similar tally chart table for the apples in Sample 2.
c) Comment on any differences between the two samples of apple.

■ ■

Hypothesis testing and bias

An *hypothesis* is a statement that could be true but has not been
proved, for example 'More people would use public transport for
going to work if the services were better.'

Questionnaires are often written to see if hypotheses are true.

We have to be careful of false results through bias. This can happen if
we do not ask a fair sample. Consider the question, 'Would you use a
new leisure centre?'. Suppose the person doing the survey asked every
tenth person coming out of a supermarket on a Tuesday morning.
The people shopping at that time would be mostly retired or mothers
with young children. They would not form a fair sample.

There are other reasons for bias. A Year 11 pupil is testing the
hypothesis 'More girls than boys in Year 11 smoke.' Year 11 pupils
may not admit to smoking if their teacher is listening. The questions
also have to be worded carefully. 'You don't smoke, do you?' is more
likely to get the answer 'No' than 'Do you smoke?'.

Sometimes we need our sample to come from a particular group. If
we want to test the hypothesis 'More people would use public
transport for going to work if the services were better', we need to
ask people who use their cars for going to work at present.

PAUSE

In each of these, write a question or questions that could be used to test the hypothesis. Suggest different samples that could be used. What precautions do you need to take against bias?

1 'More mothers than fathers take their children to school.'

2 'School dinners contain plenty of vegetables and salad.'

3 'More people would use a bicycle for short journeys if there were special routes for bicycles.'

4 'The pavements would be cleaner if there were more dog toilet bins.'

5 'Women usually have to wait longer than men when using a public toilet.'

6 'Most school pupils would prefer not to wear a uniform.'

7 'Most school pupils would like more PE lessons.'

8 'Girls have to help more with housework than boys.'

SECTION 2　Representing Data

Graphs and diagrams can be used to illustrate data. They make data easier to understand. These are the different types of graphs and diagrams that are commonly used.

Bar charts

Bar charts are used to illustrate qualitative data and discrete quantitative data.

Example

Number of children per family	0	1	2	3	4	5	6
Frequency	10	9	18	8	5	0	0

This is the completed bar graph.

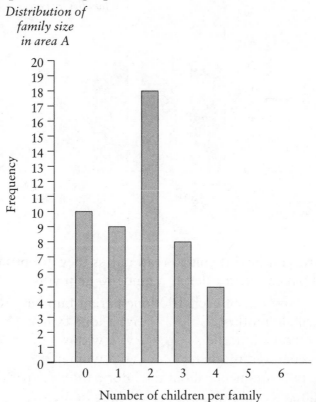

*Distribution of
family size
in area A*

1 Draw a bar chart to show the information in this table:

Number of children per family	0	1	2	3	4	5	6
Frequency	3	4	20	9	5	4	5

2 A survey in 1992 revealed that 21% of all families with children were lone parent families. This bar chart shows how each 100 of those families can be divided into different groups (source: General Household Survey 1994).

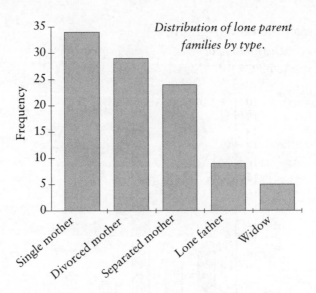

Distribution of lone parent families by type.

Bar charts can be difficult to read unless they are printed on squared paper. Use a ruler as a guide to help you.

a) How many out of each 100 lone parent families are headed by:
   i) a single mother,   iv) a lone father,
   ii) a divorced mother,   v) a widow,
   iii) a separated mother,   vi) a woman?

b) Write two comments about the information the bar chart shows.

3 This double bar graph shows the results of a survey when 2176 Year 11 pupils were asked, 'When choosing what to eat, do you consider your health?' (source: Young people in 1993).

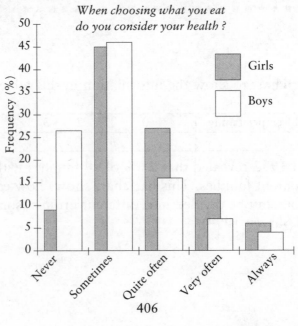

When choosing what you eat do you consider your health ?

a) When asked this question, what percentage of girls replied:
    i) never,   iv) very often,
    ii) sometimes,   v) always?
    iii) quite often,

b) When asked this question, what percentage of boys replied:
    i) never,   iv) very often,
    ii) sometimes,   v) always?
    iii) quite often,

c) Write two comments about the information the double bar chart shows.

4 This table shows the GCSE Maths grades obtained by Year 11 pupils in two different schools.

GCSE maths grade	A	B	C	D	E	F	G	U
Frequency (school A)	3	12	10	17	1	3	4	1
Frequency (school B)	10	11	27	13	5	0	3	0

a) Draw a double bar chart to show the information in this table.
b) Compare the results of the two schools.

5 This bar chart shows the number of deaths from solvent abuse each year from 1971 to 1991 (source: Re-solv 1994).

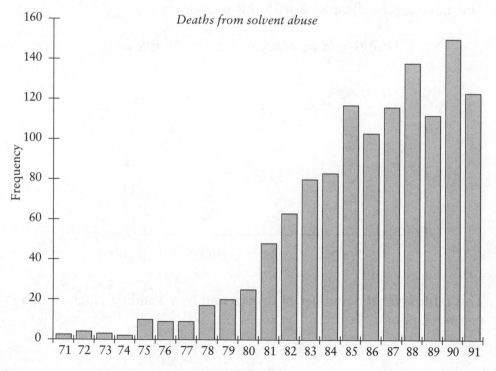

Deaths from solvent abuse

a) Estimate the number of deaths in:
   i) 1971,   iv) 1986,
   ii) 1976,   v) 1991.
   iii) 1981,
b) Write two comments about the information the bar chart shows.

■■■■■■■■■■■■■■■■■■■■■■■■■■■■■■■■■■■■■

Pictograms

A pictogram is a variation of the bar chart, using lines of symbols to replace the solid bar.

This is data on the number of drinks sold by a vending machine in a one hour period:

Type of drink	Number sold
Tea	5
Coffee	9
Soup	3
Chocolate	4

This table can be illustrated with this pictogram:

Drinks sold by vending machine in one hour

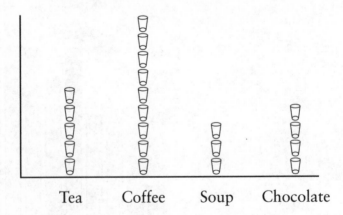

Tea   Coffee   Soup   Chocolate

This table gives the number of drinks sold by a vending machine in a one week period.

Type of drink	Number sold
Tea	120
Coffee	125
Soup	73
Chocolate	85

It would take too long to draw lines with 120 or 135 cups. We use a scale and let one cup symbol represent 10 cups of drink sold. If a number sold does not divide exactly by 10, we draw part of a cup. This is the completed pictogram, drawn horizontally this time.

Drinks sold by vending machine in one week

Tea	☕☕☕☕☕☕☕☕☕☕☕☕
Coffee	☕☕☕☕☕☕☕☕☕☕☕☕◡
Soup	☕☕☕☕☕☕☕◡
Chocolate	☕☕☕☕☕☕☕☕◡

PAUSE

Choose very simple symbols for your pictograms. You have to draw the symbols several times.

1 This table illustrates the types of pets owned by a class. Draw a pictogram to illustrate the data.

Pet	Frequency
Dogs	12
Cats	15
Rabbits	6
Fish	7
Snakes	1

2 This table illustrates the type of footware worn by a group on pupils on a school trip. Draw a pictogram to illustrate the data.

Footware	Frequency
Shoes	11
Trainers	10
Boots	8
Sandals	2

3 This table illustrates the transport used by groups of Year 7 and Year 11 pupils to come to school.

Draw two separate pictograms to illustrate the data.

Transport	Year 7	Year 10
Car	30	13
Bus	38	34
Cycle	25	29
Walk	52	66

■ ■

Pie charts

Bar charts are very useful for picking out the most or least common variable. *Pie charts* give a better display of the way that a variable is 'shared out.'

For example, this pie chart illustrates the reasons given for not truanting in a survey of Year 10 and Year 11 students (source: Truancy in English secondary schools 1994).

Reasons for not truanting

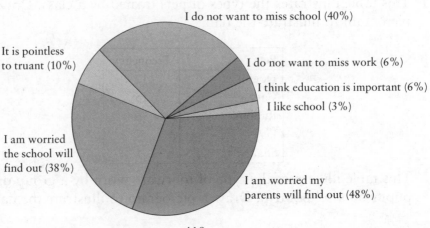

I do not want to miss school (40%)

It is pointless to truant (10%)

I do not want to miss work (6%)

I think education is important (6%)

I like school (3%)

I am worried the school will find out (38%)

I am worried my parents will find out (48%)

When drawing our own pie charts there are two methods we can use:

Method 1: Using a pie chart scale.

Example:

An international committee has 18 members, 6 from Great Britain, 7 from America, 3 from France and 2 from Japan. Illustrate this data with a pie chart.

Country	Number of members
Great Britain	6
America	7
France	3
Japan	2
Total	18

In any pie chart, we start with a full circle or 100% to share out. We add a column to the table showing the percentage represented by the members from each country.

Country	Number of members	Percentage
Great Britain	6	33.3%
America	7	38.9%
France	3	16.7%
Japan	2	11.1%
Total	18	100%

The percentages are worked out by dividing each number by the total and then multiplying by 100.

For example, for Great Britian, the percentage = $6 \div 18 \times 100 = 33.3\%$.

We can now draw the pie chart. Draw a circle of a suitable size. Start at any point to divide it up into slices. Use a pie chart scale and the percentages in the table. Give the pie chart a title and label the slices.

Nationality of Committee Members

411

Method 2: Using a protractor

Example

An international committee has 18 members, 6 from Great Britain, 7 from America, 3 from France and 2 from Japan. Illustrate this data with a pie chart.

Country	Number of members
Great Britain	6
America	7
France	3
Japan	2
Total	18

In any pie chart, we start with a full circle or 360° to share out. We add a column to the table showing the angle represented by the members from each country.

Country	Number of members	Pie chart angle
Great Britain	6	120°
America	7	140°
France	3	60°
Japan	2	40°
Total	18	360°

The angles are worked out by dividing each number by the total and then multiplying by 360.

For example, for Great Britian, the angle = $6 \div 18 \times 360 = 120°$.

We can now draw the pie chart. Draw a circle of a suitable size. Start at any point to divide it up into slices. Use a protractor and the angles in the table.

Nationality of Committee Members

412

1 This table shows how every 100 households in Great Britain can be divided up into different types of dwelling (source: General Household Survey 1994).

Type of dwelling	Frequency in every 100 dwellings
Detached House	20
Semi-Detached House	31
Terraced House	28
Flat or Maisonette (purpose built)	15
Flat or Maisonette (converted)	5
With Business or Shop	1
Total	100

Draw a pie chart to show the information in the table.

2 In a survey, Year 7 pupils were asked whether, when choosing what to eat, they thought about their health. This table shows the results for every 100 boys or girls who answered the questions.

When choosing food do you think about your health?	Boys	Girls
Always	8	7
Very Often	8	11
Quite often	24	27
Sometimes	44	47
Never	16	8
Total	100	100

a) Draw two separate pie charts to illustrate this information.
b) Write two comments about the information the pie charts show.

3 This table shows the numbers of each type of drink sold from a vending machine in a one week period. Illustrate this data with a pie chart.

Type of drink	Number sold
Tea	120
Coffee	90
Chocolate	75
Tomato soup	60
Chicken soup	15

4 A local council did a survey in their area of pre-school provision for under fives. The results are shown in the table.

Type of provision	Private nursery schools	Council nursery schools	Nursery class in infant school	Playgroup
Frequency	6	1	4	9

Draw a pie chart to illustrate the data.

5 A survey was carried out to see which television channel people watched most over Christmas. The results are shown in the pie chart.

Most watched television channel over Christmas

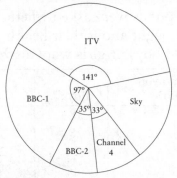

a) Calculate the percentage of people who watched Sky television most.
b) 2350 of the people in this survey watched ITV most.
   How many people were questioned altogether?

6   *Land use in Eastshire*

414

The pie chart shows land use in the county of Eastshire.

The total area of the county is 2500 km².

a) Measure and write down the size of the angle representing 'Farming'.

b) Calculate the total area of land used for 'Farming'.

c) Express, as a ratio in the form 1:*n*,

Industrial area: total area.   (ULEAC)

7 A youth club offers the following activities:

Football   Snooker   Table Tennis   Disco

All the members of the club are asked which of these activities they prefer. The pie chart represents their replies.

Activities chosen by members of youth club

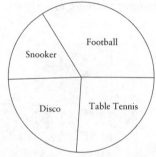

a) What is the size of the angle which represents snooker?

b) Which of these activities is the most popular?

c) The club has 180 members.
   How many members prefer the disco?

d) Show the same information as a bar chart.   (SEG)

8 720 students were asked how they travelled to school.

The pie chart shows the results of this survey.

How students travel to school

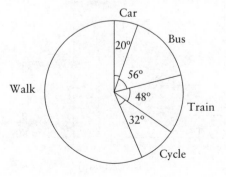

Work out:

a) how many of the students travelled to school by bus,
b) how many of the students walked to school.

(ULEAC)

■ ■

Histograms and frequency polygons

Length of earthworm (nearest cm)	1	2	3	4	5	6	7	8	9	10
Number of earthworms	3	7	12	19	18	18	11	10	2	0

Length is a continuous variable so when we draw a bar chart, the horizontal axis will be continuous. There will be no gaps between the bars.

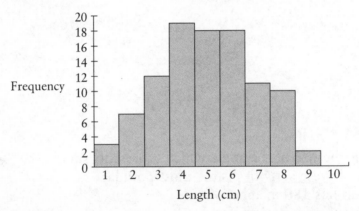

Distribution of length in 100 earthworms

The blocks of the bar chart represent groups of lengths L cm where $0.5 < L < 1.5$, $1.5 < L < 2.5$ and so on. We could label the horizontal axis like this:

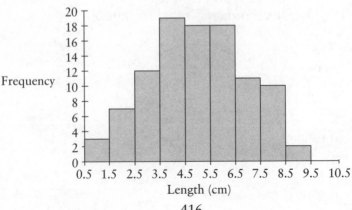

Distribution of length in 100 earthworms

416

Special bar charts like this are known as histograms. Histograms are used to illustrate grouped or continuous data.

A frequency polygon can be used instead of a histogram. The same type of axes are used, but the frequency is plotted as a single point above the mid-point of the group or interval.

These points are then connected to form a line graph. Here is a frequency polygon to illustrate the data on earthworms.

Distribution of length in a sample of 100 earthworms

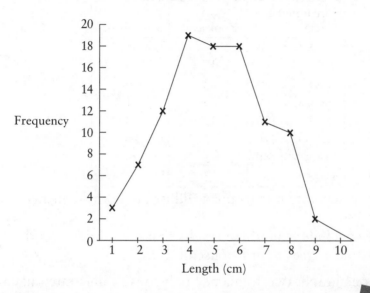

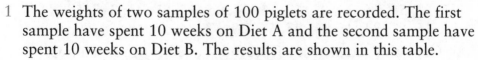

PAUSE

1 The weights of two samples of 100 piglets are recorded. The first sample have spent 10 weeks on Diet A and the second sample have spent 10 weeks on Diet B. The results are shown in this table.

Weight (nearest kg)	Frequency (Diet A)	Frequency (Diet B)
25	20	17
26	23	20
27	31	30
28	19	23
29	7	8
30	0	2

a) Draw two separate histograms to illustrate the frequency distributions.
b) Which diet do you consider is the most successful?

2 Michelle Jackson, the film star, owns a country estate to 'get away from it all'. She is interested in preserving nature and asks for a survey into the size of trees in two pieces of woodland she owns. The circumferences of 100 selected trees are measured in Rooksend and Streamside.

Circumference C (cm)	Rooksend	Streamside
$0 \leq C < 20$	3	5
$20 \leq C < 40$	17	3
$40 \leq C < 60$	22	6
$60 \leq C < 80$	25	11
$80 \leq C < 100$	19	18
$100 \leq C < 120$	7	24
$120 \leq C < 140$	5	26
$140 \leq C < 180$	2	7

a) Draw two separate histograms to illustrate both frequency distributions.
b) Can we conclude from these results that Streamside Wood contains older trees than Rooksend Wood?

3 Draw, on one graph, two frequency polygons to illustrate the data table shown in question 1.

4 Draw, on one graph, two frequency polygons to illustrate the data table shown in question 2.

5 Here are the weights, in kg, of 30 students.

45, 52, 56, 65, 34, 45, 67, 65, 34, 45, 65, 87, 45, 34, 56, 54, 45, 67, 84, 45, 67, 45, 56, 76, 57, 84, 35, 64, 58, 60.

a) Copy and complete the frequency table below using a class interval of 10 starting at 30.

Weight range (w)	Tally	Frequency
$30 \leq w < 40$		

b) Which class interval has the highest frequency?

c) Draw a histogram representing this information.

(ULEAC)

6 The table gives information about the weight of the potato crop produced by 100 potato plants of two different types.

Weight of potatoes per plant w kg	Number of plants Type X	Number of plants Type Y
$0 \leq w < 0.5$	0	0
$0.5 \leq w < 1.0$	3	0
$1.0 \leq w < 1.5$	12	6
$1.5 \leq w < 2.0$	55	39
$2.0 \leq w < 2.5$	23	32
$2.5 \leq w < 3.0$	7	23
$3.0 \leq w < 3.5$	0	0

a) Using a scale of 1 kg = 4 cm on the horizontal axis and 10 plants = 2 cm on the vertical axis, draw a frequency polygon for each type of potato.

b) Which type of potato produces the heavier crop?

c) i) Which type of potato has more variation in the weight of the crop?

   ii) Give a reason for your answer.

(SEG)

■ ■

SECTION 3 Making Comparisons Between Sets of Data

A set of values for a variable is called a *frequency distribution*. The obvious way to compare frequency distributions is to place them side by side and to look for differences. This takes a long time and various methods have been developed to summarise data. The data can then be compared more easily.

Averages and spreads

When comparing distributions, it is very useful if a single typical value can be selected to represent the distribution. Such values are called *averages*. An average gives an impression of the type of values to be found in a distribution. For example, we are told that the *average* test mark of one class is 60% and the *average* test mark for another class (on the same test) is 85%. We expect a typical student in the second class to be more able than one in the first class.

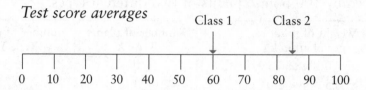

Test score averages

Averages on their own should be treated with care. It is better to make comparisons using an average and *a measure of spread*. Measures of spread tell us of how the distribution is arranged around the average value.

The simplest measure of spread is **the range**. The range is the difference between the greatest and the least values in a distribution.

For example, the greatest mark for a student in Class 1 was 98 and the least mark was 22, the range was $98-22=76$.

The greatest mark for a student in Class 2 was 95 and the least mark was 75, the range was $95-75=20$.

We can now tell that it is not true to say that a typical student in the second class is likely to be more able than a typical student in the first class. Class 1 is probably a mixed ability class and Class 2 is a class that has been set according to ability.

Test score averages

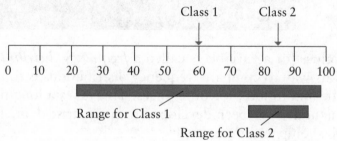

There are three types of average in common use. These are the **mean**, the **median**, and the **mode**.

The mean

The mean, is calculated by adding up all the values in a distribution and then dividing this total by the number of values in the distribution.

Find the mean and range of these values:

$$1,1,2,2,3,4,4,5,6,6,6,8.$$

First, we find the total of all the values. In this example the total is 48. Second, we divide the total by the number of values. In this example there are 10 values.

So, the mean average is:

$$\frac{48}{10} = 4.8$$

The range is $8 - 1 = 7$.

PAUSE

1 Find the mean and the range of each of the following distributions.
   a) 12,6,7,9,6
   b) 1,1,1,1,2,2,3,3,3,3,3,4,4,5,6,6
   c) £1.50, £1.50, £2.00, £3.00
   d) −2, −1, −2, 0, 0, 0, 3, 4, 1, 2, 6
   e) 2.7, 3.4, 8.9, 5.6, 6.6, 7.3, 5.2, 2.9, 8.0, 7.1
   f) 12 cm, 13 cm, 15 cm, 17 cm, 12 cm, 13 cm, 15 cm, 14 cm, 29 cm, 15 cm, 17 cm, 12 cm
   g) 450 g, 230 g, 770 g, 125 g, 431 g, 874 g, 455 g, 637 g, 888 g, 97 g
   h) 125 cm, 34 cm, 220 cm, 1.34 m, 1.90 m, 340 cm, 1.78 m, 2.55 m, 3.04 m
   i) 550 g, 650 g, 750 g, 850 g, 950 g, 1.05 kg, 1.15 kg, 1.25 kg, 1.35 kg
   j) 1716, 1785, 1745, 1745, 1780, 1765, 1731, 1734, 1754, 1700, 1735

2 The test marks of a group of ten Year 11 students are 56,72,35,47,89,98,13,45,34 and 73. The test marks of a group of ten Year 10 students who take the same test are:

58,70,40,41,42,11,94,62,43 and 60

a) Calculate the mean and the range of each set of marks.

b) Comment on your results.

3 Two darts players, Maureen and Eric each throw 9 darts. Maureen's scores are 20,40,1,5,60,20,20,1 and 60. Eric's scores are 20, 40, 40, 0, 60, 20, 0, 5 and 1.

a) Find the mean and the range of each set of scores.

b) Comment on your results.

4 A group of five trees in a garden have a mean height of 4.5 metres. A sixth tree with a height of 2.1 metres is planted.

a) What is the total height of the five original trees?

b) What is the mean height of all six trees?

5 One group of 5 students have a mean score of 45 on a test. A second group of 11 students have a mean score of 53 on the same test.

a) What was the total score of the group of 5 students?

b) What was the total score of the group of 11 students?

c) Find the mean score for the whole group of 16 students.

■ ■

Finding the mean and the range of a discrete distribution presented in a frequency table

Calculate the mean number of children per family and the range for this frequency distribution.

Number of children per family	0	1	2	3	4	5	6
Frequency	13	13	38	17	10	4	5

Remember, to calculate the mean average, we need to know
■ the total number of families,
■ the total number of children in these families.

We can find the number of families by adding all the frequencies (middle column). The third column is used to work out the total number of children.

Number of children per family	Frequency	Number x Frequency
0	13	0
1	13	13
2	38	76
3	17	51
4	10	40
5	4	20
6	5	30
Total	100	230

We now know that 100 families have a total of 230 children. We can calculate the mean:

$$\text{Mean} = \frac{230}{100} = 2.3 \text{ children.}$$

The range is the difference between the greatest value and the least value.

The range $= 6 - 0 = 6$ children per family.

PAUSE

1 This table shows the results of a survey into the number of children per family on two housing estates.

Number of children per family	0	1	2	3	4	5	6
Frequency (Estate A)	10	9	18	8	5	0	0
Frequency (Estate B)	3	4	20	9	5	4	5

a) Calculate the mean number of children per family and the range for each estate.

b) Comment on your results.

2 The cost of a litre of petrol was checked at a sample of 50 garages in two different cities. The results are shown in the table below.

Cost per litre (pence)	City A	City B
49	2	4
50	5	7
51	12	4
52	21	11
53	7	13
54	3	5
55	0	6

a) Calculate the mean price per litre and the range in each city.

b) Comment on your results.

3 In the GCSE Mathematics examination, grades A,B,C,D,E,F,G and U are awarded. This table shows the results for the students entered in the same year from two different schools.

	A	B	C	D	E	F	G	U
School A	3	14	15	23	13	20	9	3
School B	2	32	18	20	20	0	0	8

a) Calculate the mean grade for each school (Note. To do this you will need to replace grades A to U with the numbers 1 to 8. Calculate the averages and then translate them back into a grade).

b) Calculate the range for each school.

c) Comment on your results.

■ ■

Finding the mean and the range of a distribution presented in a grouped frequency table

Calculate the mean and the range for the students' test marks shown in this table.

Test mark	1–15	6–10	11–15	16–20	21–25	26–30	31–35	36–40	41–45	46–50
Frequency	1	2	11	17	25	18	13	6	3	4

Each group of marks is called a *class interval*.

Remember, to calculate the mean, we need to know:

■ the total number of students taking the test,

■ the total number of marks they scored.

It is impossible to use this data to calculate the exact total number of marks scored by the students. We know, for example, that 18 students scored a mark between 26 and 30. We have no idea of the exact marks that each of the 18 students scored.

We solve this problem by using an approximation. For example, we assume that 18 students scoring a mark between 26 and 30 is approximately the same as 18 students scoring a mark of 28 (the middle value of the class interval).

The completed table looks like this.

Test mark	Mid-value	Frequency	Mid-value × Frequency
1– 5	3	1	3
6–10	8	2	16
11–15	13	11	143
16–20	18	17	306
21–25	23	25	575
26–30	28	18	504
31–35	33	13	429
36–40	38	6	228
41–45	43	3	129
46–50	48	4	192
Total		100	2525

We now know that 100 students have scored a total of 2525 marks. Therefore:

$$\text{the mean} = \frac{2525}{100} = 25.25 \text{ marks,}$$

$$\text{the range} = 50 - 1$$
$$= 49.$$

Note: This is a more conventional orientation for tables and the one exam boards usually use.

The mid-value need not be a whole number. To find the mid-value of a class interval, add the boundary values and divide by 2.

For example, the mid-value of 0–9 is $\frac{0+9}{2} = 4.5$.

These are some more examples of class intervals and their mid-values.

Class interval	Mid-value
120–130	125
0–9	4.5
1–9	5
1–4	2.5
1–5	3
16–25	20.5
15–25	20

This example calculates the mean and the range when the mid-value is not a whole number.

This table shows the results when a player throws 60 sets of 3 darts.

Score	1–30	31–60	61–90	91–120	121–150	151–180
Frequency	8	7	9	10	9	7

These are the completed calculations.

Score	Mid-value	Frequency	Mid-value × Frequency
1– 30	15.5	8	124
31– 60	45.5	7	318.5
61– 90	75.5	9	679.5
91–120	105.5	10	1055
121–150	135.5	9	1219.5
151–180	165.5	7	1158.5
Total		60	4555

Mean = $\dfrac{4555}{60}$ = 75.9 points per set of three darts

Range = 180 – 1
= 179

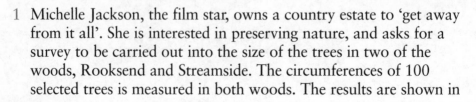

PAUSE

1 Michelle Jackson, the film star, owns a country estate to 'get away from it all'. She is interested in preserving nature, and asks for a survey to be carried out into the size of the trees in two of the woods, Rooksend and Streamside. The circumferences of 100 selected trees is measured in both woods. The results are shown in

this table.

Circumference C (cm)	Rooksend	Streamside
$0 \leq C < 20$	3	5
$20 \leq C < 40$	17	3
$40 \leq C < 60$	22	6
$60 \leq C < 80$	25	11
$80 \leq C < 100$	19	18
$100 \leq C < 120$	7	24
$120 \leq C < 140$	5	26
$140 \leq C < 160$	2	7

a) Calculate the mean circumference of the trees in each wood and the range.

b) Comment on your results.

2 Two darts players, Maureen and Eric both throw their three darts 50 times. The 50 total scores for each player are shown in this table.

Score	1–30	31–60	61–90	91–120	121–150	151–180
Frequency (Maureen)	7	8	10	9	8	8
Frequency (Eric)	0	16	18	14	2	0

a) Calculate the mean score (per three dart throw) for each player and the range.

b) Comment on your results.

3 The weights of two samples of 100 piglets are recorded after the first sample have spent 10 weeks on Diet A and the second sample have spent 10 weeks on Diet B. The results are shown in this table.

Weight (kg)	Frequency (Diet A)	Frequency (Diet B)
24.5–26.5	20	17
26.5–28.5	23	20
28.5–30.5	31	30
30.5–32.5	19	23
32.5–34.5	7	8
34.5–36.5	0	2

a) Calculate the mean weight for each sample of piglets and the range.

b) Comment on your results.

427

4 One hundred people complete a sponsored slim for charity. The distribution of their weights before and after their diets are shown below.

Weight (kg)	Frequency (before)	Frequency (after)
31–50	0	2
51–55	2	4
56–60	3	6
61–65	5	9
66–70	8	12
71–75	13	19
76–80	18	23
81–90	24	14
91–100	21	11
101–120	6	0

a) Calculate the mean weight of the group both before and after the sponsored slim and the range of each distribution.

b) Comment on your results.

■ ■

The median

The mean is a commonly used average. Sometimes it is even confusingly refered to as '*the average*', as if no other type of average existed. There are if fact other types of average, one of which is the *median*.

The median of a set of values is simply the *middle value when the values are placed in order*.

Find the median and the range of 1, 4, 7, 5, 6, 3, 3, 2, 1, 5, 1.

First, we arrange the values into order.

$$1, 1, 1, 2, 3, 3, 4, 5, 5, 6, 7$$

Then we pick out the *middle* value.

$$\text{The median} = 3$$
$$\text{The range} = 7 - 1 = 6$$

If there are an even number of values, we select a value which is half way between the two middle values.

Find the median and the range of £4.00, £5.00, £4.50, £6.00, £5.50, £6.00.

First, we arrange the values into order.

£4.00, £4.50, £5.00, £5.50, £6.00, £6.00

The *middle* position is half way between £5.00 and £5.50.

The median = £5.25

The range = £6.00 − £4.00

= £2.00

PAUSE

1 Find the median and the range of each of the following distributions.

a) 12, 6, 7, 9, 6

b) 1, 2, 1, 3, 4, 5, 5, 3, 4, 1, 3, 5, 2, 2, 5

c) £5.00, £1.50, £4.00, £5.50, £4.50, £6.00, £2.00

d) £5.00, £1.50, £4.00, £5.50, £4.50, £6.00, £2.00, £1.50

e) −2, 3, 0, −1, 0, 2, 0, 0, 0, −3, 1, 1, 1, 1, 2, −1, −2, 3, 3, 0, −1, −1

f) 2.7, 3.4, 8.9, 5.6, 6.6, 7.3, 5.2, 2.9, 8.0, 7.1

g) 12 cm, 13 cm, 15 cm, 17 cm, 12 cm, 13 cm, 15 cm, 14 cm, 29 cm, 15 cm, 17 cm, 12 cm

h) 450 g, 230 g, 770 g, 125 g, 431 g, 874 g, 455 g, 637 g, 888 g, 97 g

i) 125 cm, 34 cm, 220 cm, 1.34 m, 1.90 m, 340 cm, 1.78 m, 2.55 m, 3.04 m

j) 550 g, 650 g, 750 g, 850 g, 950 g, 1.05 kg, 1.15 kg, 1.25 kg, 1.35 kg

k) 1716, 1785, 1745, 1745, 1780, 1765, 1731, 1734, 1754, 1700, 1735

2 The test marks of a group of ten Year 11 students are 56, 72, 35, 47, 89, 98, 13, 45, 34 and 73. The test marks of a group of ten Year 10 students who take the same test are: 58, 70, 40, 41, 42, 11, 94, 62, 43 and 60.

a) Find the median and the range of each set of marks.

b) Comment on your results.

3 Two players, Maureen and Eric both throw 9 darts. Maureen's scores are: 20, 40, 1, 5, 60, 20, 20, 1 and 60. Eric's scores are: 20, 40, 40, 0, 60, 20, 0, 5 and 1.

a) Find the median and the range of each set of scores.

b) Comment on your results.

4 The set of 5 values: 1, 2, 5, 7, 20 has a median of 5 and a mean of 7.
Find a set of values which meets each of the following conditions.
a) Number of values = 5, median = 9, mean = 10.
b) Number of values = 7, median = 5, mean = 4.
c) Number of values = 10, median = 7.5, mean = 8.
d) Number of values = 10, median = 10, mean = 10.

■ ■

Finding the median and the range of a discrete distribution presented in a frequency table

Calculate the median and the range for this frequency distribution:

Number of children per family	0	1	2	3	4	5	6
Frequency	19	21	36	14	7	3	0

Remember, to find the median we need to know the value which represents the mid-point of the distribution.

To discover this value we add an extra line to the table, called the *cumulative frequency* or running total. So for example, in this table:

Number of children per family	Frequency	Cumulative frequency
0	19	19
1	21 ← + =	40
2	36 ← + =	76 etc
3	14	90
4	7	97
5	3	100
6	0	100

With 100 values, the median is between the 50th and 51st values. We use the cumulative frequencies to pick out the values that are in these positions.

For example, the cumulative frequencies tell us that there are 40 families with 0 or 1 child and 76 families with 0, 1 or 2 children. This means that the 50th and 51st families must both have 2 children. So:

the median = 2 children per family,

the range is 5–0 = 5 children per family.

PAUSE

1 A survey was conducted into the number of eggs in 100 nests of two different bird species. This table shows the results.

Number of eggs	0	1	2	3	4	5	6
Number of nests (species A)	12	13	22	28	18	7	0
Number of nests (species B)	10	31	29	25	5	0	0

a) Find the median number of eggs per nest for species A and the range.
b) Find the median number of eggs per nest for species B and the range.

2 Trading Standards Officers checked the claims of two brands of matches to have 'average contents 40 matches'. They counted the contents of 200 boxes of each brand. This table shows their results.

Number of matches per box	36	37	38	39	40	41	42	43	44
Number of boxes (Flares Matches)	12	19	23	46	50	30	17	3	0
Number of boxes (Squibs Matches)	0	3	13	24	27	82	31	16	4

a) Find the median number of matches per box for Flares Matches and the range.
b) Find the median number of matches per box for Squibs Matches and the range.
c) Comment on your answers.

3 A television factory has two different production lines. They find that far too many televisions are being produced with faults. They call in a quality control inspector who checks 50 television sets produced by each production line. These are her results.

Number of faults per set	0	1	2	3	4
Number of televisions (Line A)	24	23	2	1	0
Number of televisions (Line B)	11	12	27	0	0

a) Find the median number of faults per television for Line A and the range.

b) Find the median number of faults per television for Line B and the range.

c) Comment on your answers.

4 One of the questions in a survey on school meals asked Year 11 students how many days they had chips with their school meal during the previous week. These are the results.

Number of days	0	1	2	3	4	5
Frequency (girls)	12	20	21	17	6	4
Frequency (boys)	4	13	21	16	12	14

a) Find the median number of days that girls ate chips with their meal and the range.

b) Find the median number of days that boys ate chips with their meal and the range.

c) Comment on your answers.

■ ■

Finding the median and the interquartile range of a distribution presented in a grouped frequency table

Find the median for the student's test marks shown in this table.

Test mark	1–5	6–10	11–15	16–20	21–25	26–30	31–35	36–40	41–45	46–50
Frequency	1	2	11	17	25	18	13	6	3	4

We can only *estimate* the median when data is presented in a grouped table. We make our estimates by drawing a graph called a *cumulative frequency curve* (sometimes called an *ogive*). We always plot the cumulative frequencies over the *upper limit* of the class interval.

We start by adding two columns to our table.

Test mark	Frequency	Upper limit of class interval	Cumulative frequency
1–5	1	5	1
6–10	2	10	3
11–15	11	15	14
16–20	17	20	31
21–25	25	25	56
26–30	18	30	74
31–35	13	35	87
36–40	6	40	93
41–45	3	45	96
46–50	4	50	100

We now draw a graph, plotting each cumulative frequency against the upper limit of its class interval.

A cumulative frequency curve of marks obtained by 100 students in a maths test out of 50

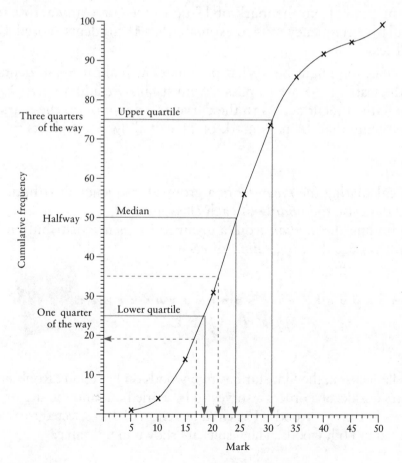

The median is the middle value of the distribution. We should really use halfway between the 50th and 51st mark but this is only an estimate so we will take the median as the 50th mark.

Median = 24 marks.

The range is a rather rough measure of spread. We can get a better measure from this graph. We take the spread of the typical results in the middle half of the distribution called the *interquartile range*.

The *lower quartile* is one quarter of the way (25 marks) and the *upper quartile* is three quarters of the way (75 marks).

Lower quartile = 18½ marks
Upper quartile = 30½ marks
Interquartile range = 30½ – 18½
= 12 marks

We can answer other questions from the graph, for example, 'How many pupils scored 17 marks or less?'.

We add an arrow from the mark of 17 up to the curve and across to the cumulative frequency axis to estimate that 19 students scored 17 marks or less.

Another question might be, 'What pass mark will allow 65 students to pass the test?'. If 65 are to pass, 35 must fail. We add an arrow from the 35th student across to the curve and down on to the marks axis to estimate that, 'A pass mark of 21 will allow 65 students to pass.'.

Remember:

■ when calculating the *mean* from a grouped frequency distribution you always use the *middle* of each class interval,
■ when finding the *median* from a grouped frequency distribution you always use the *upper limit* of each class interval.

PAUSE

1 Michelle Jackson, the film star has two woods on her estate, Rooksend and Streamside. She orders a survey to be carried out into the size of the trees in the two woods. The circumferences of 100 selected trees is measured in both woods. The results are shown in this table:

Circumference C(cm)	Rooksend	Streamside
$0 \leq C < 20$	3	5
$20 \leq C < 40$	17	3
$40 \leq C < 60$	22	6
$60 \leq C < 80$	25	11
$80 \leq C < 100$	19	18
$100 \leq C < 120$	7	24
$120 \leq C < 140$	5	26
$140 \leq C < 160$	2	7

a) Draw separate, cumulative frequency curves for the data from each wood.
b) Estimate the median for both distributions.
c) Estimate the quartiles and find the interquartile range for each distribution.
d) How many trees in the Rooksend sample were less than 50cm in diameter?
e) How many trees in the Streamside sample were between 70cm and 130cm in diameter?

2 Two darts players, Maureen and Eric both throw their three darts 50 times. The 50 total scores for each player are shown in this table.

Score	1–30	31–60	61–90	91–120	121–150	151–180
Frequency (Maureen)	7	8	10	9	8	8
Frequency (Eric)	0	16	18	14	2	0

a) Draw separate cumulative frequency curves for the data on each player.
b) Estimate the median for both distributions.
c) Estimate the quartiles and the interquartile range for both distributions.
d) How many times did Maureen score 100 or less with her three darts?
e) How many times did Eric score 100 or more with his three darts?

3 One hundred people complete a sponsored slim for charity. The distribution of their weights before and after their diets are shown below.

Weight (kg)	Frequency (before)	Frequency (after)
$30 \leq W < 50$	0	2
$50 \leq W < 55$	2	4
$55 \leq W < 60$	3	6
$60 \leq W < 65$	5	9
$65 \leq W < 70$	8	12
$70 \leq W < 75$	13	19
$75 \leq W < 80$	18	23
$80 \leq W < 90$	24	14
$95 \leq W < 100$	21	11
$100 \leq W < 120$	6	0

a) Draw separate, cumulative frequency curves for the 'before' and 'after' data.
b) Estimate the median for each distribution.
c) Estimate the quartiles and interquartile range for each distribution.
d) How many of the slimmers weighted over 85 kg before they started their diet?
e) How many of the slimmers weighted over 85kg after they completed their diet?

4 The time taken by two groups of 60 runners to complete a race are shown in this table.

Time (minutes)	Frequency (Group A)	Frequency (Group B)
$18 \leq T < 19$	3	0
$19 \leq T < 20$	8	22
$20 \leq T < 21$	21	18
$21 \leq T < 22$	14	12
$22 \leq T < 23$	6	7
$23 \leq T < 24$	5	1
$24 \leq T < 25$	3	0

a) Draw a cumulative frequency curve for each distribution using the same axes.
b) Find the range and the interquartile range of each distribution.
c) Compare the spreads of the two distributions.

5 Two groups of 60 students sat the same Mathematics text, marked out of 60. The results of test are shown in the table.

Mark	Group A	Group B
0–10	1	1
11–20	2	2
21–30	23	18
31–40	28	20
41–50	4	11
51–60	2	1

a) Draw a cumulative frequency curve for each distribution, using the same axes.
b) Find the median, range and interquartile range for each distribution.
c) Comment on your answers.

6 The Headteacher of Minastroney High School decided to make a survey of the time spent talking to each set of parents at a parents' evening. The interview times of two teachers, Ms Salt and Ms Pepper were recorded as they each spoke with 30 sets of parents. This table shows the data that was collected.

Interview Time (Minutes)	Frequency	
	Ms Salt	Ms Pepper
$0 \leq T < 1$	0	0
$1 \leq T < 2$	2	0
$2 \leq T < 3$	3	0
$3 \leq T < 4$	6	1
$4 \leq T < 5$	9	7
$5 \leq T < 6$	5	10
$6 \leq T < 7$	3	8
$7 \leq T < 8$	2	4

a) Draw on the same graph, cumulative frequency curves for the data on each teacher.
b) Estimate the median and the quartiles for both distributions.
c) Compare the spreads of the two distributions.

■ ■

The mode

The third type of average which we will study is called the *mode* (or *modal average*). The mode of a set of values is *the value that occurs most frequently*.

Find the mode and the range of: 1, 4, 7, 5, 6, 3, 3, 2, 1, 5, 1.

First, we sort the values into groups.

$$1, 1, 1, 2, 3, 3, 4, 5, 5, 6, 7$$

Then we pick out the value which occurs most frequently. In this case, 1 occurs most frequently so:

$$\text{the mode} = 1.$$

In small distributions, where many (or all) values occur the same number of times there is no sensible modal average.

Find the mode of: 11, 11, 12, 12, 13, 13, 14, 14, 15, 16, 16.

There is no value that can be selected as a sensible mode.

If data is organised into a frequency table, we select the value of the variable which has the greatest frequency.

What is the modal number of children per family for the data in this frequency distribution?

Number of children per family	0	1	2	3	4	5	6
Frequency	13	13	38	17	10	4	5

The modal number of children per family is 2, because it occurs with the greatest frequency (38).

In a grouped table, the modal average becomes a *modal group*, and we select the *group of values* which occurs most frequently. In a large distribution, if two or more values of the variable each occur with the same frequency, they may be stated as alternative values for the mode.

Find the modal group and the range for the students' test marks shown in this table.

Test mark	1–5	6–10	11–15	16–20	21–25	26–30	31–35	36–40	41–45	46–50
Frequency	1	2	11	17	25	18	25	6	3	4

Two groups each occur with a frequency of 25, which is greater than any other group's frequency.

So, the modal group is either 21–25 or 31–35.

1 Find the mode of each of the following distributions.

a) 12, 6, 7, 9, 6

b) 1, 1, 1, 1, 2, 2, 3, 3, 3, 3, 3, 4, 4, 5, 6, 6

c) £1.50, £1.50, £2.00, £3.00, £3.00, £2.00, £2.50, £1.50, £1.50, £1.50

d) –2, –1, –2, 0, 0, 0, 3, 4, 1, 2, 6

e) 2.7, 3.4, 8.9, 5.6, 6.6, 7.3, 5.2, 2.9, 8.0, 7.1

f) 1, 0, 0, 1, 0, 0, 1, –1, –1, –1, 0, 0, 0, –1, 1, 1, 1, 1, 1, 0, 1, –1, 0, –1

g) 12 cm, 13 cm, 15 cm, 17 cm, 12 cm, 13 cm, 15 cm, 14 cm, 29 cm, 15 cm, 17 cm, 12 cm

h) 10 g, 11 g, 10 g, 12 g, 11 g, 11 g, 12 g, 10 g, 11 g, 12 g, 12 g, 12 g, 12 g, 13 g, 12 g, 11 g, 11 g

i) 125 cm, 340 cm, 220 cm, 1.34 m, 1.90 m, 340 cm, 1.78 m, 2.55 m, 3. 40 m

j) 550 g, 650 g, 750 g, 850 g, 950 g, 1.05 kg, 1.15 kg, 1.25 kg, 1.35 kg

2 The cost of a litre of petrol was checked at a sample of 50 garages in two different cities. The results are shown in the table below.

Cost per gallon (pence)	City A	City B
49	2	4
50	5	7
51	12	4
52	21	11
53	7	13
54	3	5
55	0	6

Find the modal price in each of the cities.

3 Michelle Jackson, the film star has two woods in her estate, Rooksend and Streamside. She orders a survey to be carried out into the size of the trees in the two woods. The circumferences of

100 selected trees is measured in both woods. The results are shown in this table. Find the modal group for each wood.

Circumference C (cm)	Rooksend	Streamside
$0 \leq C < 20$	3	5
$20 \leq C < 40$	17	3
$40 \leq C < 60$	22	6
$60 \leq C < 80$	25	11
$80 \leq C < 100$	19	18
$100 \leq C < 120$	7	24
$120 \leq C < 140$	5	26
$140 \leq C < 160$	2	7

■ ■

A summary of the ways to compare frequency distributions

In order to condense and summarise the data in a frequency distribution, we can find:

- the mean average,
- the median average,
- the quartiles,
- the modal average,
- the range,
- the interquartile range.

In three different types of distribution:

- a simple list of values,
- a frequency distribution table,
- a grouped frequency distribution table.

■ ■
PAUSE

1 Find the mean, median, mode, and range for each list of values:

a) 1, 1, 2, 3, 3, 4, 4, 5, 8, 7, 6, 5, 4, 4, 3, 3, 3, 2, 1
b) 12, 23, 11, 19, 11, 12, 12, 17, 18, 19, 13
c) 36, 36, 37, 36, 38, 41, 39, 40, 37, 37, 38, 39, 37, 38, 40
d) 102, 101, 102, 101, 102, 100, 100, 100, 102, 100, 101, 100, 99
e) 1, 2, 4, 5, 3, 3, 3, 1, 1, 6, 7, 10, 4, 1, 1, 1, 6, 1, 1, 2, 2, 2, 2

2 In order to test the manufacturers' claims of 'average content 40 matches', two samples of 100 match boxes are opened and their contents counted. The first sample are 'Burners' matches and the second sample are 'Flames'. The table below shows the results.

Number of matches per box	Frequency 'Burners'	Frequency 'Flames'
36	2	0
37	5	0
38	7	12
39	13	24
40	24	38
41	26	17
42	13	9
43	8	0
44	2	0

a) Find the mean for each frequency distribution.
b) Find the median for each frequency distribution.
c) Find the mode for each frequency distribution.
d) Find the range for each frequency distribution.
e) Comment on your results, comparing and contrasting the two distributions.

3 Each year, a game warden in East Anglia checks on the contents of 60 nests of the Egyptian Goose, a bird introduced from Africa in the eighteenth century. The table below shows her results for two successive years.

Number of eggs	0	1	2	3	4	5	6	7	8	9
Number of nests (Year 1)	12	0	0	3	0	12	17	12	4	0
Number of nests (Year 2)	17	0	1	5	0	19	12	6	0	0

a) Find the mean for each frequency distribution.
b) Find the median for each frequency distribution.
c) Find the mode for each frequency distribution.
d) Find the range for each frequency distribution.
e) Comment on your results, comparing and contrasting the two distributions.

4 A games teacher starts a ten week training programme for a health and fitness club. She times the students over a one mile race both before and after they complete the programme. The table below shows her results.

Time (seconds)	Frequency (before)	Frequency (after)
$260 \leq T < 280$	0	6
$280 \leq T < 300$	2	12
$300 \leq T < 320$	4	15
$320 \leq T < 340$	14	16
$340 \leq T < 360$	16	8
$360 \leq T < 380$	11	3
$380 \leq T < 400$	8	0
$400 \leq T < 420$	5	0

a) Find the mean for each frequency distribution.
b) Find the median and the quartiles for each frequency distribution.
c) Find the mode for each frequency distribution.
d) Find the range and interquartile range for each frequency distribution.
e) How many club members ran the race in less than 350 seconds before the training program?
f) How many club members ran the race in less than 350 seconds after the training program?
g) Comment on your results, comparing and contrasting the two distributions.

5 A comparison test is carried out between the batteries produced by two rival companies. Fifty pairs of batteries from each company are tested, each pair being used to power a personal stereo until they fail. The results obtained are shown in the table below.

Time before failing (minutes)	Frequency (Duraready)	Frequency (Evercell)
$40 \leq T < 50$	0	0
$50 \leq T < 60$	2	0
$60 \leq T < 70$	4	0
$70 \leq T < 80$	6	0
$80 \leq T < 90$	8	10
$90 \leq T < 100$	12	28
$100 \leq T < 110$	17	11
$110 \leq T < 120$	1	1

a) Find the mean for each frequency distribution.
b) Find the median and the quartiles for each frequency distribution.

c) Find the mode for each frequency distribution.
d) Find the range and interquartile range for each frequency distribution.
e) What percentage of the Evercell batteries lasted for more than 100 minutes?
f) What percentage of the Duraready batteries lasted for more than 100 minutes?
g) What percentage of the Evercell batteries lasted for less than 80 minutes?
h) What percentage of the Duraready batteries lasted for less than 80 minutes?
i) Comment on your results, comparing and contrasting the two distributions, which batteries would you recommend and why?

■■■■■■■■■■■■■■■■■■■■■■■■■■■■■■■■■■■■

Correlation and scatter diagrams

All the work so far has involved values of a *single* variable. Many of the questions that are asked involve *more than* one variable. For example, we might ask:

- do people with large incomes tend to live in large houses?
- do tall parents tend to have tall children?
- how is the amount that is spent on advertising a new product related to its sales success?
- do taller than average people tend to be heavier than average?
- do people who are good at Maths tend to be also good at Geography?

All these questions involve the idea that the values of two variables may be *correlated*. This means that the values are linked in some way. For example, if we say that height and shoe size in children are *correlated*, we mean that taller children tend to have bigger feet.

Scatter diagrams are used to detect correlations.

Jamie and Jessica wonder if there is a correlation between a person's homework marks in Maths and Science. They collect the following data from 20 friends who have just had homeworks from both subjects marked out of 10.

443

Individual	A	B	C	D	E	F	G	H	I	J	K	L	M	O	N	P	Q	R	S	T
Maths mark	1	9	3	6	5	10	7	4	8	5	5	3	3	4	10	7	7	4	5	6
Science mark	2	3	4	6	3	10	5	1	8	6	5	2	3	3	8	6	5	5	4	5

To construct a *scatter* diagram from this data, each individual plots as a single point on the graph. Their two marks are used as *x* and *y* coordinates.

This is a scatter diagram drawn from Jamie's and Jessica's results. The arrow on the scatter diagram points to individual E, plotted at the point (5,3) because his Maths mark is 5 and his Science mark is 3.

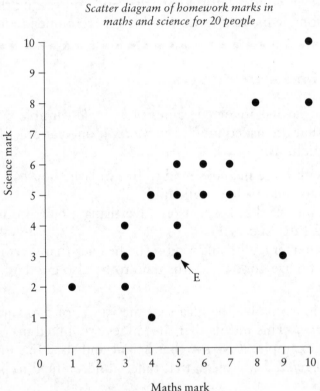

Scatter diagram of homework marks in maths and science for 20 people

The scatter diagram shows that there is a *positive correlation* between the homework marks in Maths and Science. This means that the diagram shows that a student who gets a good mark for Maths tend to also get a good mark for Science. Students who get a poor mark for Maths tend to also get a poor mark for Science.

There are exceptions to this rule, Individual B for example scores 9 for Maths but only 3 for Science.

We can have *negative correlation* as well as *positive correlation*.

For example, this scatter diagram shows the relationship between the engine sizes of 20 cars and their average fuel consumption in miles per gallon. It is clear that as the engine size *increases*, the fuel consumption tends to *decrease*. We say that the two variables in this case are *negatively correlated*.

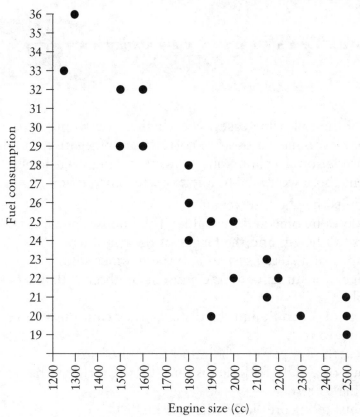

A scatter diagram to show the relationship between engine size and average fuel consumption for 20 cars

We can of course have instances where two variables are not correlated at all. In this case we would expect to see a random distribution of dots in the scatter diagram.

These diagrams illustrate variables with positive correlation, negative correlation and no correlation.

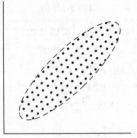
Positive correlation

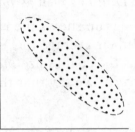
Negative correlation

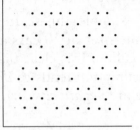
No correlation

PAUSE

1 In each of the following cases, say whether you would expect the two variables to have a positive correlation, a negative correlation or no correlation. Explain your answers, saying what likely exceptions there would be to any expected correlation.

a) An adults' height and weight.
b) A person's income and the value of the house they live in.
c) A person's height and the height of their mother.
d) The price of a stereo system and the number sold.
e) The age of a student and the amount of absence they have from school.
f) The age of an adult and the distance they can run without stopping to rest.
g) The length of a person's hair and their shoe size.
h) The age of a child and the distance they can run without stopping to rest.
i) Students' test scores in Maths and English.
j) The age of a used car and its value.

2 The records of 20 pupils are checked and the number of days absence they had in the past year recorded, together with their year group.

Pupil	A	B	C	D	E	F	G	H	I	J
Year group	7	7	7	7	8	8	8	8	9	9
Days absent	0	3	2	4	2	5	3	6	4	7

Pupil	K	L	M	N	O	P	Q	R	S	T
Year group	9	9	10	10	10	10	11	11	11	11
Days absent	5	9	4	7	7	11	6	8	9	15

a) Draw a scatter diagram.

b) Comment on any correlation indicated in the scatter diagram.

3 The length of 20 girls' hair (measured in inches) and their shoe size is recorded in this table.

Girl	A	B	C	D	E	F	G	H	I	J
Hair length	6	7	9	15	11	12	3	7	9	12
Shoe size	3	4	4	3	5	3	6	5	5	4

Girl	K	L	M	N	O	P	Q	R	S	T
Hair length	8	12	15	11	12	2	18	12	11	15
Shoe size	5	3	3	3	6	4	4	5	6	4

a) Draw a scatter diagram.

b) Comment on any correlation indicated in the scatter diagram.

4 A shopkeeper makes a record of the temperature each day for the first two weeks in September. She also records the number of ice-creams she sells. The table below shows her results.

Date	Number sold	Temp (C°)
1	20	21
2	15	16
3	25	24
4	35	27
5	30	27
6	20	18
7	15	18
8	25	24
9	20	24
10	35	29
11	15	16
12	20	18
13	30	21
14	25	18

a) Draw a scatter diagram.

b) Comment on any correlation indicated in the scatter diagram.

5 This table shows the latitude of 15 cities and their average annual temperature.

City	Latitude (nearest degree)	Average temperature (nearest C°)
Algiers	37	58
Amsterdam	52	12
Berlin	53	13
Bombay	19	31
Copenhagen	56	11
Hong Kong	22	25
Karachi	25	31
Leningrad	60	8
London	52	14
Madrid	40	19
Manila	15	32
Paris	49	15
Phnom Penh	12	32
Rangoon	17	32
Saigon	11	32

a) Draw a scatter diagram.

b) Comment on any correlation indicated in the scatter diagram.

■■■■■■■■■■■■■■■■■■■■■■■■■■■■■■■■■■■■■

Adding a line of best fit to a scatter diagram

If a correlation is strong, the points in the scatter diagram will be tightly grouped.

These diagrams show strong positive and negative correlations.

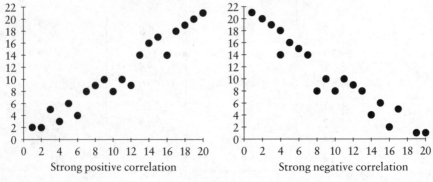

Strong positive correlation   Strong negative correlation

448

If the correlation is moderate, the points in the scatter diagram will be loosely grouped.

These diagrams show moderate positive and negative correlations.

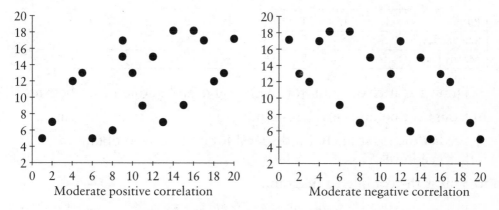

Moderate positive correlation   Moderate negative correlation

Remember, if there is no correlation the dots will be randomly spread over the scatter diagram, with no noticeable pattern.

If a scatter diagram shows a strong positive or negative correlation, we can add *a line of best fit* to the diagram.

A line of best fit is drawn by eye, using a transparent ruler and carefully balancing points above the line with points below the line.

These diagrams show lines of best fit added to diagrams with strong positive and strong negative correlations.

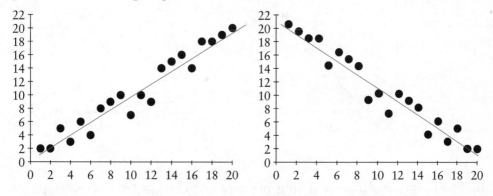

We can use a line of best fit to make predictions based on the correlation shown in the scatter diagram.

Example
This table shows the Maths and Science levels obtained by a a group of twenty Year 9 students in the National SAT examinations.

Pupil	A	B	C	D	E	F	G	H	I	J
Science level	3	3	3	4	4	4	4	5	5	5
Maths level	4	3	3	4	5	4	6	4	6	5

Pupil	K	L	M	N	O	P	Q	R	S	T
Science level	5	5	5	6	6	6	7	7	8	8
Maths level	6	4	5	6	6	5	7	8	6	7

a) Draw a scatter diagram for these results and add a line of best fit.

b) Comment on any correlation indicated in the scatter diagram.

c) Predict the most likely Maths level for a pupil who obtains a Science Level 5.

a) This is the completed diagram.

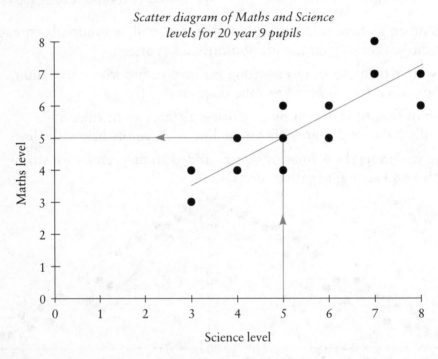

Scatter diagram of Maths and Science levels for 20 year 9 pupils

b) The diagram shows a strong positive correlation between a pupil's Science level and their Maths level.

c) The arrowed line shows how the line of best fit can be used to predict that the most likely Maths level is also a Level 5.

1 A cafe owner makes a record of the temperature each day in October and the number of cups of soup she sells. The table below shows her results.

Date	Temp (C°)	Number sold	Date	Temp (C°)	Number sold
1	24	3	16	19	7
2	24	4	17	14	10
3	25	3	18	14	9
4	23	4	19	13	10
5	22	5	20	14	11
6	20	5	21	12	13
7	17	8	22	11	11
8	16	7	23	11	14
9	17	9	24	11	13
10	17	7	25	10	13
11	18	8	26	11	12
12	19	6	27	10	12
13	21	5	28	10	13
14	21	4	29	9	15
15	20	6	30	9	14

a) Draw a scatter diagram for the table and add a line of best fit.
b) Comment on any correlation indicated in the scatter diagram.
c) Predict the most likely number of cups of soup that will be sold on a day when the temperature is 15°C.
d) Predict the most likely temperature on a day when 9 cups of soup are sold.

2 A research scientist is experimenting with the effect of watering on a type of seedling. She has 10 test beds of seedlings and supplies them with different quantities of water and measures the growth of the seedlings. These are her results.

Water (1000 litres)	1	2	3	4	5	6	7	8	9	10
Average growth (centimetres)	3	4	4	5	7	7	8	9	8	10

a) Draw a scatter diagram for the table and add a line of best fit.

b) Comment on any correlation indicated in the scatter diagram.
c) Predict the most likely average growth for seedlings watered with:
   i) 3500 litres of water
   ii) 4500 litres of water
   iii) 7500 litres of water
d) Predict the most likely quantity of water that has been supplied to seedling with an average growth of:
   i) 6 cm
   ii) 9.5 cm
   iii) 3.5 cm

3 A survey is made into the time that twenty Year 11 students spend each week watching television or doing homework. These are the results.

Watching TV (hours)	0	2	2	3	4	4	5	5	6	8
Doing homework (hours)	14	12	13	11	12	10	8	12	9	8

Watching TV (hours)	10	10	12	12	12	14	14	16	17	18
Doing homework (hours)	5	7	5	4	6	1	3	2	1	0

a) Draw a scatter diagram for the table and add a line of best fit.
b) Comment on any correlation indicated in the scatter diagram.
c) Predict the most likely time spent doing homework by somebody who each week watches television for:
   i) 7 hours
   ii) 13 hours
   iii) 15 hours
d) Predict the most likely time spent watching television by somebody who each week does homework for:
   i) 3 hours
   ii) 6 hours
   iii) 11 hours

4 A trainer records the heights and weights of the 15 players in a team. These are the results.

Height (nearest cm)	170	175	175	175	180	180	180	180	185	185
Weight (nearest kg)	70	75	80	85	80	85	90	95	90	100

Height (nearest cm)	190	195	195	200	205
Weight (nearest kg)	105	105	110	115	120

a) Draw a scatter diagram for the table and add a line of best fit. Start the height axis at 160 cm and the weight axis at 60 kg.
b) Comment on any correlation indicated in the scatter diagram.
c) Predict the most likely weight for a player with a height of:
   i) 187 cm
   ii) 192 cm
   iii) 177 cm
d) Predict the most likely height for a player with a weight of:
   i) 72 kg
   ii) 91 kg
   iii) 117 kg

5 This table shows the circulation figures of five Sunday newspapers and the cost of a full page advertisement in these newspapers (source: BRAD 1994).

Paper	Cost (£ 10 000)	Circulation (1 000 000)
Mail on Sunday	2.8	2.0
News of the World	3.0	4.8
Observer	2.4	0.5
Sunday Mirror	2.9	2.6
Sunday Times	4.7	1.2

a) Draw a scatter diagram from the table.
b) Comment on any correlation indicated in the scatter diagram.
c) Do you think it is sensible to add a line of best fit to the scatter diagram? Explain your answer.

■■■■■■■■■■■■■■■■■■■■■■■■■■■■■■■■■

SECTION 4 Probability

Probability is the name given to the branch of statistics which allows us to predict how *likely* a given result is. This means we can compare our experimental results with theoretical 'perfect' results and draw conclusions about their reliability. For example, probability predicts that when you roll a normal die you have the same chance of scoring a 6 as of scoring a 1,2,3,4 or 5. If, you rolled a die 600 times and recorded 300 sixes, you would suspect that something was wrong with either the die or the experiment.

Simple probability (single events with equally likely outcomes)

An **event** is something which happens, like tossing a coin or rolling a die.

Most events have more than one possible outcome, for example, when a die is rolled, there are six possible outcomes, 1,2,3,4,5 or 6.

If an event is *fair*, all these outcomes will be **equally likely**.

If an event is **biased** all these outcomes will not be **equally likely**.

Several of these outcomes may give the same result. For example, the scores 2,4 and 6 on a die all give the result 'an even number'.

If the die is *fair*, the *probability* that any particular result will be obtained in an event is the fraction:

$$\frac{\text{the number of outcomes that contains the desired result}}{\text{the total number of all possible outcomes}}$$

So, when a fair die is rolled, the probability of obtaining an even score is:

$$\frac{3}{6} = \frac{1}{2}$$

Probabilities can be written as fractions, decimals or percentages. Remember:

- to turn a fraction into a decimal, divide the top number by the bottom number,
- to turn a decimal into a percentage, multiply by 100,
- you will often need to approximate your answers to a sensible number of decimal places.

For example, the probability of obtaining an even score when a fair die is rolled can be written as:

$$\frac{1}{2} = 0.5 = 50\%$$

The probability of rolling a fair die and obtaining a 6 can be written as:

$$\frac{1}{6} = 0.167 \text{ (to 3 decimal places)} = 16.7\%$$

Remember, these rules can only be applied if an event is *fair* and *all outcomes are equally likely*.

Events may involve a selection, for example taking a bead from a bag or selecting a number for a National Lottery card. It this case, we can only apply the rules if the selection is made *at random*, with no personal preference involved in the selection.

Examples

1. What is the probability of cutting a deck of cards and obtaining an ace? Give your answer as a fraction, decimal and percentage.

   The number of outcomes that includes the result is 4 (ace of hearts, diamonds, clubs and spades).

   The total number of all possible outcomes is 52.

   The probability of obtaining an ace is:

   $$\frac{4}{52} = \frac{1}{13} = 0.077 \text{ (to 3 decimal places)} = 7.7\%.$$

2. What is the probability of selecting a letter at random from the word 'probability' and obtaining a vowel? Give your answer as a fraction, decimal and percentage.

   The number of outcomes that includes the result is 4 (o, a, i, i).

   The total number of all possible outcomes is 11.

   The probability of obtaining a vowel is:

   $$\frac{4}{11} = 0.364 \text{ (to 3 decimal places)} = 36.4\%.$$

If any result is certain to happen because it is included in all the possible outcomes it has a probability of 1. For example, the probability of cutting a deck of cards and obtaining a red or black card is 1 (52 out of 52).

If any result cannot happen because it is not included in any of the possible outcomes it has a probability of 0. For example, the probability of cutting a deck of cards and obtaining a green card is 0 (0 out of 52).

PAUSE

Give all numerical answers in this exercise as a fraction, decimal and percentage.

1 List three different ways that somebody could select National Lottery numbers 'at random'.

2 A fair die is rolled. What is the probability that the result is:
a) a 4,
b) an odd number,
c) a prime number,
d) a 1 or a 6?

3 A card is cut from a normal deck of 52 cards. What is the probability that the card is:
a) a picture card,
b) not a picture card,
c) a 5 of spades,
d) a 5,
e) a red 5,
f) a red picture card,
g) a red card or a king,
h) a club or a queen,
i) a diamond or a club,
j) a card higher than a 7?

4 A bag contains 3 red, 5 yellow and 4 green beads. If one bead is selected at random from the bag, what is the probability that it is:
a) red,
b) yellow,
c) green,
d) red or green,
e) red or yellow,
f) yellow or green,
g) not green,
h) purple,
i) red yellow or green,
j) not yellow?

5 In a maths class of 30 students, 15 hate the subject, 7 dislike it, 5 like it and 3 love it. If a student is selected at random from the group, what is the probability that they will:
a) hate maths,
b) dislike maths,
c) like maths,
d) love maths,

e) either hate or dislike maths,
f) either like or love maths,
g) either love or hate maths?

6 This table shows the number of children per family in 50 families on each of two housing estates.

Number of children per family	0	1	2	3	4	5	6
Frequency (Estate A)	10	9	18	8	5	0	0
Frequency (Estate B)	3	4	20	9	5	4	5

a) What is the probability of selecting a family at random on Estate A and obtaining one with:
   i) no children,
   ii) 4 children,
   iii) less than the median average number of families for that estate?
b) What is the probability of selecting a family at random on Estate B and obtaining one with:
   i) 2 children,
   ii) 1 child,
   iii) more than the mean average number of families for that estate?

7 The weights of two samples of 100 piglets are recorded after the first sample have spent 10 weeks on Diet A and the second sample have spent 10 weeks on Diet B. The results are shown in this table.

Weight (kg)	Frequency (Diet A)	Frequency (Diet B)
$24.5 \leq W < 26.5$	20	17
$26.5 \leq W < 28.5$	23	20
$28.5 \leq W < 30.5$	31	30
$30.5 \leq W < 32.5$	19	23
$32.5 \leq W < 34.5$	7	8
$34.5 \leq W < 36.5$	0	2

a) What is the probability that a pig selected at random from the distribution for diet A will:
   i) weigh less than 28.5 kg,
   ii) weigh 28.5 kg or more,
   iii) weigh more than 26.5 but less than 30.5 kg?

b) What is the probability that a pig selected at random from the distribution for diet B will:
   i) weigh less than 28.5 kg,
   ii) weigh 28.5 kg or more,
   iii) weigh more than 26.5 but less than 30.5 kg?

8 Two darts players, Maureen and Eric both throw their three darts 50 times. The 50 total scores for each player are shown in this table.

Score	1–30	31–60	61–90	91–120	121–150	151–180
Frequency (Maureen)	7	8	10	9	8	8
Frequency (Eric)	0	16	18	14	2	0

a) If we assume that these scores are typical of the players' performances, what is the probability that on her next throw Maureen will score:
   i) more than 90,
   ii) less than 61,
   iii) between 31 and 150?
b) If we assume that these scores are typical of the players' performances, what is the probability that on his next throw Eric will score:
   i) more than 90,
   ii) less than 61,
   iii) between 31 and 150?
c) Comment on your answers to this question. Can we be certain that the probabilities are accurate or are they just estimates?

■ ■

The probability scale

A probability of 0 means an outcome cannot happen, it is *impossible*.

A probability of 1 means an outcome must happen, it is *certain*.

All probabilities must be between 0 and 1.

Probabilities can be illustrated on a *probability scale*.

A probability scale is a number line from 0 to 1, divided into either fractions, decimals or percentages. This diagram shows fraction, decimal and percentage probability scales.

If outcomes are *mutually exclusive,* this means that if one happens the other(s) cannot happen.

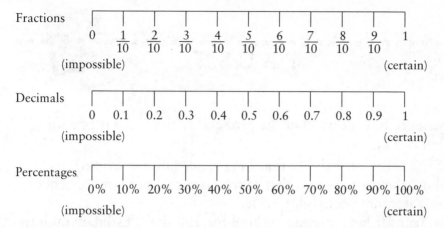

Fractions

$0 \quad \frac{1}{10} \quad \frac{2}{10} \quad \frac{3}{10} \quad \frac{4}{10} \quad \frac{5}{10} \quad \frac{6}{10} \quad \frac{7}{10} \quad \frac{8}{10} \quad \frac{9}{10} \quad 1$

(impossible)   (certain)

Decimals

$0 \quad 0.1 \quad 0.2 \quad 0.3 \quad 0.4 \quad 0.5 \quad 0.6 \quad 0.7 \quad 0.8 \quad 0.9 \quad 1$

(impossible)   (certain)

Percentages

$0\% \quad 10\% \quad 20\% \quad 30\% \quad 40\% \quad 50\% \quad 60\% \quad 70\% \quad 80\% \quad 90\% \quad 100\%$

(impossible)   (certain)

For example, it may rain tomorrow or it may stay dry all day. The two outcomes are *mutually exclusive* because it cannot both rain and stay dry all day.

The probability of all the mutually exclusive outcomes of an event *must add to 1.*

Example

The probability that it will rain on a June day in Hemerton is 0.15.

a) What is the probability that it will stay dry all day on a June day in Hemerton?

b) Show the probabilities that it will rain or stay dry on a decimal probability scale.

c) Estimate on how many days it will stay dry during a typical June in Hemerton.

a) The probability that it will stay dry all day $= 1 - 0.15 = 0.85$.

b)

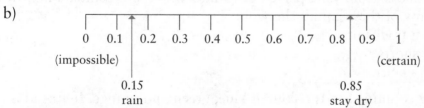

$0 \quad 0.1 \quad 0.2 \quad 0.3 \quad 0.4 \quad 0.5 \quad 0.6 \quad 0.7 \quad 0.8 \quad 0.9 \quad 1$

(impossible)   (certain)

0.15
rain

0.85
stay dry

c) Estimate of number of dry days during June
   = probability that it will stay dry all day × total number of days in June
   $= 0.85 \times 30$
   $= 25.5$ days or 26 days to the nearest day.

459

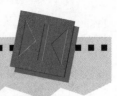

PAUSE

1 Experience has shown that the probability that Jane is late for school is 0.1.
   a) What is the probability that Jane is not late for school?
   b) Show the probabilities that Jane is late and not late for school on a decimal probability scale.
   c) In Year 10 Jane attended school for 190 days. Estimate on how many of these days she was late.

2 Experience has shown that the probability that Simon wears white socks to school is 0.6.
   a) What is the probability that Simon does not wear white socks to school?
   b) Show the probabilities that Simon does or does not wear white socks on a decimal probability scale.
   c) In Year 11 Simon attended school for 185 days. Estimate on how many of these days he wore white socks.

3 June always wears either a tracksuit, shorts or a skirt to play netball. Experience has shown that the probability that she wears a tracksuit is 0.25 and the probability that she wears shorts is 0.4.
   a) What is the probability that June wears a skirt to play netball?
   b) Show the probabilities that June wears a tracksuit, shorts or a skirt to play netball on a decimal probability scale.
   c) In one season, June played 60 games of netball. Estimate in how many of these games she wore:
      i) a tracksuit,
      ii) shorts,
      iii) a skirt.

4 A bag contains beads coloured blue, green and orange. If a bead is selected at random, the probability that it is blue is $\frac{1}{12}$ and the probability that it is green is $\frac{7}{12}$.
   a) What is the probability that a bead selected at random from the bag is orange?
   b) Show all three probabilities on a probability scale divided into twelfths.

460

c) If there are 36 beads in the bag, estimate how many are:
   i) blue,
   ii) green,
   iii) orange.

5 A bag contains a mixture of mints, toffees and eclairs. If a sweet is selected at random, the probability that it is a mint is $\frac{3}{16}$ and the probability that it is a toffee is $\frac{5}{16}$.
   a) What is the probability that a sweet selected at random from the bag is an eclair?
   b) Show all three probabilities on a probability scale divided into sixteenths.
   c) If there are 64 sweets in the bag, estimate how many are:
      i) mints,
      ii) toffees,
      iii) eclairs.

6 The probability that a milling machine breaks at least one cutter during an 8 hour working shift has been found to be $\frac{7}{20}$.
   a) What is the probability that the machine does not break a cutter during the shift?
   b) Show both probabilities on a probability scale divided into twentieths.
   c) Bill Yates works the machine for 100 shifts. Estimate in how many of these shifts Bill did not need to change the cutter because it had broken.

7 In a Year 11 class, 56% of the students are boys.
   a) What is the probability that a student selected at random from the class will be a girl?
   b) Show the probabilities that a selected student will be a boy or a girl on a percentage probability scale.
   c) One student is selected at random from the class every day to do litter duty. Estimate how many girls will be picked during a school term lasting for 75 school days.

8 A bag contains beads coloured grey, green and purple. If a bead is selected at random, the probability that it is grey is 37.5% and the probability that it is green is 12.5%.
   a) What is the probability that a bead selected at random from the bag is purple?
   b) Show all three probabilities on a percentage probability scale.

461

c) If there are 120 beads in the bag, estimate how many are:
   i) grey,
   ii) green,
   iii) purple.
d) Explain why there cannot be 60 beads in the bag.

9 An ice cream van sells four flavours of ice cream, vanilla, strawberry, mint and chocolate. The owner worked out that the probability a customer chose vanilla was 40%, the probability that a customer chose strawberry was 25%, and the probability that a customer chose mint was 10%.
   a) What was the probability that a customer chose chocolate?
   b) Show all four probabilities on a percentage probability scale.
   c) If the van sold 460 ice creams during the day, estimate how many were:
      i) vanilla,
      ii) strawberry,
      iii) mint,
      iv) chocolate.

■■■■■■■■■■■■■■■■■■■■■■■■■■■■■■■■■

Research and experimental probability

There are many real life situations where equally likely outcomes cannot be assumed.

For example, when a drawing pin is dropped on a table, it will either land point down or point up. We cannot assume that these two outcomes are equally likely.

The way to find the probability of the drawing pin landing point up or point down is to do an experiment.

Suppose we dropped the drawing pin 100 times and obtained these results.

	Point up	Point down
Frequency	45	55

We estimate the probability of the drawing pin landing point up as $\frac{45}{100}$ or 0.45, and the probability of it landing point down as $\frac{55}{100}$ or 0.55.

The fractions $\frac{45}{100}$ and $\frac{55}{100}$ are called the ***relative frequencies*** with which the drawing pin landed point up or point down.

Relative frequencies become more accurate measures of probability if an experiment is repeated a large number of times.

Example

James, Carl, Vali and Jodie each toss two coins 100 times and record the number of times they obtain two heads, two tails or one head with one tail. These are their results.

	One head, one tail	Two tails	Two heads
James	45	31	24
Carl	56	17	27
Vali	61	20	19
Jodie	42	33	25

a) Write down the relative frequencies of one head with one tail, two tails and two heads for each student's results.

b) Write down the relative frequencies of one head with one tail, two tails and two heads for the students' combined results.

c) Which relative frequencies are likely to be the most accurate estimates of the probabilities of one head with one tail, two tails or two heads?

d) If the experiment is repeated 2000 times, estimate how many of the outcomes will be two heads.

a) This table shows the relative frequencies.

	One head, one tail	Two tails	Two heads
James	$\frac{45}{100}$	$\frac{31}{100}$	$\frac{24}{100}$
Carl	$\frac{56}{100}$	$\frac{17}{100}$	$\frac{27}{100}$
Vali	$\frac{61}{100}$	$\frac{20}{100}$	$\frac{19}{100}$
Jodie	$\frac{42}{100}$	$\frac{33}{100}$	$\frac{25}{100}$

b) The combined relative frequencies are:

One head, one tail	Two tails	Two heads
$\frac{204}{400}$	$\frac{101}{400}$	$\frac{95}{100}$

c) The combined relative frequencies are likely to be the most accurate estimates of the probabilities of one head with one tail, two tails or two heads.

d) Estimate of number of two heads outcomes $= \frac{95}{400} \times 2000 = 475$.

1 A bag contains a very large number of mixed red and green seeds. Stefan, Wayne and Surbajit each select 100 seeds and note the seed's colour, returning each seed to the bag before another is selected. These are their results.

	Green	Red
Stefan	58	42
Wayne	63	37
Surbajit	59	41

a) Write down the relative frequencies of green and red seeds for each student's results.
b) Write down the relative frequencies of green and red seeds for the students' combined results.
c) Which relative frequencies are likely to be the most accurate estimates of the probabilities of selecting a green seed or a red seed?
d) If there are 5000 seeds in the bag, estimate how many of the seeds are red.

2 Four plant researchers each open 30 pods of a new type of broad bean and count the contents. These are their results.

Number of beans	4	5	6	7	8	9
Number of pods (Researcher A)	3	4	7	6	5	5
Number of pods (Researcher B)	6	7	6	5	6	0
Number of pods (Researcher C)	0	11	0	12	5	2
Number of pods (Researcher D)	1	4	6	9	8	2

a) Write down the relative frequencies 4, 5, 6, 7, 8 or 9 beans per pod for each researcher's results.
b) Write down the relative frequencies of 4, 5, 6, 7, 8 or 9 beans per pod for the researchers' combined results.
c) Which relative frequencies are likely to be the most accurate estimates of the probabilities of 4, 5, 6, 7, 8 or 9 beans per pod?
d) If 1000 pods were opened, estimate how many would contain 9 beans.

3 A market researcher asks four groups of 50 people to select their favourite ice cream flavour from the list strawberry, vanilla, chocolate and mint. These are his results.

	Strawberry	Vanilla	Chocolate	Mint
Group A	4	16	24	6
Group B	5	25	15	5
Group C	8	27	10	5
Group D	7	23	8	12

a) Write down the relative frequencies of selecting strawberry, vanilla, chocolate or mint for each group.
b) Write down the relative frequencies of selecting strawberry, vanilla, chocolate or mint for the groups' combined results.
c) Which relative frequencies are likely to be the most accurate estimates of the probabilities that a person will select strawberry, vanilla, chocolate or mint?
d) If a group of 2000 people were asked the same question, estimate how many would select vanilla.

■■■■■■■■■■■■■■■■■■■■■■■■■■■■■■■■■■■■

Deciding when sufficient trials have been made in a probability experiment

It can be very difficult to decided when to stop a probability experiment. We want to be confident that any relative frequencies we have calculated are close to the real probabilities for those outcomes.

One way to judge this is illustrated in the following example.

Example
A student decided to experiment with a single die.

She knew the die should produce exactly one sixth of its outcomes as a score of one but wondered whether it actually would do this.

She threw the die 4200 times and recorded the number of ones scored after every 600 throws.

Number of ones	91	208	292	382	505	605	709
Total throws	600	1200	1800	2400	3000	3600	4200

She then worked out the relative frequencies by dividing the number of ones by the total number of throws (to 2 decimal places).

465

Number of ones	91	208	292	382	505	605	709
Total throws	600	1200	1800	2400	3000	3600	4200
Relative frequency	0.15	0.17	0.16	0.16	0.17	0.17	0.17

She drew a graph showing the total number of throws and the relative frequency.

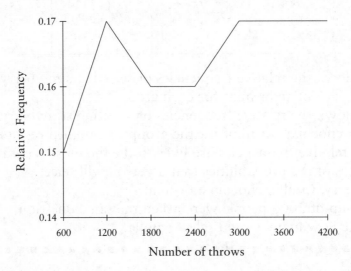

She decided that the relative frequency had settled to a value close to 0.17 and stopped the experiment.

PAUSE

1 A sack contains a large number of mixed red and white seeds. A group of 10 students each select 500 seeds from the bag and count the number which are red. These are their cumulative results.

Red seeds counted	206	412	608	791	1000
Total number counted	500	1000	1500	2000	2500

Red seeds counted	1181	1373	1561	1765	1962
Total number counted	3000	3500	4000	4500	5000

a) Complete a table showing the relative frequency of red seeds after every 500 seeds counted.

b) Draw a graph showing the relative frequency of red seeds after every 500 seeds counted.

c) Do you think the students have produced a reliable estimate of the probability that a seed selected from the sack will be red?

d) There are 26 000 seeds in the sack, estimate the number that are red.

2 A group of students flipped two coins 4000 times. They recorded their results after every 400 flips. These are their results.

One head, one tail	214	407	611	823	1025
Total number of flips	400	800	1200	1600	2000

One head, one tail	1228	1433	1632	1830	2022
Total number of flips	2400	2800	3200	3600	4000

a) Complete a table showing the relative frequency of one head and one tail after every 400 flips.

b) Draw a graph showing the relative frequency of one head and one tail after every 400 flips.

c) Do you think the students have a reliable estimate of the probability that the two coins will land as one head and one tail?

■ ■

Probability problems with two events

We may be asked to find the probability of obtaining a particular result when two or more events are combined. For example, we may be asked to find the probability of obtaining two heads when two coins are tossed, or of obtaining 5 sixes when five dice are rolled. When *two* events are combined we often draw a small graph or *sample space* to illustrate clearly all the possible outcomes that can be obtained.

Examples

1. Two coins are tossed, find the probability that the result obtained is two heads.

To illustrate all the outcomes that are possible, we draw this diagram.

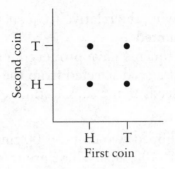

A diagram like this is called a *sample space*. It is a graph with all the outcomes of one event listed on the *x*-axis and all the outcomes of the other event listed on the *y*-axis. All the possible *combined outcomes are illustrated by the dots on the graph.*

Our sample space shows that there are four possible combined outcomes, HH, HT, TH and TT.

Of these, only one gives the required result of two heads.

So the probability of obtaining two heads = $\frac{1}{4}$ = 0.25 = 25%.

2. Two dice are rolled, find the probability that the score will be greater than 9.

We can draw this sample space.

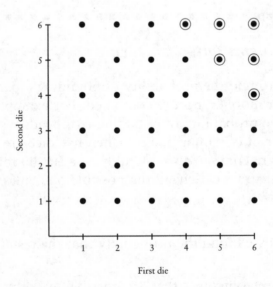

From the sample space we see that there are 36 possible outcomes.

Of these, 6 (circled in the diagram) give a score greater than 9.

So, the probability of a score greater than $9 = \frac{6}{36} = \frac{1}{6} = 0.167$ (to 3 decimal places) = 16.7%.

3. An ordinary pack of 52 playing cards is shuffled and a card is chosen. The card is replaced, the pack is shuffled and a second card is chosen. What is the probability that both cards chosen are red?

We can draw this sample space:

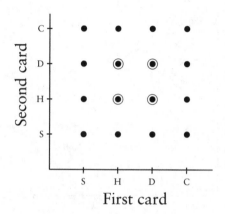

There are 16 possible outcomes of these, 4 give two red cards.

The probability of two red cards is $\frac{4}{16}$.

PAUSE

Give all numerical answers in this exercise as a fraction, decimal and percentage.

1 Find the probability that when two coins are tossed the result will be:
   a) two tails,
   b) a head and a tail.

2 If two coins are tossed 2000 times, how many of the results would you predict will be:
   a) two heads,

b) two tails,

c) a head and a tail?

Would you expect an actual experiment to exactly match your predictions?

3 Two dice are rolled together, and their total score recorded.

Copy and complete the sample space diagram.

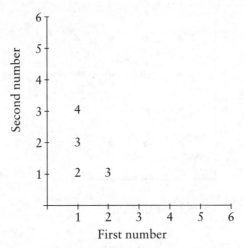

Use the diagram to answer these questions. Find the probability of the result being:

a) an 8, f) a 12,

b) a 7, g) an even number,

c) a 4, h) a prime number,

d) a 5, i) greater than 1,

e) a 10, but less than 13.

4 If two dice are rolled 1800 times, how many of the total scores would you predict will be:

a) a 9, d) a 6,

b) a 2, e) an 11,

c) a 3, f) a square number?

5 An ordinary pack of 52 playing cards is shuffed and a card is chosen. The card is replaced, the pack is shuffled and a second card is chosen. What is the probability that the cards cut will be:

a) two spades,

b) two black cards,

c) one red and one black card,
d) both of the same suit,
e) one club and one diamond?

6 Two spinners are made, one numbered from 1 to 5 and the other from 6 to 10. Find the probability that when the spinners are spun together the total score will be greater than 12.

7 Two special dice are made, one has the set of numbers 0,1,1,2,2,3 and the other the set of numbers 0,0,1,2,4,5. Bill rolls the first dice and Anthea rolls the second dice.

a) Copy and complete the sample space diagram:

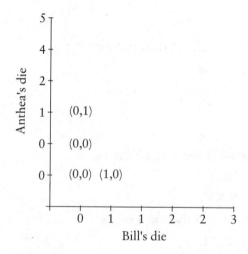

b) What is the probability that Anthea beats Bill?

■ ■

Probability problems with more than two events

If more than two events are combined, a sample space cannot be drawn. In simple cases the required combined probability can be found by *multiplying* the probabilities for each of the separate events.

Examples

1. Find the probability of rolling three dice and scoring 18.

To score 18, we must score a 6 on each die. The separate probabilities of doing this are:

$$\frac{1}{6}, \frac{1}{6} \text{ and } \frac{1}{6}.$$

The probability of obtaining three sixes is $\frac{1}{6} \times \frac{1}{6} \times \frac{1}{6} = \frac{1}{216} = 0.038$ (to 3 decimal places) = 3.8%.

2. A bag contains 4 red and 5 black beads. A bead is chosen at random from the bag. Its colour is noted and the bead is replaced. This experiment is repeated three times. Find the probability that the three beads chosen are:

a) all black,
b) all red,
c) a red followed by a black followed by a red.

a) The probability of three black beads is:

$$\frac{5}{9} \times \frac{5}{9} \times \frac{5}{9} = \frac{125}{729} = 0.171$$
$$= 17.1\%.$$

b) The probability of three red beads is:

$$\frac{4}{9} \times \frac{4}{9} \times \frac{4}{9} = \frac{64}{729} = 0.088$$
$$= 8.8\%.$$

c) The probability of red, black, red is:

$$\frac{4}{9} \times \frac{5}{9} \times \frac{4}{9} = \frac{80}{729} = 0.11$$
$$= 11.0\%.$$

PAUSE

Give all numerical answers in this exercise as a fraction, decimal and percentage.

1 Find the probability of rolling four dice and scoring four sixes.

2 An ordinary pack of 52 playing cards is shuffled and a card is chosen. The card is then replaced. This experiment is repeated four times. Find the probability that all the cards cut are clubs.

3 If three coins are tossed, find the probability that three tails will be obtained.

4 A bag contains 3 red and 4 green beads. A bead is chosen at random, its colour is noted, then the bead is replaced. This is done three times. Find the probability that the three beads are:

a) all red,
b) all green,
c) a red followed by a green followed by a red,
d) a red followed by a red followed by a green.

5 A bag contains 3 red, 4 green and 5 black beads. A bead is chosen at random, its colour is noted, then the bead is replaced. This is done three times. Find the probability that the three beads are:

a) all red,
b) all green,
c) all black,
d) a red followed by a green followed by a black,
e) a black followed by a red followed by a green.

■ ■

Tree diagrams

In more complicated cases, we draw a ***tree diagram***. A tree diagram starts with a dot to represent the first event. From this dot, 'branches' are drawn to represent all the possible outcomes from the event. The probability of each outcome is written on it's 'branch'. For example, if we are drawing a tree diagram to illustrate flipping a coin three times, we start with this diagram which shows that the first event can result in a head or a tail.

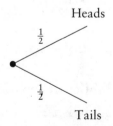

Since the second event can follow either of these outcomes, dots are placed at the end of both the branches to represent the second event. From these dots branches are drawn to represent all the outcomes of the second event. Adding the branches for the second coin to our tree diagram gives us this diagram.

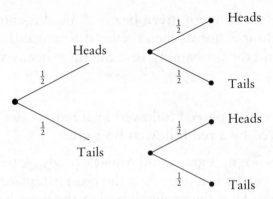

This process continues until all the successive events have been dealt with. The probabilities on the branches are multiplied together to obtain the probabilities of obtaining each possible combined outcome. These probabilities are added to the tree diagram. This is our completed tree diagram.

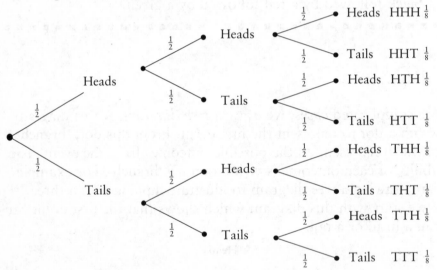

These are examples of some questions that can be answered from the diagram.

Notice that we *add* the probabilities of separate outcomes.

The probability of obtaining 2 heads is:

$$\frac{1}{8} + \frac{1}{8} + \frac{1}{8} = \frac{3}{8} = 0.375 = 37.5\%.$$

The probability of obtaining at least 2 heads is:

$$\frac{1}{8} + \frac{1}{8} + \frac{1}{8} + \frac{1}{8} = \frac{1}{2} = 0.5 = 50\%.$$

The probability of obtaining at least 1 tail is:

$$\frac{1}{8} + \frac{1}{8} + \frac{1}{8} + \frac{1}{8} + \frac{1}{8} + \frac{1}{8} + \frac{1}{8} = \frac{7}{8} = 0.875 = 87.5\%.$$

Example

A bag contains 3 red, 4 green and 5 black beads. A bead is chosen at random and replaced, then a second bead is chosen. Find the probability that at least one of the beads is green.

This is the tree diagram.

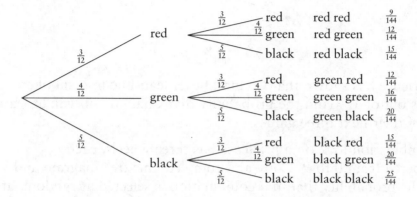

The probability that at least one bead is green is:

$$\frac{12}{144} + \frac{12}{144} + \frac{16}{144} + \frac{20}{144} + \frac{20}{144} = \frac{80}{144} = \frac{5}{9} = 0.556 \text{ (to 3 decimal places)}$$
$$= 55.6\%.$$

■ ■
PAUSE

Give all numerical answers in this exercise as a fraction, decimal and percentage.

1 A bag contains 3 red and 2 white beads. Two beads are selected from the bag, the first being replaced before the second is selected. Draw a tree diagram and find the probability that:
a) at least one red bead is selected,
b) two beads of the same colour are selected.

2 A bag contains 3 red and 2 white beads. Three beads are selected
   from the bag, each one being replaced before the next is selected.
   Draw a tree diagram and find the probability that:

   a) at least one red bead is selected,
   b) three beads of the same colour are selected.

3 Three dice are rolled. Copy and complete this tree diagram and
   find the probability that at least two sixes are obtained.

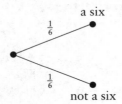

4 Experience has shown that Chantal beats Jean-Luc in 6 snooker
   games out of 10. Find the probability that Chantal will win at least
   two of their next three games.

5 A combination lock on a briefcase has three wheels, each
   numbered from 0 to 9. Copy and complete this tree diagram and
   find the probability that, if a combination is selected at random, at
   least one digit is correct.

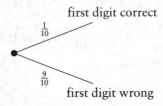

6 Cindy has a bag of sweets with two mints and two toffees left.
   Davinder has a bag of sweets with three mints and two toffees left.
   They both offer Errol a sweet and he selects one at random from
   each bag. Draw a tree diagram and find the probability that:

   a) one toffee is selected,
   b) at least one toffee is selected,
   c) two mints are selected.

7 An electronics factory buys television tubes from suppliers in
   batches of ten. A quality control inspector checks each batch of ten
   by selecting one tube at random from the batch and testing it. If
   the inspector finds the tube is faulty, the whole batch is sent back
   to the supplier. A dishonest supplier sends three batches of tubes.

She knows that the first batch has 2 faulty tubes, the second batch has 1 faulty tube and the third batch has 3 faulty tubes. The quality control inspector checks all three batches. Draw a tree diagram and find the probability that:

a) only one batch is returned to the supplier,
b) at least one batch is returned to the supplier,
c) all three batches are returned to the supplier,
d) all three batches are accepted.

REWIND

1 Explain the difference between qualitative variables, discrete variables and continuous variables, giving some examples of each type.

2 These are the test scores of 50 students in a maths test. Use a tally chart to organise the results into a table.

2	7	8	9	3	4	1	9	10	2
6	6	5	7	8	9	2	6	4	9
3	4	9	10	6	7	8	4	5	3
2	5	6	7	8	7	6	5	9	10
9	8	7	4	5	6	7	2	8	7

3 Illustrate the 50 test marks in Question 2 with a bar chart.

4 This bar chart illustrates a serious problem that existed in British prisons. Estimate from the bar chart the average daily overcrowding for each year from 1977 to 1987.

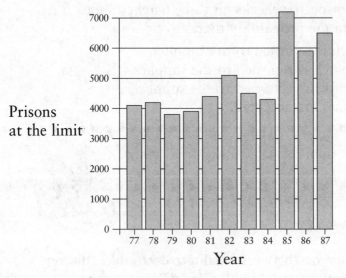

Overcrowding year by year: average daily total of prisoners above recommended level

5 Draw a pie chart to illustrate the data in this table:

Activity	Time spent in one day (hours)
Working	9
Travel	1
Leisure	4
Sleeping	8
Meals	2

6 Draw a histogram to illustrate the data in this table.

Height of seedling (mm)	$0 \leq H < 5$	$5 \leq H < 10$	$10 \leq H < 15$	$15 \leq H < 20$	$20 \leq H < 25$	$25 \leq H < 30$
Frequency	5	11	20	26	12	6

7 Find the mean, median, mode and range of each of these sets of numbers.

a) 2,2,4,6,7,4,2,6,1,2,6,8,2,3,1,3,5,7,7

b) 21,22,23,24,22,22,21,23,25,23,22,21,21,24,21

c) −1,0,−1,2,3,−3,2,1,1,1,1

d) 200,300,200,400,500,600,200,300,300,400,500,200,200

8 Find the mean, median, mode and range of the data shown in each of the following tables.

478

a)

Number of matches per box	36	37	38	39	40	41	42	43
Frequency	3	12	18	27	23	12	5	0

b)

Number of children per family	0	1	2	3	4	5	6
Frequency	45	67	124	33	17	12	2

9 Find the mean, median, quartiles, modal group, range and interquartile range of the data shown in each of the following tables.

a)

Height of seedling (mm)	$0 \leq H < 5$	$5 \leq H < 10$	$10 \leq H < 15$	$15 \leq H < 20$	$20 \leq H < 25$	$25 \leq H < 30$
Frequency	5	11	20	26	12	6

b)

Length of vehicle (metres)	Frequency
$3.0 \leq L < 4.0$	36
$4.0 \leq L < 4.5$	85
$4.5 \leq L < 5.0$	49
$5.0 \leq L < 6.0$	21
$6.0 \leq L < 7.0$	6
$7.0 \leq L < 12.0$	3

10 There is growing concern about the difficulty of recruiting teachers, particularly in Science subjects and in areas of the country where houses are very expensive. This diagram appeared in the 'Times Educational Supplement', a special newspaper for teachers.

a) Write a summary of the information illustrated by the diagram.

b) Estimate the average number of replies that would be received in an area where the average price of a semi-detached house was £40 000.

The average number of replies to TES advertisements for physics teachers received by local authorities against the average price of a semi-detached house in that region.

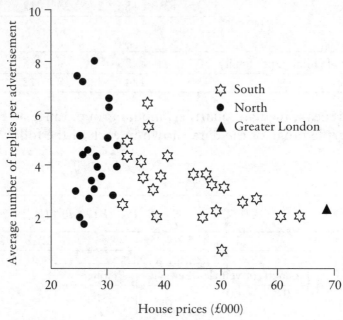

11 If a single die is rolled, what is the probability that the score will be:

a) even,

b) a factor of 12,

c) less than 5,

d) a 3?

12 If a single card is cut from a deck of 52 cards, what is the probability that it will be:

a) a club,   e) not a spade,

b) a red card,   f) a queen or a red card,

c) the king of diamonds,   g) a queen and a red card,

d) a picture card,   h) a five or a black card?

13 Hannah is an animal researcher studying a large colony of bats. She wishes to know the relative frequencies of males and females in the colony. She traps ten groups of twenty bats, determines their gender and then releases them. These are her results:

Males	9	15	24	34	43	51	55	63	73	78
Females	11	25	36	46	57	69	85	97	107	122
Total	20	40	60	80	100	120	140	160	180	200

a) Write down the relative frequencies of females after each of the ten samples has been checked.

b) Which relative frequency is likely to be the most accurate estimates of the probabilities that a bat selected at random will be female?

c) If there are 1200 bats in the colony, estimate the number of these bats which are female.

14 Two dice are rolled together. Draw a sample space for the results and find the probability that the final score will be:

a) a prime number,

b) 12,

c) greater than 8,

d) 7,

e) a factor of 12,

f) 13,

g) 1,

h) 2 or 12.

15 Find the probability of:

a) cutting 4 red cards in succession from a normal pack of 52 cards,

b) tossing 15 pennies and getting 15 heads,

c) rolling 10 dice and getting a total score of 60.

16 There are two sets of traffic lights on the route that Mr Grimby takes on his way to work each morning. Experience has shown that the first lights are red on 4 mornings out of 10 and the second lights are red on 6 mornings out of 10. Draw a tree diagram and find the probability that:

a) Mr Grimby meets at least one red light on his way to work,

b) Mr Grimby meets two green lights on his way to work.

17 An operation has a 70% success rate the first time it is attempted. If it fails, it can be repeated but with only a 30% chance of success. If the operation fails twice it is not repeated. Draw a tree diagram and find the probability that a patient about to have the operation for the first time will eventually have a successful operation after either one or two attempts.

1 Lorraine is writing a questionnaire for a survey about her local Superstore.

She thinks that local people visit the store more often than people from further away. She also thinks that local people spend less money per visit.

Write three questions which would help her to test these ideas.

Each question should include at least three responses from which people choose one. (SEG)

2 The bar chart shows the numbers of customers shopping in a department store one week.

a) On which day of that week did the store have the least number of customers?

b) How many customers did the store have:
   i) on Wednesday,
   ii) on Thursday?

On Saturday, each customer spent an average of £14.20.

c) Work out the total amount of money spent in the store on Saturday. (ULEAC)

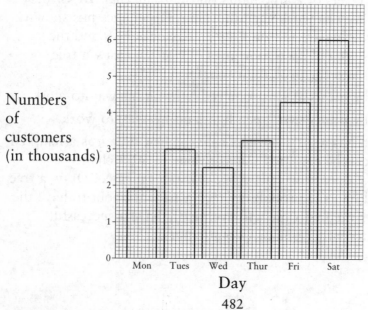

3 Sally did a survey of car colours. The notebook shows all her results.

w	w	r	w	g
r	w	r	b	r
r	r	w	r	r
b	w	r	r	w
r	g	r	w	r
g	w	w	b	r

Key:
w white
b blue
r red
g green

a) Copy and complete her frequency table.

Colour	Tally	Frequency
White		
Blue		
Red		
Green		

b) Show this information as a bar chart.   (SEG)

4 Each morning in September, Liz picked mushrooms from her field. The bar chart below shows the number of mushrooms that she picked in the 30 days of September last year.

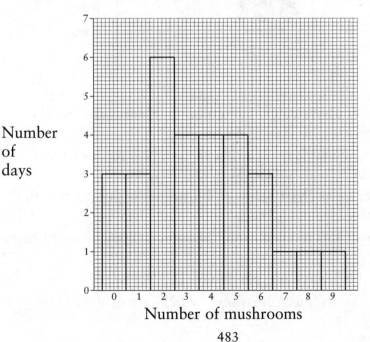

Number of mushrooms

483

a) Write down the number of days on which Liz picked:
   i) two mushrooms,
   ii) no mushrooms.
b) Find the total number of mushrooms Liz picked in the month.
c) Find the modal number of mushrooms Liz picked in the month.
d) Find the median number of mushrooms Liz picked in the month.
e) Calculate the mean number of mushrooms Liz picked in the month.

These days in September are typical.

f) Find the probability that Liz will pick at least four mushrooms on 23rd September next year. (ULEAC)

5 The table below shows the number of records sold by a small record shop in Megtown.

MEGTOWN RECORD SHOP	represents 20 records	
JANUARY	⦿ ⦿ ⦿	60
FEBRUARY	⦿ ⦿ ⦿ ⦿ ◖	95
MARCH	⦿ ⦿ ⦿ ⦿	
APRIL	⦿ ⦿ ◖	
MAY		100
JUNE		45

a) In the table above enter the number of records sold in March and April.
b) Complete the table by drawing the symbols to show the number of records sold in May and June.
c) Complete a bar chart below to illustrate the information in the above table. (MEG)

6 Out of every £180 received from the rates a district council spent as follows:

$$\text{Highways and planning} \quad £13$$
$$\text{Sports and recreation} \quad £29$$

	Environmental health	£37
	Housing	£24
	Administration	£49
	Emergency fund	£28

Illustrate this expenditure by means of a pie chart of radius 5 cm.

(SEG)

7

Type of crisp	Tally	Frequency	Angle
Plain	ⅢⅡ IIII		
Salt and vinegar	ⅢⅡ ⅢⅡ ⅢⅡ I		
Cheese and onion	ⅢⅡ ⅢⅡ I		
Beef	ⅢⅡ I		
Crispy bacon	III		
	Total	45	

The table shows the result of a survey taken on the sale of crisps one day in a School Tuckshop.

a) Copy the table and complete the Frequency column.

The information is to be shown on a pie chart.

b) Calculate the size of the angle of each sector of this pie chart.

Write your answers in the Angle column in the table.

c) Draw an accurate pie chart to show this information.

Label the sectors clearly. (ULEAC)

8 Tracey asked 50 pupils how they travelled to school. Here are the results.

Type of travel	Number	Percentage
Special bus	23	
Public transport	14	
Walked	8	
Car	5	
Total	50	

She wants to draw a pie chart.

a) Work out the percentages and fill them in on a copy of the table.

b) Draw the pie chart. (MEG)

9 The three main costs of a factory are shown in the pie chart as wages, overheads and raw materials. Two angles at the centre of the circle are given.

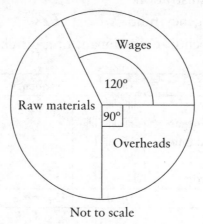

Not to scale

a) Calculate the angle which represents 'raw materials'.
b) What fraction of the total cost is 'wages'? Give your fraction in its lowest terms.   (NEAB)

10

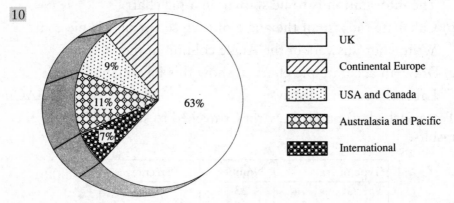

The chart above shows where a major manufacturing company does business.

a) What percentage of its business is done with Continental Europe?

b) Give the ratio of business done in U.K. to business done in USA and Canada. Give your answer in its lowest terms.

The company does business worth £22 000 000 in Australasia and Pacific.

c) What is the value of business done in UK?

(ULEAC)

486

11 Thirty-five people took part in an ice-skating competition.
The points they scored are shown below.

18 24 19   3 24 11 25 10 25 14 25   9 16 26
21 27 13 23   5 26 22 12 27 20   7 28 21 20
22 16 12 25   7 25 19

a) Work out the range of points scored.   (ULEAC)

b) Complete the frequency table.

Class interval	Frequency
1-5	2
6-10	4
11-15	
16-20	
21-25	
26-30	

c) Draw a histogram to represent this information.   (ULEAC)

12 The table below gives information about the expected lifetimes, in hours, of 200 light bulbs.

Lifetime (t)	$0<t \leq 400$	$400<t \leq 800$	$800<t \leq 1200$	$1200<t \leq 1600$	$1600<t \leq 2000$
Frequency	32	56	90	16	6

a) Mr Jones buys one of the light bulbs.
  i) What is the probability that it will not last more than 400 hours?
  ii) What is the probability that it will last at least 800 hours but not more than 1600 hours?
b) Using a horizontal scale of 400 hours = 2 cm and a vertical scale of 20 light bulbs = 1 cm, draw a frequency polygon to illustrate the information in the table.

(MEG)

13 The number of goals scored in each of the eleven First Division Football League matches one Saturday last season was:

0, 0, 1, 1, 2, 2, 4, 5, 6, 6, 6.

Find:

a) the mode,

b) the median,

c) the mean,

d) the range. (MEG)

14 Samantha and Teresa both did ten Mathematics homeworks. Here are their marks out of ten.

Samantha	10	10	7	4	9	8	5	2	9	8
Teresa	5	7	6	8	6	8	7	8	7	6

a) Work out the mean mark for each of them.

b) Work out the range for each of them.

c) Say who you think was better at maths. Give a reason for your answer. (MEG)

15 Richard saved some of his pocket money each week to go on holiday. He made a list of how much he saved each week.

£2   £1   £1.50 £1.30 £1.75 £2 £2   £1 £1   £2

£1.30 £1.30 £1.75 £1.50 £1.30 £2 £1.30 £1 £1.30 £1.50

Calculate:

a) the total number of weeks in which he saved money,

b) the total amount of money saved,

c) the mean amount of money saved per week,

d) the median,

e) the mode. (NICCEA)

16 Eunice measures the lengths of runner beans in a gardening competition. The lengths in centimetres of the longest ten runner beans are given.

26, 34, 27, 28, 24, 36, 30, 28, 25, 32

a) What is the range in the lengths of these runner beans?

b) What is the mean length of these runner beans?

c) In last year's competition the mean length of the longest ten runner beans was 29 cm and the range was 8 cm. Explain how the lengths of the runner beans differ this year from last year. (SEG)

Monthly rainfall (millimetres)

	Jan	Feb	Mar	Apr	May	June	July	Aug	Sept	Oct	Nov	Dec
Great Britain	74	44	40	48	50	29	48	37	61	75	84	70
The Gambia	0	0	0	0	1	2	84	352	185	81	27	0

a) The mean rainfall per month in Great Britain is 55 mm.
   Calculate the mean rainfall per month in The Gambia.

b) Find the range of the monthly rainfall:
   i) in The Gambia,
   ii) in Great Britain.

c) In which of these two countries are water shortages more likely?

d) Explain your answer to part c), using the means *and* the ranges. (MEG)

18 Ian and David are both batsmen in a local cricket team.

In May and June 1992, Ian completed ten innings and David completed eight innings.

Their scores are shown below.

Ian	0	44	17	56	2	80	13	42	1	45
David	25	42	37	31	22	48	N	10		

At the end of June both batsmen had the same batting average (mean score).

a) How many runs, N, did David score in his seventh innings?

b) Using other information you could deduce from Ian's and David's scores, compare their batting skills. (NEAB)

19 The frequency table shows the temperature at midnight for a ski resort in December 1992.

Temperature °C	−3	−3	−1	0	1	2	3	4	5	6	7
Number of nights	2	3	4	3	4	2	3	6	3	0	1

a) State the modal temperature.

b) Find the median temperature.

c) Calculate the mean temperature. Give your answer correct to one decimal place. (ULEAC)

20 Ian looked at a passage from a book. He recorded the number of words in each sentence in a frequency table using class intervals of 1–5, 6–10, 11–15, etc.

Class interval	Frequency f	Mid-interval value x	$f \times x$
1–5	16	3	48
6–10	28	...	...
11–15	26	13	338
16–20	14	...	...
21–25	10	23	230
26–30	3	...	...
31–35	1	33	33
36–40	0	...	...
41–45	2	43	86
	Total		Total

a) Complete the table.

b) Write down:

   i) the modal class interval,

   ii) the class interval in which the median lies.

c) Work out an estimate of the mean number of words in a sentence. (ULEAC)

21 Cole's sells furniture and will deliver up to a distance of 20 miles. The diagram shows the delivery charges made by Cole's.

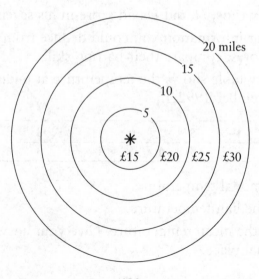

The table shows the information in the diagram and also the number of deliveries made in the first week of May 1994.

Distance (d) from Cole's in miles	Delivery charge in pounds	Number of deliveries			
$0 < d \le 5$	15	27			
$5 < d \le 10$	20	11			
$10 < d \le 15$	25	8			
$15 < d \le 20$	30	4			

a) Copy the table.

b) Calculate the mean charge per delivery for these deliveries.

c) Calculate an estimate for the mean distance of the customers' homes from Cole's.   (ULEAC)

22 The heights of 100 men are shown in the distribution table.

Height in cm	168–170	171–173	174–176	177–179	180–182	183–185	186–188	189–191
Frequency	5	10	14	20	16	19	10	6
Upper class-boundary	170.5							
Cumulative frequency	5							

a) Complete the upper class-boundaries and cumulative frequencies.

b) Draw a cumulative frequency polygon.

c) Use your diagram to estimate:

   i) the median height,

   ii) the interquartile range,

   iii) the number of men taller than 1.82 metres.   (NEAB)

23 A greengrocer takes delivery of 200 grapefruit. The graph shows the cumulative frequency distribution of their weights.

a) From the graph estimate:

   i) the median,

   ii) the interquartile range.

b) The greengrocer classifies the fruit as follows:

less than 200 grams: small
between 200 and 250 grams: medium
over 250 grams: large

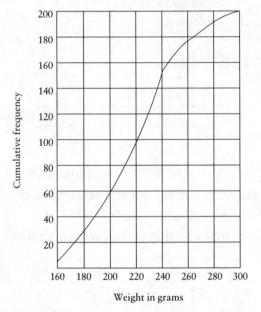

Estimate the number of medium-sized grapefruit in the consignment.

(MEG)

24 The speeds, in miles per hour (mph), of 200 cars travelling on the A320 road were measured.

The results are shown in the table.

Speed (mph)	Cumulative frequency
not exceeding 20	1
not exceeding 25	5
not exceeding 30	14
not exceeding 35	28
not exceeding 40	66
not exceeding 45	113
not exceeding 50	164
not exceeding 55	196
not exceeding 60	200
Total	200

a) Draw a cumulative frequency graph to show these figures.

b) Use your graph to find an estimate for:

   i) the median speed (in mph),

   ii) the interquartile range (in mph).   (ULEAC)

25 The table below shows the distribution of the marks obtained by 250 students in a recent examination.

Mark	0 to 19	20 to 39	40 to 59	60 to 79	80 to 99
Frequency	20	41	64	84	41

a) Calculate an estimate of the mean mark.

b) Draw a cumulative frequency diagram to display this information.

c) Use your diagram to answer the following questions.

   i) 12% of these students got a grade A.

What was the minimum mark needed to get a grade A?

   ii) Students who scored between 45 and 60 marks inclusive got a grade C.

How many of them got a grade C?

26 This table gives the marks scored by pupils in a French and in a German test.

French	15	35	34	23	35	27	36	34	23	24	30	40	25	35	20
German	20	37	35	25	33	30	39	36	27	20	33	35	27	32	28

a) Draw a scatter graph of the marks scored in the French and German tests.

b) Describe the correlation between the marks scored in the two tests.   (ULEAC)

27 The table gives information about the age and value of a number of cars of the same type.

Age (years)	1	3	4½	6	3	5	2	5½	4	7
Value (£)	8200	5900	4900	3800	6200	4500	7600	2200	5200	3200

a) Use this information to draw a scatter graph. Use a scale of 2cm = 1 year on the horizontal axis and 2cm = £1000 on the vertical axis.

b) What does the graph tell you about the value of these cars as they get older?

c) The information is correct but the age and value of one of these cars looks out of place. Give a possible reason for this.

d) Draw a line of best fit.

e) John has a car of this type which is $3\frac{1}{2}$ years old and is in average condition. Use the graph to estimate its value.   (SEG)

28 Janine throws an ordinary dice.

a) What is the probability she gets a 3?

To win a game she needs a 3 or a 5.

b) What is the probability that she wins?   (MEG)

29 Chris taped five pop broadcasts, each on a separate tape. The broadcasts lasted for 40 minutes, 55 minutes, 30 minutes, 42 minutes and 48 minutes.

a) Chris picks a tape at random.

What is the probability that the chosen broadcast will last for more than $\frac{3}{4}$ hour?

b) Find the mean length of time of the five recordings.

(ULEAC)

30 To play a game you spin the pointer. You win the prize on which the pointer stops.

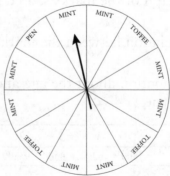

Richard has one spin.

a) Which prize is Richard most likely to win?

b) Explain your answer to part a).

494

Donna has one spin.

c) On the line below mark with a P the probability that Donna will win a pen.

d) On the line below mark with a W the probability that Donna will win a watch.

31 A game is played using the two boards shown below.

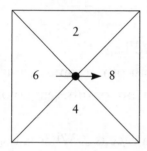
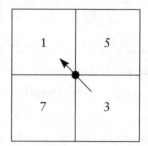

In one 'go' a player spins both arrows.

The sum of the numbers where the two arrows point is the score for the 'go'.

In the diagram the score for the 'go' is 9.

a) Copy and complete the table below showing all possible scores for one 'go'.

+	1	3	5	7
2				
4				
6				
8	9			

b) Write down the most likely score in one 'go'.

c) Write down the probability that a player will score 11 in one 'go'.

d) Calculate the probability that a player will score 13 in each of two 'go's'. (ULEAC)

32 a) A bag contains 2 red marbles, 1 blue marble and 1 yellow marble.

495

A second bag contains 1 red marble, 2 blue marbles and 1 yellow marble.

A marble is drawn from each bag.

Complete the table showing all the possible pairs of colours.

Marble from second bag

		R	B	B	Y
Marble from first bag	R	RR	RB	RB	RY
	R	RR			
	B	BR			
	Y	YR			

b) 2 marbles are drawn from a third bag.

The probability that they are both of the same colour is $\frac{5}{9}$.

What is the probability that they are of different colours?

(MEG)

33 a) The probability that Jane wins a race is $\frac{1}{4}$. What is the probability that she does not win the race?

b) The probability that Claire wins a race is $\frac{1}{3}$. Which of Jane or Claire is more likely to win a race? Explain your answer.

c) The probability that Andrew wins a race is 1. Explain what this means. (MEG)

34 A machine makes compact discs.

The probability that a perfect compact disc will be made by this machine is 0.85.

Work out the probability that a compact disc made by this machine will not be perfect. (ULEAC)

35 A bag contains a total of 20 beads.

There are 6 red beads, 9 blue beads and 5 white beads.

a) A bead is taken at random from the bag.

The probability that it is red is 0.3.

i) What is the probability that it is white?

ii) What is the probability that it is *not* white?

All the beads are taken from the bag, numbered 1, 2, 3 or 4 and then replaced. When a bead is taken from the bag at random the probability of each number is:

Number on bead	Probability
1	0.3
2	0.4
3	0.2
4	0.1

The red beads are numbered,

b) A bead is taken at random from the bag.

   i) What is the probability that it is red or numbered 4?

   ii) Explain why the probability of getting a red bead or a bead numbered 2 is *not* 0.3 + 0.4. (SEG)

36 Alan, Barbara and Chris usually travel on the same bus to school.

a) Copy and complete the following table by listing all the possible outcomes of Alan (A), Barbara (B) and Chris (C) trying to catch the bus. Use a tick to indicate 'catches the bus' and a cross to indicate 'does not catch the bus'.

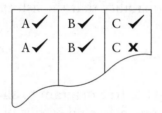

b) On any one day, the probability that Alan catches the bus is 0.8, the probability that Barbara catches the bus is 0.9 and the probability that Chris catches the bus is 0.75.

   i) Calculate the probability that only Alan and Barbara catch the bus on any one day.

   ii) Calculate the probability that at least two of the three pupils catch the bus on any one day. (MEG)

37 David oversleeps one day in ten.

On days when he oversleeps the probability that he is late for work is $\frac{1}{2}$.

When he does not oversleep the probability that he is late for work is $\frac{1}{6}$.

a) Copy and complete the tree diagram.

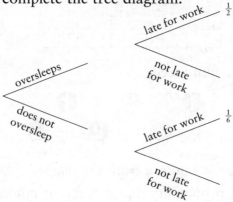

b) What is the probability that David oversleeps and is late for work?

c) What is the probability that David is late for work?　(NEAB)

38 There are two fish tanks in a pet shop.
In tank A there are four white fish and one black fish.
In tank B there are three white fish and four black fish.
One fish is to be taken out of each tank at random.

a) Write down the probability that the fish taken from tank A will be:

i) black,

ii) white.

b) Copy and complete the tree diagram to show all the probabilities when one fish is taken out of each tank.

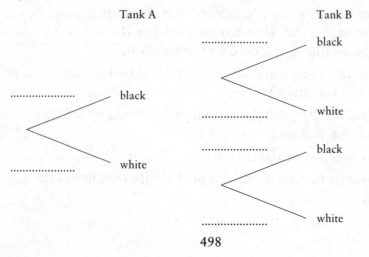

One fish is to be taken out of each tank at random.

c) Work out the probability that:

   i) the two fish will both be white,

   ii) the two fish will be of different colours.

(ULEAC)

39 Ahmed and Kate play a game of tennis.

The probability that Ahmed will win is $\frac{5}{8}$.

Ahmed and Kate play a game of snooker.

The probability that Kate will win is $\frac{4}{7}$.

a) Copy and complete the probability tree diagram below.

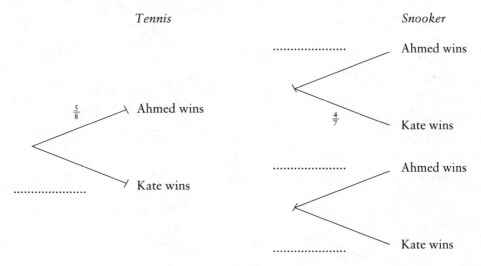

b) Calculate the probability that Kate will win both games.

(ULEAC)

How to do Coursework

This tells you how to do a piece of coursework. It is for students who do a task over a period of time rather than in school like an exam. However most of the suggestions here will help with a timed task.

It is important that you understand the *starting point* for your task. If you are unsure, ask for this to be made clear.

Once you understand what the task is about, think of the *simplest case* and build up from there.

Work *methodically*. This means:

1. looking for a path through the investigation,
2. organising your information,
3. looking for patterns and relationships,
4. finding a conclusion,
5. possibly extending the work,
6. writing a report.

When you are working on the task you may get other ideas you want to explore. Always make sure you complete the part of the problem you are working on first. Some of your ideas may not work out. Do not discard them. All the work you do is important.

1. Looking for a path through the investigation

Once you feel that you know what you are looking for, you need to plan how to find it. Your plan might change as you proceed, but that does not matter. Make a note of the changes that you have made and your reasons. Include these in your write up.

Example
'I have a mug rack with four pegs on it. I have two yellow mugs and two blue ones. How many different ways are there to arrange the mugs on the rack?'

In this case you would look for one answer before going on to investigate other combinations of colours and pegs.

Can I simplify the problem in any way?

Example
'The maximum number of straight lines that connect five dots is ten. How many lines will connect 20 dots?'

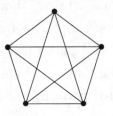

Start by considering how many lines connect two dots, then three, then four.

Can I break the problem up into smaller problems?

Example
'I am 40 and 40 can be written 6 + 7 + 8 + 9 + 10.
6, 7, 8, 9 and 10 are consecutive numbers.
Will I be able to write my age as the sum of consecutive numbers when I am 42?
Are there any ages that cannot be written in this way?'

6 + 7 + 8 + 9 + 10 is the sum of **five** consecutive numbers. Break the problem into smaller problems by looking at the sums of two consecutive numbers, then three consecutive numbers etc.

Can I already see a pattern?

If you see a pattern at any time, then point this out. It may be something very simple like 'the answers are all odd numbers'. Remember there may be several patterns to find.

Do I need to investigate lots of similar situations?

If your answer is yes, you need to give the method careful thought. In the example with the mugs you need to choose a different number of colours or a different number of pegs. Changing yellow and blue to red and green does not change the problem.

2. Organising your information

You need to do this methodically and accurately. You cannot look for patterns if you are missing information or if some of your results are wrong.

The way you record your results is important. You gain marks from well-organised tables and diagrams. They also help you to see patterns.

The problem on joining dots would have sketches showing different numbers of dots connected. It would have its results recorded in a table like this:

No of dots	1	2	3	4	5
No of lines	0	1	3	6	10

Suppose you have designed a questionnaire for a statistical project and have collected responses. You need to put the results in a table (or database) so that you can start to process them. You need to draw statistical diagrams to illustrate your data.

3. Looking for patterns and relationships

It is important to have accurate results. Suppose you got this result to the joining dots problem:

The second +2 spoils the pattern so the 5 needs to be checked.
The corrected pattern can be used to predict the next term:

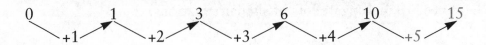

This prediction can be checked by drawing a diagram with 6 dots.
The correct numbers show a pattern so there must be a connection.
The rule is: 'multiply two adjacent numbers and divide by 2'.

No of dots	1	2	3	4	5	6
No of lines		$\frac{1 \times 2}{2} = 1$	$\frac{2 \times 3}{2} = 3$	$\frac{3 \times 4}{2} = 6$	$\frac{4 \times 5}{2} = 10$	$\frac{5 \times 6}{2} = 15$

The nth term, where n is the number of dots, is given by: $\frac{n(n+1)}{2}$
You should try to explain the pattern:

There are 4 lines joining one dot to all the
other dots and there are 5 dots.

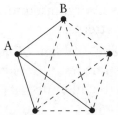

4×5 gives twice the answer, as the line joining
A to B is the same as the line joining
B to A, and so on.

4. Finding a conclusion

You will get to the stage where you feel that you have got an answer.
This may be the rule for a numerical pattern. You may have proved
or disproved a statistical hypothesis or made a model.

Can I see the reasons why my relationships come out the way they do?

Do my conclusions fit the task I started with or the ones I have developed?

If your answer to either of these questions is 'no' then you are not
ready to write a conclusion.
If you finish your work quite quickly, then you could consider

extending it. This should never be done until you have reached a conclusion for the original task.

5. Extending the work

Avoid doing the same type of work again, unless it helps you get a deeper understanding of the task.

It may be that you have been using a square grid for a task. A suitable extension might be to work on other shapes of grid. You might change a space and shape problem from one dimension to two or three dimensions or a numerical problem to sets of more difficult numbers. If the original problem has lots of rules, try breaking the rules one at a time.

6. Writing a report

Remember that you are not just writing about your work for your own teacher. Moderators and teachers from other schools may read your work. Show the reader exactly what you understand from the work that you have been doing. You could do this at different points throughout the work or altogether at the end.

Start with an introduction that states the initial problem. If it is a statistical investigation state your hypothesis clearly.

Describe your path through the investigation. Show your results with clear diagrams and tables and say what you did to get them.

Write about anything you have noticed or any ideas of your own that you have developed. Explain any predictions you have made.

Give a precise conclusion. Include any rough work at the end of your report.

Answers

NUMBER
SECTION 1
TYPES OF NUMBER

Pause – page 3

1 1, 2, 4, 8   **2** 1, 5, 25   **3** 1, 13
4 a) i) 1, 2, 3, 6, 9, 18   iii) 1, 23
       ii) 1, 3, 7, 21   iv) 1, 2, 4, 8, 16
    b) i) a prime number has only two factors
        – itself and 1.
        ii) 23
5 42, 170, 1500
6 2, 3, 5, 7, 11, 13, 17, 19, 23, 29
7 6, 12, 18, 24, 30, 36, 42
8 7, 14, 21, 28, 35, 42
9 a) 12, 15, 18   b) 12, 18
10 a) 7, 21   b) 7, 21   c) 7, 13, 19

Pause – page 4

1 a) 4   c) 6   e) 15
    b) 5   d) 12   f) 5
2 a) 24   c) 26   e) 70
    b) 30   d) 28   f) 108

Pause – page 5

1 1, 2, 5, 10, 25, 50
2 2, 5
3 $50 = 2 \times 5 \times 5 = 2 \times 5^2$
4 $400 = 10 \times 40$
       $= 2 \times 5 \times 40$
       $= 2 \times 5 \times 5 \times 2 \times 2 \times 2$
       $= 2^4 \times 5^2$

5 a) $36 = 2 \times 2 \times 3 \times 3 = 2^2 \times 3^2$
    b) $100 = 2 \times 2 \times 5 \times 5 \times = 2^2 \times 5^2$
    c) $48 = 2 \times 2 \times 2 \times 2 \times 3 = 2^4 \times 3$
    d) $25 = 5 \times 5 = 5^2$
    e) $125 = 5 \times 5 \times 5 = 5^3$
    f) $162 = 2 \times 3 \times 3 \times 3 \times 3 = 2 \times 3^4$
    g) $2700 = 2 \times 2 \times 3 \times 3 \times 3 \times 5 \times 5$
               $= 2^2 \times 3^3 \times 5^2$

Pause – page 6

1 1, 4, 9, 16, 25, 36, 49, 64, 81, 100, 121,
    144
2 4, 6, 8, 9, 10, 12, 14, 15, 16, 18, 20, 21,
    22, 24, 25, 26, 27, 28, 30, 32

3 1, 3, 6, 10, 15, 21, 28, 36, 45, 55
4 pentagonal numbers

Pause – page 7

1 a) $\sqrt{81} = 9$ g) $\sqrt{400} = 20$
    b) $\sqrt{49} = 7$ h) $\sqrt{900} = 30$
    c) $\sqrt{121} = 11$ i) $\sqrt{1600} = 40$
    d) $\sqrt{100} = 10$ j) $\sqrt{2500} = 50$
    e) $\sqrt{144} = 12$ k) $\sqrt{6400} = 80$
    f) $\sqrt{64} = 8$ l) $\sqrt{10\,000} = 100$

SECTION 2
ESTIMATION AND CALCULATION

Pause – page 9

	Estimate	Accurate answer
1	$20 \times 6 = 120$	108
2	$30 \times 9 = 270$	306
3	$7 \times 60 = 420$	413
4	$2 \times 90 = 180$	184
5	$8 \times 80 = 640$	608
6	$600 \times 9 = 5400$	5157
7	$60 \times 90 = 5400$	5130
8	$5 \times 600 = 3000$	3010
9	$60 \times 50 = 3000$	3150
10	$90 \times 400 = 36\,000$	34\,000
11	$80 \times 800 = 64\,000$	63\,200
12	$20 \times 300 = 6000$	6900
13	$500 \times 70 = 35\,000$	31\,500
14	$700 \times 3 = 2100$	2025
15	$9 \times 600 = 5400$	5004
16	$400 \times 60 = 24\,000$	22\,440
17	$1000 \times 90 = 90\,000$	89\,100
18	$800 \times 400 = 320\,000$	332\,800
19	$400 \times 600 = 240\,000$	226\,800
20	$2000 \times 500 = 1\,000\,000$	1\,170\,500

Pause – page 9

	Estimate	Accurate answer
1	$200 \times 50 = 10\,000$	10 530
2	$600 \times 30 = 18\,000$	14 846
3	$700 \times 40 = 28\,000$	29 568
4	$1000 \times 50 = 50\,000$	51 940
5	$700 \times 30 = 21\,000$	21 663
6	$800 \times 40 = 32\,000$	31 242
7	$200 \times 50 = 10\,000$	11 024
8	$1000 \times 20 = 20\,000$	24 244
9	$700 \times 30 = 21\,000$	16 875
10	$300 \times 90 = 27\,000$	29 993
11	$400 \times 200 = 80\,000$	84 777
12	$300 \times 400 = 120\,000$	140 193
13	$400 \times 200 = 80\,000$	82 418
14	$300 \times 300 = 90\,000$	91 980
15	$900 \times 300 = 270\,000$	253 344
16	$400 \times 200 = 80\,000$	70 034
17	$300 \times 800 = 240\,000$	233 496
18	$300 \times 100 = 30\,000$	35 643
19	$400 \times 800 = 32\,000$	345 978
20	$600 \times 200 = 120\,000$	145 314

Pause – page 10

1. a) 6 c) 79 e) 14
   b) 23 d) 10
2. a) 4 francs
   b) 20 francs
   c) 42 francs
   d) 101 francs
   e) 278 francs
3. a) 400
   b) 2484
   c) Financial Times
   d) Daily Mail
   e) i) Independent ii) 284

Pause – page 12

1	3	11	60	21	30
2	5	12	90	22	9
3	2	13	40	23	7
4	8	14	20	24	20
5	6	15	80	25	40
6	7	16	4	26	50
7	6	17	5	27	200
8	6	18	8	28	3000
9	7	19	8	29	600
10	8	20	50	30	90

Pause – page 13

	Estimate	Accurate answer
1	$700 \div 10 = 70$	55
2	$300 \div 30 = 10$	11
3	$900 \div 40 = 22.5$	22

	Estimate	Accurate answer
4	$800 \div 20 = 40$	43
5	$500 \div 20 = 25$	22
6	$400 \div 40 = 10$	11
7	$400 \div 30 = 13\cdot3$	13
8	$400 \div 40 = 10$	9
9	$400 \div 30 = 13\cdot3$	16
10	$800 \div 10 = 80$	56
11	$900 \div 30 = 30$	27 rem 5
12	$800 \div 40 = 20$	21 rem 21
13	$700 \div 20 = 35$	32 rem 17
14	$600 \div 30 = 20$	17 rem 33
15	$800 \div 10 = 80$	63 rem 7
16	$100 \div 20 = 5$	4 rem 15
17	$300 \div 20 = 15$	14 rem 8
18	$600 \div 40 = 15$	15 rem 17
19	$1000 \div 50 = 20$	21 rem 32
20	$1000 \div 40 = 25$	26 rem 21

Pause – page 14

1. a) £4, nothing left over
   b) 4 sweets, 2 left over
   c) 14 roses, 2 left over
   d) 2 mince pies, 4 left over
   e) 250 exercise books, none left over
   f) £25 000, nothing left over
   g) 19 marbles, 1 left over
   h) 9 mice, 30 left over
   i) 8 sandwiches, 2 left over
   j) 11 sandwiches, 16 left over

SECTION 3
DECIMALS

Pause – page 15

1. a) $2 + \frac{7}{10}$
   b) $3 + \frac{9}{10}$
   c) $70 + 5 + \frac{2}{10}$
   d) $7 + \frac{5}{10} + \frac{2}{100}$
   e) $100 + 30 + 8$
   f) $10 + 3 + \frac{8}{10}$
   g) $1 + \frac{3}{10} + \frac{8}{100}$
   h) $\frac{1}{10} + \frac{3}{100} + \frac{8}{1000}$
   i) $8 + \frac{7}{10} + \frac{5}{100}$
   j) $\frac{8}{10} + \frac{7}{100} + \frac{5}{1000}$
   k) $80 + 7 + \frac{5}{10}$
   l) $30 + 4 + \frac{1}{10} + \frac{5}{100}$
   m) $\frac{1}{1000}$
   n) $1 + \frac{1}{100}$
   o) $100 + 1 + \frac{1}{10} + \frac{1}{100}$

2. a) $4\cdot5$ g) $6\cdot15$
   b) $13\cdot2$ h) $7\cdot37$
   c) $348\cdot3$ i) $7\cdot653$
   d) $1\cdot9$ j) $750\cdot01$
   e) $17\cdot6$ k) $604\cdot5$
   f) $33\cdot1$ l) $530\cdot371$

3. a) 74, 0·47 c) 654, 0·456
   b) 91, 0·19 d) 872, 0·278

Pause – page 16

1 a)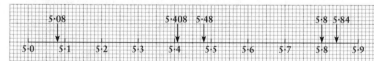

b) 1·08, 1·15, 1·23, 1·32, 1·48, 1·53, 1·63, 1·70, 1·85, 1·97

2 a)

b) 5·06, 5·17, 5·23, 5·33, 5·4, 5·51, 5·57, 5·61, 5·8, 5·9

3 a)

b) 40·07, 40·19, 40·23, 40·32, 40·4, 40·5, 40·68, 40·7, 40·86, 40·91

Pause – page 17

1 a)

b) 1·009, 1·013, 1·02, 1·032, 1·037, 1·041, 1·047, 1·052, 1·068, 1·07

2 a)

b) 4·956, 4·962, 4·972, 4·981, 4·99, 5·007, 5·015, 5·027, 5·048

3 a)

b) 15·308, 15·319, 15·328, 15·339, 15·340, 15·35, 15·363, 15·374, 15·38, 15·392

Pause – page 18

1

5·08, 5·408, 5·48, 5·8, 5·84

2

2·06, 2·1, 2·106, 2·16, 2·61

3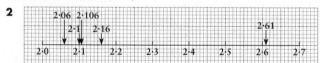

8·5, 8·517, 8·57, 8·705, 8·75

4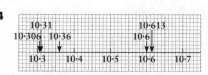

10·306, 10·31, 10·36, 10·6, 10·613

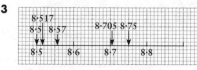

5

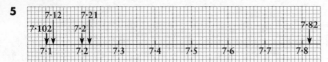

7·102, 7·12, 7·2, 7·21, 7·82

6

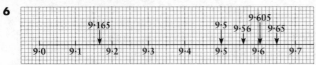

9·165, 9·5, 9·56, 9·605, 9·65

7

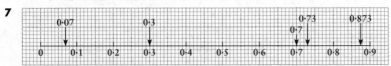

0·07, 0·3, 0·7, 0·73, 0·873

8

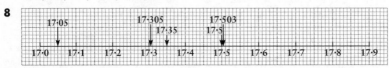

17·05, 17·305, 17·35, 17·5, 17·503

Pause – page 19

1 6, 2, 8, 7, 4, 5, 6, 2, 10, 1
2 3, 5, 9, 1, 5, 3, 7, 8, 7, 9
3 8, 1, 6, 3, 8, 10, 5, 6, 3, 2
4 2, 0, 2, 4, 4, 6, 7, 8, 8, 10
5 2, 2, 1, 4, 5, 6, 7, 9, 5, 10

Pause – page 20

1 1·6, 2·0, 1·2, 1·5, 1·7, 1·5, 1·1, 1·3, 1·9, 1·2
2 5·2, 5·6, 5·5, 5·8, 5·1, 5·3, 5·4, 5·6, 5·9, 5·2
3 5·3, 5·5, 5·6, 5·9, 5·0, 5·4, 5·4, 5·7, 6·0, 5·1
4 8·1, 8·9, 8·7, 8·7, 8·2, 8·2, 8·5, 8·3, 8·5, 8·8
5 8·0, 8·9, 8·8, 8·7, 8·2, 8·3, 8·4, 8·4, 8·5, 8·9
6 15·7, 15·1, 15·4, 15·9, 15·3, 15·2, 15·6,
15·8, 15·4, 15·5
7 15·8, 15·1, 15·5, 16·0, 15·2, 15·1, 15·7,
15·9, 15·3, 15·6

Pause – page 21

1 5·59, 5·51, 5·58, 5·52, 5·57, 5·53, 5·56,
5·54, 5·55, 5·55
2 15·40, 15·30, 15·39, 15·32, 15·38, 15·32,
15·36, 15·33, 15·36, 15·34
3 1·01, 1·07, 1·03, 1·05, 1·05, 1·07, 1·02,
1·01, 1·04, 1·04
4 5·01, 4·96, 4·98, 5·03, 5·05, 4·96, 4·97,
4·99, 5·02, 5·03
5 a) 1·73   e) 6·48   i) 4·24
b) 2·74   f) 12·25   j) 4·37
c) 4·47   g) 1·80   k) 4·76
d) 2·90   h) 9·49   l) 14·14

Pause – page 23

1 a) 12·45   c) 3·138   e) 47·05
b) 27·235   d) 0·6512   f) 45·04
2 a) 1·89   d) 3·465   g) 6·6
b) 3·16   e) 0·398   h) 0·39
c) 4·17   f) 0·014   i) 1·285

Pause – page 24

1 a) 49·6   c) 36·204   e) 112·554
b) 57·68   d) 51·4   f) 1·944
2 a) 7·35   c) 4·7125   e) 2·346
b) 5·8   d) 0·03175   f) 1·501166…
3 a) 0·2   c) 0·6   e) 0·0048
b) 0·48   d) 0·056   f) 0·00093
4 a) 4 c) 6   e) 0·3 g) 20 i) 0·5 k) 40
b) 2 d) 0·3 f) 30 h) 20 j) 20 l) 8

Pause – page 25

1 a) 170 mm   f) 78·5 mm
b) 17 mm   g) 89·3 mm
c) 1·7 mm   h) 12·43 mm
d) 89 mm   i) 14·7 mm
e) 93·2 mm   j) 10·47 mm
2 a) 245 cm   f) 155·7 cm
b) 200 cm   g) 120 cm
c) 45 cm   h) 570 cm
d) 1237·5 cm   i) 730 cm
e) 89·5 cm   j) 76·5 cm
3 a) 1237 g   f) 7892·3 g
b) 1657 g   g) 16 666·6 g
c) 7045 g   h) 100 g
d) 8230 g   i) 456 g
e) 7600 g   j) 7031·4 g

4 a) 6400 ml f) 1 ml
b) 6470 ml g) 0·1 ml
c) 6473 ml h) 1000·1 ml
d) 6473·5 ml i) 2·34 ml
e) 10 ml j) 0·25 ml

Pause – page 27
1 a) 1·7 cm f) 0·785 cm
b) 0·17 cm g) 7·85 cm
c) 0·017 cm h) 78·5 cm
d) 0·89 cm i) 0·0785 cm
e) 9·32 cm j) 0·06 cm
2 a) 2·45 m f) 0·01557 m
b) 2 m g) 0·012 m
c) 0·45 m h) 0·057 m
d) 0·123 m i) 73·55 m
e) 0·0895 m j) 86·52 m
3 a) 1·237 kg f) 78·923 kg
b) 1·657 kg g) 0·0166 kg
c) 7·045 kg h) 0·1 kg
d) 0·823 kg i) 0·01 kg
e) 0·076 kg j) 0·0073 kg
4 a) 0·0064 *l* f) 0·000 001 *l*
b) 0·00647 *l* g) 0·3 *l*
c) 0·006 *l* h) 0·07 *l*
d) 6·4735 *l* i) 5 *l*
e) 0·000 01 *l* j) 0·65 *l*

Pause – page 28
1 £116·20 **6** £99
2 £168 **7** £179·20
3 £56 **8** £81
4 £52·20 **9** £158·84
5 £212·40 **10** £114·21

Pause – page 29
1 £201·60 **6** £225·60
2 £220·80 **7** £288
3 £213·60 **8** £283·20
4 £259·20 **9** £290·40
5 £235·20 **10** £259·20

Pause – page 30
1 10 **6** 4 **11** 10 **16** 0·4 **21** 0·03
2 20 **7** 2 **12** 10 **17** 0·7 **22** 20
3 90 **8** 4 **13** 30 **18** 0·004 **23** 3000
4 300 **9** 10 **14** 50 **19** 0·003 **24** 30
5 800 **10** 6 **15** 50 **20** 0·002 **25** 0·01

Pause – page 31
1 $10 \times 20 = 200$ **6** $10 \times 0·4 = 4$
2 $90 \times 4 = 360$ **7** $0·4 \times 0·7 = 0·28$
3 $300 \times 10 = 3000$ **8** $0·004 \times 20 = 0·08$
4 $6 \times 2 = 12$ **9** $0·003 \times 10 = 0·03$
5 $0·7 \times 50 = 35$ **10** $20 \times 0·01 = 0·2$

11 $20 \div 10 = 2$ **16** $0·03 \div 0·002 = 15$
12 $30 \div 2 = 15$ **17** $10 \div 0·4 = 25$
13 $50 \div 10 = 5$ **18** $30 \div 300 = 0·1$
14 $3000 \div 10 = 300$ **19** $300 \div 800 = 0·375$
15 $0·7 \div 0·4 = 1·75$ **20** $0·7 \div 20 = 0·035$

Pause – page 31
1 a) £48 b) 422
2 a) £90·16 b) £3426·08
3 4·73 *l*
4 a) 313·5 hours b) 765 *l*

Pause – page 34
1 2·25 lb, 4·5 lb, 11·25 lb, 33·75 lb, 45 lb
2

Pints	1	2	3	4	5	6	7	8	9	10
Litres	0·57	1·14	1·71	2·29	2·86	3·43	4	4·57	5·14	5·71

Pints	10	20	30	40	50	60	70	80	90	100
Litres	5·71	11·43	17·14	22·86	28·57	34·29	40	45·71	51·43	57·14

3 a) 100 miles c) 60 miles e) 92·5 miles
b) 130 miles d) 95 miles f) 128·75 miles
4

Centimetres	5	10	15	20	25	30
Inches	2·0	3·9	5·9	7·9	9·8	11·8

Centimetres	30·5	61·0	91·4	121·9	152·4	182·9
Feet	1	2	3	4	5	6

Pause – page 36
1 6·111 **6** 0·592
2 12·667 **7** 9·705
3 50·625 **8** 2·279
4 1·152 **9** 15·242
5 3·913 **10** 0·274

Pause – page 38
1 £16·62, £19·06, £17·52, £28·27, £21·30
2 £39·96, £3·64, £27·20, £38·68, £84·31
3 £118·80, £126, £139·50, £160·74, £178·61
4 0·5 kg, 0·6 kg, 0·9 kg, 1·3 kg, 0·83 kg
5

Parts	Parts ÷ 2	Labour	Labour + parts ÷ 2	Call out fee (£25)	VAT (17·5%)	Special reduction (£12)
£32	£16	£80	£96	£121	£142·18	£130·18
£48	£24	£89	£113	£138	£162·15	£150·15
£64	£32	£32	£64	£89	£104·58	£92·58
£12	£6	£50	£56	£81	£95·18	£83·18
£86	£43	£65	£108	£133	£156·28	£144·28
£25	£12·50	£37	£49·50	£74·50	£87·54	£75·54
£59	£29·50	£95	£124·50	£149·50	£175·66	£163·66
£102	£51	£86	£137	£162	£190·35	£178·35
£155	£77·50	£43	£120·50	£145·50	£170·96	£158·96
£99	£49·50	£99	£148·50	£173·50	£203·86	£191·86
£128	£64	£102	£166	£191	£224·43	£212·43

SECTION 4
FRACTIONS

Pause – page 40

1 a) $\frac{7}{10}$ b) $\frac{4}{10}=\frac{2}{5}$ c) $\frac{7}{10}$ d) $\frac{7}{10}$ e) $\frac{3}{10}$

2 a) shaded $\frac{1}{8}$ d) shaded $\frac{1}{9}$
    unshaded $\frac{7}{8}$ unshaded $\frac{8}{9}$

    b) shaded $\frac{1}{5}$ e) shaded $\frac{5}{8}$
    unshaded $\frac{4}{5}$ unshaded $\frac{3}{8}$

    c) shaded $\frac{1}{7}$ f) shaded $\frac{1}{6}$
    unshaded $\frac{6}{7}$ unshaded $\frac{5}{6}$

Pause – page 42

1 $\frac{1}{3},\frac{2}{6}$ **6** $\frac{1}{4},\frac{4}{16}$

2 $\frac{1}{2},\frac{4}{8}$ **7** $\frac{1}{10},\frac{10}{100}$

3 $\frac{2}{3},\frac{6}{9}$ **8** $\frac{1}{2},\frac{50}{100}$

4 $\frac{1}{4},\frac{2}{8}$ **9** $\frac{75}{100},\frac{15}{20},\frac{3}{4}$

5 $\frac{4}{16},\frac{2}{8}$ **10** $\frac{3}{5},\frac{6}{10},\frac{12}{20},\frac{60}{100}$

Pause – page 43

1 a) $4\frac{3}{5}$ e) $4\frac{3}{4}$ i) $2\frac{9}{12}=2\frac{3}{4}$

    b) $1\frac{2}{7}$ f) $5\frac{2}{5}$ j) $9\frac{4}{5}$

    c) $7\frac{1}{2}$ g) $1\frac{8}{11}$

    d) $5\frac{2}{3}$ h) $3\frac{3}{8}$

2 a) $\frac{15}{4}$ c) $\frac{26}{3}$ e) $\frac{31}{4}$ g) $\frac{45}{16}$ i) $\frac{35}{3}$

    b) $\frac{19}{2}$ d) $\frac{5}{4}$ f) $\frac{77}{8}$ h) $\frac{165}{32}$ j) $\frac{43}{4}$

Pause – page 44

1 a) $\frac{2}{5}=\frac{4}{10}$ e) $\frac{12}{17}=\frac{36}{51}$ i) $\frac{5}{9}=\frac{25}{45}$

    b) $\frac{5}{8}=\frac{15}{24}$ f) $\frac{1}{9}=\frac{3}{27}$ j) $\frac{8}{11}=\frac{40}{55}$

    c) $\frac{7}{6}=\frac{21}{18}$ g) $\frac{11}{12}=\frac{44}{48}$ k) $\frac{7}{10}=\frac{70}{100}$

    d) $\frac{3}{7}=\frac{9}{21}$ h) $\frac{4}{5}=\frac{28}{35}$ l) $\frac{1}{9}=\frac{11}{99}$

Pause – page 45

1 a) $\frac{5}{7}$ c) $\frac{2}{3}$ e) $\frac{4}{5}$ g) $\frac{13}{18}$ i) $\frac{1}{3}$

    b) $\frac{5}{12}$ d) $\frac{4}{5}$ f) $\frac{7}{13}$ h) $\frac{5}{7}$ j) $\frac{2}{3}$

Pause – page 46

1 a) $\frac{1}{3}$ d) $\frac{5}{9}$ g) $\frac{7}{10}$ j) $\frac{5}{11}$

    b) $\frac{4}{5}$ e) $\frac{4}{6}$ h) $\frac{1}{2}$ k) $\frac{9}{11}$

    c) $\frac{5}{6}$ f) $\frac{5}{7}$ i) $\frac{2}{3}$ l) $\frac{3}{5}$

2 Maths   **6** 5B
3 Biology   **7** Arsenal
4 The second bay $\left(\frac{5}{30}\right)$ **8** The first batch $\left(\frac{3}{10}\right)$
5 Susan   **9** John

Pause – page 47

1 a) $\frac{1}{6}$ b) $\frac{5}{12}$ c) $\frac{1}{5}$ d) $\frac{2}{3}$ e) $\frac{1}{2}$ f) $\frac{1}{4}$

2 a) $\frac{1}{10}$ b) $\frac{5}{6}$ c) $\frac{7}{10}$ d) $\frac{7}{12}$ e) $\frac{3}{4}$ f) $\frac{1}{3}$

3 a) $\frac{1}{24}$ c) $\frac{1}{4}$ e) $\frac{1}{12}$ g) $\frac{2}{3}$ i) $\frac{9}{16}$

    b) $\frac{1}{6}$ d) $\frac{1}{3}$ f) $\frac{1}{2}$ h) $\frac{3}{4}$ j) $\frac{7}{32}$

4 a) $\frac{1}{4}$ c) $\frac{1}{10}$ e) $\frac{17}{20}$ g) $\frac{12}{25}$ i) $\frac{1}{50}$

    b) $\frac{2}{5}$ d) $\frac{1}{2}$ f) $\frac{3}{4}$ h) $\frac{1}{25}$ j) $\frac{27}{100}$

Pause – page 48

1 a) 45   c) i) 50   ii) 100
    b) i) 12   ii) 36   d) i) 13   ii) 39
    e) 80   g) 9   i) 140
    f) 16   h) 63   j) 121·5

2 3125   **3** about 40

Pause – page 50

1 a) $\frac{9}{10}$ c) $\frac{5}{8}$ e) $\frac{9}{10}$ g) $\frac{11}{12}$

    b) $\frac{7}{12}$ d) $\frac{11}{12}$ f) $\frac{29}{35}$ h) $1\frac{17}{24}$

2 a) $\frac{1}{6}$ c) $\frac{1}{12}$ e) $\frac{1}{8}$ g) $\frac{1}{6}$

    b) $\frac{1}{12}$ d) $\frac{1}{2}$ f) $\frac{4}{35}$ h) $1\frac{3}{8}$

Pause – page 51

1 a) $\frac{1}{6}$ c) $\frac{1}{9}$ e) $\frac{1}{2}$ g) $\frac{1}{2}$

    b) $\frac{3}{8}$ d) $\frac{1}{3}$ f) $\frac{5}{8}$ h) $1\frac{1}{6}$

2 a) 2   c) 2   e) $\frac{1}{6}$ g) 6

    b) 2   d) 3   f) 4   h) 28

SECTION 5
PERCENTAGES

Pause – page 52

1 a) $\frac{1}{4}$ b) $\frac{99}{100}$ **2** a) 50% b) 10%

    b) $\frac{3}{5}$ g) $\frac{2}{5}$ b) 85% g) 24%

    c) $\frac{17}{25}$ h) $\frac{3}{20}$ c) 70% h) 12%

    d) $\frac{9}{10}$ i) $\frac{8}{25}$ d) 55% i) 50%

    e) $\frac{16}{25}$ j) $\frac{19}{20}$ e) 48% j) 44%

3 $\frac{3}{10}$

Pause – page 53

1 a) 39   d) 769·5   g) £22·72
    b) 87   e) 40·96   h) £6·75
    c) 187   f) 2475   i) 14·4 kg

2 £347·20   **4** £10 260
3 12 320   **5** £32.20

Pause – page 55

1 a) 32%   d) 35%   g) 34%   j) 40%
    b) 40%   e) 15%   h) 75%   k) 12·5%
    c) 2·5%   f) 75%   i) 80%   l) 75%

2 36%   47%   65%   74%   78%
    80%   85%   93%   54%   88%
    70%   18%   14%   57%   60%
    98%   73%   59%   49%   50%

3 a) 15%, 16·7%, 17·5%
    b) The first one (3 out of 20)
4 The first school (6 out of 10)
5 a) 6·9%   b) 17·9%   c) 18·2%

6 a) $\frac{1}{8}$ b) 12·5%

7 a) 5% b) 3 extra oranges c) $\frac{3}{18} = \frac{1}{6}$

8 a) £80 b) 15%

9 a) 95 p b) £10·54 c) 9% d) 5·2%

10 a) $12 880 b) £7235·96 c) 24·4%

Pause – page 60

1 a) 0·75 e) 0·5 i) 0·66…
  b) 0·8 f) 0·2 j) 0·833…
  c) 0·875 g) 0·25
  d) 0·9 h) 0·1875

2 a) 75% e) 50% i) 66·66…%
  b) 80% f) 20% j) 83·33…%
  c) 87·5% g) 25%
  d) 90% h) 18·75%

3 a) 10% e) 64% i) 50%
  b) 80% f) 8% j) 37·5%
  c) 30% g) 0·5%
  d) 35% h) 5%

4 a) $\frac{1}{10}$ c) $\frac{3}{10}$ e) $\frac{16}{25}$ g) $\frac{1}{200}$ i) $\frac{1}{2}$
  b) $\frac{4}{5}$ d) $\frac{7}{20}$ f) $\frac{2}{25}$ h) $\frac{1}{20}$ j) $\frac{3}{8}$

5 a) $80\% = \frac{4}{5} = 0·8$ f) $55\% = \frac{11}{20} = 0·55$
  b) $32\% = \frac{8}{25} = 0·32$ g) $40\% = \frac{2}{5} = 0·4$
  c) $25\% = \frac{1}{4} = 0·25$ h) $48\% = \frac{12}{25} = 0·48$
  d) $70\% = \frac{7}{10} = 0·7$ i) $65\% = \frac{13}{20} = 0·65$
  e) $98\% = \frac{49}{50} = 0·98$ j) $50\% = \frac{1}{2} = 0·5$

6 37·5% 7 a) $\frac{54}{1000} = \frac{27}{500}$ b) 0·054

8 a) i) 0·125 ii) 12·5% b) £3783·81

Pause – page 61

1 a) $\frac{1}{2} = 0·5$ e) $\frac{1}{16} = 0·063$
  b) $\frac{1}{4} = 0·25$ f) $\frac{1}{20} = 0·05$
  c) $\frac{1}{6} = 0·167$ g) $\frac{1}{40} = 0·025$
  d) $\frac{1}{11} = 0·091$ h) $\frac{1}{30} = 0·033$

Pause – page 62

1 a) £595·51 c) £848·72 e) £2121·80
  b) £1166·40 d) £377·91 f) £795·91

2 a) £145·80 c) £105·47 e) £109·85
  b) £256 d) £17·5 f) £4423·68

3 a) £88·20 b) £1822·50

4 a) £43·18 b) £598·03

5 £29 866·88

Pause – page 64

1 a) £200 e) £400 i) £12·50
  b) £300 f) £50 j) £45·63
  c) £450 g) £56·25
  d) £80 h) £112·50

2 a) £200 e) £180 i) £428·57
  b) £400 f) £210 j) £366·67
  c) £150 g) £250
  d) £140 h) £285·71

3 £3 050 000

4 £428·57

5 a) 28·9% b) £7·99

Pause – page 65

1 a) £341 b) £61
2 a) £221.80 b) £31·80
3 a) £371 b) £21
4 a) £427·80 b) £7·80
5 a) £3625 b) £125
6 a) £1373·54 b) £123·54
7 a) £419·90 b) £20·90
8 a) £198 b) £18
9 a) £1355·19 b) £106·19

SECTION 6
RATIO

Pause – page 66

1 a) 2:3 e) 6:7 i) 1:10 m) 5:7
  b) 3:4 f) 7:4 j) 8:17 n) 8:73
  c) 1:5 g) 3:5 k) 1:100 o) 1:4
  d) 5:6 h) 2:7 l) 5:6 p) 1:5

2 11:9 3 41:9 4 7:1

Pause – page 67

1 a) 12 kg, 24 kg g) 3 kg, 33 kg
  b) 27 kg, 9 kg h) 15 kg, 21 kg
  c) 30 kg, 6 kg i) 34 kg, 2 kg
  d) 20 kg, 16 kg j) 10 kg, 26 kg
  e) 8 kg, 28 kg k) 6 kg, 12 kg, 18 kg
  f) 32 kg, 34 kg l) 8 kg, 12 kg, 16 kg

2 a) £75, £25 g) £10, £90
  b) £40, £60 h) £70, £30
  c) £20, £80 i) £95, £5
  d) £62·50, £37·50 j) £15, £85
  e) £87·50, £12·50 k) £30, £30, £40
  f) £50, £50 l) £10, £25, £65

3 £9375, £15 625

4 £6·20, £9·30. The shares should take account of the amount of time each puts in working on the stall, as well as the amount of money.

Pause – page 69

1	a)	b)	c)	d)	e)
flour	400 g	600 g	800 g	300 g	500 g
lard	100 g	150 g	200 g	75 g	125 g
butter	100 g	150 g	200 g	75 g	125 g
apples	1000 g	1500 g	2000 g	750 g	1250 g
	f)	g)	h)	i)	j)
flour	100 g	150 g	700 g	900 g	50 g
lard	25 g	37·5 g	175 g	225 g	12·5 g
butter	25 g	37·5 g	175 g	225 g	12·5 g
apples	250 g	375 g	1750 g	2250 g	125 g

2 a) 448 km c) 720 km e) 99·2 km
  b) 1120 km d) 56 km

3 a) 150 miles c) 25 miles e) 202.5 miles
   b) 500 miles d) 60 miles
4 a) 15 books d) 30 books g) 24 books
   b) 20 books e) 35 books h) 28 books
   c) 25 books f) 18 books
5   pineapple   lemonade
   a) 2 l   8 l
   b) 3 l   12 l
   c) 4 l   16 l
   d) 1·5 l   6 l
   e) 0·5 l   2 l
   f) 0·6 l   2·4 l
   g) 1·7 l   6·7 l
   h) 2·3 l   9·3 l
   i) 1·3 l   5·3 l
   j) 1·4 l   5·7 l

SECTION 7
DIRECTED NUMBERS

Pause – page 70

1 a $= -43°C$ e $= 5°C$ i $= 28°C$
   b $= -27°C$ f $= 12°C$ j $= 36°C$
   c $= -14°C$ g $= 19°C$
   d $= -8°C$ h $= 21°C$

Pause – page 72

1 25°C   **4** −15°C   **7** 25°C
2 35°C   **5** 25°C   **8** 10°C
3 10°C   **6** 5°C
9 a) −30°C, −4°C, −2°C, 14°C, 25°C
   b) −10°C, −1°C, 7°C, 16°C, 18°C
   c) −12°C, −5°C, 0°C, 8°C, 9°C
   d) −14°C, −8°C, −5°C, 4°C, 7°C
   e) −20°C, −13°C, −10°C, −6°C, 0°C
10 a) 12°C   **11** a) 18°C
   b) 8°C   b) 6°C
   c) 12°C   c) −3°C
   d) 14°C   d) −4°C
   e) 25°C   e) −12°C
   f) 12°C   f) −7°C

Pause – page 73

1 $8 + 5 = 13$ $8 - 5 = 3$
   $8 + -5 = 3$ $-8 - 5 = -13$
   $-8 + 5 = -3$ $8 - -5 = 13$
   $-8 + -5 = -13$ $-8 - -5 = -3$
2 $3 + 7 = 10$ $3 - 7 = -4$
   $3 + -7 = -4$ $3 - -7 = 10$
   $-3 + 7 = 4$ $-3 - 7 = -10$
   $-3 + -7 = -10$ $-3 - -7 = 4$

Pause – page 74

1 $13 + 9 = 22$ $13 - 9 = 4$
   $13 + -9 = 4$ $13 - -9 = 22$
   $-13 + 9 = -4$ $-13 - 9 = -22$
   $-13 + -9 = -22$ $-13 - -9 = -4$

2 $5 + 23 = 28$ $5 - 23 = -18$
   $5 + -23 = -18$ $5 - -23 = 28$
   $-5 + 23 = 18$ $-5 - 23 = -28$
   $-5 + -23 = -28$ $-5 - -23 = 18$

Pause – page 75

1 $10 \times 5 = 50$ $10 \times -5 = -50$
   $10 \div 5 = 2$ $10 \div -5 = -2$
   $-10 \times 5 = -50$ $-10 \times -5 = 50$
   $-10 \div 5 = -2$ $-10 \div -5 = 2$
2 $12 \times 8 = 96$ $12 \times -8 = -96$
   $12 \div 8 = 1·5$ $12 \div -8 = -1·5$
   $-12 \times 8 = -96$ $-12 \times -8 = 96$
   $-12 \div 8 = -1·5$ $-12 \div -8 = 1·5$
3 $6·5 \times 5 = 32·5$ $6·5 \times -5 = -32·5$
   $6·5 \div 5 = 1·3$ $6·5 \div -5 = -1·3$
   $-6·5 \times 5 = -32·5$ $-6·5 \times -5 = 32·5$
   $-6·5 \div 5 = -1·3$ $-6·5 \div -5 = 1·3$
4 $1·2 \times 0·4 = 0·48$ $1·2 \times -0·4 = -0·48$
   $1·2 \div 0·4 = 3$ $1·2 \div -0·4 = -3$
   $-1·2 \times 0·4 = -0·48$ $-1·2 \times -0·4 = 0·48$
   $-1·2 \div 0·4 = -3$ $-1·2 \div -0·4 = 3$

SECTION 8
POWERS AND STANDARD FORM

Pause – page 76

1 243   **5** 2401   **9** 0·81
2 125   **6** 16384   **10** 1·61051
3 64   **7** 729
4 256   **8** 15·625

Pause – page 77

1 a) 81   e) 400   i) 0·36
   b) 144   f) 225   j) 0·0625
   c) 324   g) 56·25
   d) 121   h) 68·89
2 a) 64   e) 10 000 i) 4096
   b) 64   f) 343   j) 15 625
   c) 243   g) 6561
   d) 216   h) 1331

Pause – page 78

1 a) 15 625   f) 128   k) $\frac{1}{64}$
   b) 7776   g) $\frac{1}{16}$ l) 1
   c) 100   h) $\frac{1}{5}$ m) 1
   d) 1000   i) 10   n) 35
   e) 10 000   j) 1   o) $\frac{1}{256}$
2 a) 2^5 e) 10^5 i) 3^0 m) 10^5
   b) 3^9 f) 10^6 j) 4^0 n) 10^0
   c) 4^{10} g) 3^4 k) 2^3 o) 10^{-2}
   d) 2^{11} h) 5^5 l) 10^1 p) 10^5

Pause – page 79

1 a) $3{\cdot}7 \times 10^4$ e) $6{\cdot}5 \times 10^8$ i) $6{\cdot}5 \times 10^4$
   b) $3{\cdot}7 \times 10^3$ f) $6{\cdot}5 \times 10^7$ j) $6{\cdot}5 \times 10^3$
   c) $3{\cdot}7 \times 10^5$ g) $6{\cdot}5 \times 10^6$ k) $6{\cdot}5 \times 10^2$
   d) $1{\cdot}5 \times 10^5$ h) $6{\cdot}5 \times 10^5$ l) $6{\cdot}5 \times 10$

2 a) 200 000   g) 8300
   b) 300 000 000   h) 472 000
   c) 17 000   i) 6 720 000
   d) 3 450 000 000   j) 1 990 000 000 000
   e) 1 600 000 000   k) 730
   f) 3800   l) 80 900 000 000

3 a) $3{\cdot}7 \times 10^{-1}$ g) $6{\cdot}5 \times 10^{-1}$
   b) 4×10^{-3} h) $6{\cdot}5 \times 10^{-2}$
   c) $4{\cdot}5 \times 10^{-6}$ i) $6{\cdot}5 \times 10^{-3}$
   d) $5{\cdot}6 \times 10^{-8}$ j) $6{\cdot}5 \times 10^{-4}$
   e) $5{\cdot}78 \times 10^{-4}$ k) $6{\cdot}5 \times 10^{-5}$
   f) 1×10^{-7} l) $6{\cdot}5 \times 10^{-6}$

4 a) 0·000 02
   b) 0·000 000 03
   c) 0·000 17
   d) 0·000 000 003 45
   e) 0·000 000 001 6
   f) 0·0038
   g) 0·0083
   h) 0·000 047 2
   i) 0·000 006 73
   j) 0·000 000 000 001 99
   k) 0·073
   l) 0·000 000 000 809

5 a) The lifetime of an omega particle is $1{\cdot}1 \times 10^{-10}$ seconds.
   b) The mass of the Earth is $5{\cdot}967 \times 10^{24}$ kg.
   c) The average distance of Uranus from the Sun is $2{\cdot}869 \times 10^9$ km.
   d) A large orange contains $1{\cdot}6 \times 10^{-2}$ grams of vitamin C.
   e) The moon orbits at an average distance of $3{\cdot}84 \times 10^5$ km from the Earth.
   f) A light year, which is the distance travelled by light in one year is equal to $9{\cdot}4605 \times 10^{15}$ metres.

Pause – page 81

1 a) 4×10^2, 400
   b) 7×10^5, 700 000
   c) $8{\cdot}4 \times 10^6$, 8 400 000
   d) 9×10^{-3}, 0·009
   e) $1{\cdot}2 \times 10^{-4}$, 0·000 12
   f) 3×10^{-10}, 0·000 000 000 3
   g) $6{\cdot}4 \times 10^{-1}$, 0·64
   h) $7{\cdot}2 \times 10^{-5}$, 0·000 072
   i) $1{\cdot}9 \times 10^{-2}$, 0·019
   j) $5{\cdot}6 \times 10^2$, 560
   k) $6{\cdot}7 \times 10$, 67
   l) $1{\cdot}3 \times 10^{-5}$, 0·000 013

Pause – page 82

1 a) 85 000
   b) 225 000 000
   c) 48 750 000
   d) 0·0112
   e) 0·000 461 5
   f) 60 800 000 000 000
   g) 345 320 000
   h) 4750
   i) 830 000
   j) 0·000 000 000 368 55
   k) 16·8
   l) 9175·3
   m) 13 157 895
   n) 0·000 000 000 083
   o) 0·000 096 9

Rewind – page 82

1 a) 1, 2, 3, 4, 6, 8, 12, 24
   b) 1, 3, 17, 51
   c) 1, 2, 4, 5, 8, 10, 20, 25, 40, 50, 100, 200

2 a) 2, 3, 5   b) 2, 3, 7   c) 3, 5, 7

3 a) $2 \times 2 \times 3 \times 5 = 2^2 \times 3 \times 5$
   b) $2 \times 3 \times 5 \times 7$
   c) $2 \times 2 \times 2 \times 3 \times 3 \times 3 \times 13 = 2^3 \times 3^3 \times 13$

4 1, 4, 9, 25, 36, 49, 64, 81, 100, 121, 144

5 4, 6, 8, 9, 10, 12, 14, 15, 16, 18, 20, 21

6 1, 3, 6, 10, 15, 21, 28, 36, 45, 55, 66, 78

7 a) 9   b) 7   c) 11

8 a) 80, 130, 270, 350, 480, 590, 610, 770, 840, 950
   b) 100, 100, 300, 300, 500, 600, 600, 800, 800, 900

9 a) 900, 1260, 2500, 3790, 4520, 5630, 6170, 7380, 8910, 9170
   b) 900, 1300, 2500, 3800, 4500, 5600, 6200, 7400, 8900, 9200
   c) 1000, 1000, 3000, 4000, 5000, 6000, 6000, 7000, 9000, 9000

	Estimate	Accurate answer
10 a)	200 + 300 = 500	579
b)	700 + 300 = 1000	960
c)	300 − 300 = 0	77
d)	500 − 100 = 400	352
e)	20 × 60 = 1200	1288
f)	300 × 90 = 27 000	28 449
g)	1000 ÷ 30 = 33	54
h)	700 ÷ 10 = 70	52·2

11 a) 2·6, 2·6, 3·4, 3·7, 4·0, 5·4, 5·5, 6·7, 7·8, 9·1
   b) 2·2, 2·28, 3·21, 3·22, 3·25, 3·26, 3·27, 3·29, 3·3
   c) 1·5, 1·56, 1·562, 1·563, 1·564, 1·567, 1·567, 1·568, 1·569, 1·57

12 a) 92 d) 38·9 g) 10·9
  b) 11·9 e) 3·3 h) 2·5
  c) 9·2 f) 6·5

13 a) 118·53 c) 190·63 e) 621·44
  b) 620·9 d) 20·49 f) 4954·89

14 a) 807 c) 7287·066
  b) 1053·9 d) 51·75

15 a) 2·1 e) 0·1 i) 965·3
  b) 3·2 f) 2 j) 9·04
  c) 0·36 g) 20 k) 197
  d) 1·8 h) 500 l) 99·988

16 a) 166·5 c) 1412·35 e) 9·5
  b) 444·6 d) 12·478 f) 4·653

17 a) 797·41 c) 94·6192 e) 16·79
  b) 2551·5 d) 31·5 f) 20·08

18 a) i) 33·1 ii) 19·4 iii) 3·6 iv) 1·8
  b) i) 30 ii) 20 iii) 4 iv) 2

19 a) 1561·15 c) 1·06
  b) 33·14 d) 0·73

20 a) $2\frac{2}{5}$ b) $3\frac{1}{10}$ c) $4\frac{6}{7}$ d) $2\frac{1}{3}$ e) $1\frac{3}{4}$

21 a) $\frac{7}{4}$ b) $\frac{11}{3}$ c) $\frac{15}{2}$ d) $\frac{133}{11}$ e) $\frac{41}{6}$

22 a) $\frac{2}{5}=\frac{6}{15}$ d) $\frac{3}{7}=\frac{12}{28}$ g) $\frac{11}{12}=\frac{55}{60}$
  b) $\frac{5}{8}=\frac{10}{16}$ e) $\frac{12}{17}=\frac{48}{68}$ h) $\frac{4}{5}=\frac{52}{65}$
  c) $\frac{7}{6}=\frac{28}{24}$ f) $\frac{1}{9}=\frac{2}{18}$

23 a) $\frac{5}{6}$ c) $\frac{2}{5}$ e) $\frac{2}{3}$ g) $\frac{13}{15}$
  b) $\frac{5}{14}$ d) $\frac{2}{3}$ f) $\frac{7}{15}$

24 a) $\frac{3}{8}$ b) $\frac{3}{5}$ c) $\frac{5}{60}$

25 Biology

26 a) $\frac{1}{8}$ d) $\frac{3}{8}$ g) $\frac{7}{12}$
  b) $\frac{5}{24}$ e) $\frac{5}{12}$ h) $\frac{19}{48}$
  c) $\frac{1}{4}$ f) $\frac{1}{2}$ i) $\frac{25}{72}$

27 a) $\frac{1}{4}$ c) $\frac{19}{20}$ e) $\frac{21}{25}$
  b) $\frac{1}{2}$ d) $\frac{3}{4}$ f) $\frac{2}{25}$

28 a) 150 c) 108 e) 200 g) 36
  b) 35 d) 9 f) 32 h) 84

29 a) $\frac{1}{4}$ b) $\frac{1}{2}$ c) $\frac{14}{25}$ d) $\frac{47}{50}$ e) $\frac{8}{25}$

30 a) 50% c) 30% e) 72%
  b) 75% d) 85%

31 a) 36 c) 221 e) 51·84
  b) 138 d) 540 f) 9900

32 a) 360 c) £341·55 e) £19 972·44
  b) £92 d) £26 250 f) £1950

33 a) 304 c) £248 e) £43 240
  b) £80·75 d) £37 050 f) £968

34 a) 64% e) 7% i) 40%
  b) 60% f) 93·75% j) 60%
  c) 24·5% g) 45·2%
  d) 30% h) 25%

35 £612·52

36 £44

37

Fraction	Percentage	Decimal
$\frac{1}{2}$	50%	0·5
$\frac{1}{4}$	25%	0·25
$\frac{3}{4}$	75%	0·75
$\frac{1}{5}$	20%	0·2
$\frac{1}{10}$	10%	0·1
$\frac{3}{20}$	15%	0·15
$\frac{3}{5}$	60%	0·6
$\frac{13}{25}$	52%	0·52
$\frac{1}{25}$	4%	0·04
$\frac{1}{20}$	5%	0·05

38 a) 2 : 5 c) 1 : 6 e) 4 : 7 g) 11 : 21
  b) 2 : 3 d) 5 : 7 f) 1 : 2 h) 1 : 10

39 a) 10 kg, 30 kg d) 35 kg, 5 kg
  b) 32 kg, 8 kg e) 28 kg, 12 kg
  c) 25 kg, 15 kg

40 a) £90, £30 d) £75, £45
  b) £48, £72 e) £75, £30, £15
  c) £24, £96

41 a) 75 books b) 80 cm

42 $8 + 15 = 23$ $8 + -15 = -7$
  $-8 + 15 = 7$ $-8 + -15 = -23$
  $8 - 15 = -7$ $8 - -15 = 23$
  $-8 - 15 = -23$ $-8 - -15 = 7$
  $7 + 12 = 19$ $7 + -12 = -5$
  $-7 + 12 = 5$ $-7 + -12 = -19$
  $7 - 12 = -5$ $7 - -12 = 19$
  $-7 - 12 = -19$ $-7 - -12 = 5$

43 $2 \times 5 = 10$ $2 \div 5 = 0\cdot4$
  $-2 \times 5 = -10$ $-2 \div 5 = -0\cdot4$
  $2 \times -5 = -10$ $2 \div -5 = -0\cdot4$
  $-2 \times -5 = 10$ $-2 \div -5 = 0\cdot4$
  $2\cdot5 \times 10 = 25$ $2\cdot5 \div 10 = 0\cdot25$
  $-2\cdot5 \times 10 = -25$ $-2\cdot5 \div 10 = -0\cdot25$
  $2\cdot5 \times -10 = -25$ $2\cdot5 \div -10 = -0\cdot25$
  $-2\cdot5 \times -10 = 25$ $-2\cdot5 \div -10 = 0\cdot25$

44 a) 4096 e) $\frac{1}{10}$ i) 10
  b) 32 f) $\frac{1}{128}$ j) 1
  c) 100 000 g) $\frac{1}{9}$ k) $\frac{1}{64}$
  d) 10 000 h) $\frac{1}{10000}$ l) 1

45 a) 2^7 c) 2^{-2} e) 10^8 g) 10^0
  b) 3^9 d) 10^9 f) 5^0 h) 10^1

46 a) 3×10^3 d) $3\cdot679 \times 10^2$
  b) $4\cdot5 \times 10^4$ e) $2\cdot34 \times 10^{10}$
  c) $3\cdot46 \times 10^2$

47 a) 3 100 000 c) 191 000 000
  b) 56 700 d) 50 670 000 000

48 a) 3×10^{-5} c) 1×10^{-6}
  b) $1\cdot7 \times 10^{-2}$ d) $9\cdot009 \times 10^{-3}$

49 a) 0·000 003 1   c) 0·000 000 019 1
    b) 0·000 567   d) 0·000 000 000 506 7
50 a) 0·0069   d) 0·000 000 005 143 68
    b) 0·006 321   e) 0·000 000 000 058 182
    c) 570 000   f) 0·000 017 2

Fast Forward – page 87

1 a) i) $2 \times 3 \times 3 \times 7 = 2 \times 3^2 \times 7$
      ii) $2 \times 2 \times 3 \times 5 \times 7 = 2^2 \times 3 \times 5 \times 7$
   b) 1260
2 a) 1, 2, 3, 4, 6, 8, 12, 24
   b) 1, 2, 3, 4, 5, 6, 10, 12, 15, 20, 30, 60
   c) 12
3 a) 9   b) 9   c) 12   d) 11
4 a) 19   c) 21   e) 15
   b) 16   d) 8   f) 18
5

Daily sales

	April	May
The Times	386 258	388 196
The Independent	389 523	386 227
The Guardian	429 062	415 426

6 £15 000
7 39
8 490
9 Cost of flour $205 \times 48\,\text{p} \approx 200 \times 50\,\text{p}$
                        $= 10\,000\,\text{p}$
                        $= £100$

14 people pay £0·72
This is less than £14.
Answer is not the right size.
10 a)

   b) 142 bottles at 39 p
      $\approx 100 \times 40$
      $+ 40 \times 40$
      $= 5600\,\text{p}$
      $= £56$
   c) £55·38

11 a) £55   c) £45
    b) £367   d) £360
12 a) 57·803 882   b) 57·8
13 a) £15·80 b) i) $6\frac{1}{2}$ hours ii) £2·92
14 a) £4·50   b) £6·60   c) £6
15 50 kg
16 a) 1·75   b) 8·75
17 a) 0·875   b) 12·5%

18 a) 252   b) 25%
19 a) $\frac{5}{8}$ b) $\frac{3}{8}$ c) 6
20 1 hour 13 minutes 20 seconds
21 35 miles per gallon
22 a) 48 km per hour b) 37·5 miles per hour
23 a) £164.50 b) £136   c) £208
24 12.5 minutes
25 £27
26 a) $\frac{3}{8}$ b) $\frac{1}{8}, \frac{1}{4}, \frac{3}{8}, \frac{1}{2}$
27 a) £3   b) 50 p
28 a) £76   b) £1036
29 a) $\frac{7 \cdot 2 \times 2 \cdot 9}{14 \cdot 4} = \frac{20 \cdot 88}{14 \cdot 44} = 1 \cdot 45$
   b) $(8 \cdot 1)^2 \div 24 = 65 \cdot 61 \div 24 = 2 \cdot 733\,75$
30 a) 73·75   b) £14·24
31 a) $\frac{1}{15}$ b) 90 p
32 a) £21·25 b) 200
33 a) i) 12 000   ii) 20 000
   b) i) $\frac{12\,000}{20\,000}$ ii) 60%
34 £360
36 a) £105   b) £76·65   c) 7·665%
36 £120
37 a) £7·99   b) 20%
38 £55 935
39 a) £8640   b) 5240   c) £1414.80
40 a) £286   c) £59   e) £274·94
   b) £26   d) £236   f) £24·99
   g) The Beovision as the Flan works out
      nearly as much because of the interest.
41 a) £60   b) £180   c) £30
42 £23·76
43 a) 1 400 000, 60 000 b) 1 million hectares
44 6
45 a) 57·1   b) 36%
46 £1771·56
47 a) £276   b) 26   c) £6·91
48 12 cm, 24 cm
49 75
50 80
51 Jill 12, Dave 18
52 a) 3 m   b) 4
53 a) $\frac{1}{15}$ b) $\frac{1}{5}$
54 a) 2 ounces b) 3 ounces c) 31·5 ounces
55 a) 7°C   b) 6° C
56 a) -2°C   b) 10°C
57 a) -10°C c) 8°C   e) 3°C
   b) Thurs   d) 1°C
58 a) 123 million km b) $1 \cdot 23 \times 10^6$
59 a) 5×10^{101} b) 5×10^{-7}
60 52 000 000
61 4·299 817
62 36 100 000
63 $1 \cdot 496 \times 10^8$
64 a) $n = 7$ b) i) $\frac{1}{64}$ ii) $1 \cdot 5625 \times 10^{-2}$

517

ALGEBRA

SECTION 1
ESTABLISHING RULES

Pause – page 103
1 a) $y - 21$ b) $y + 11$ c) $y - 1$ d) $y - 5$
2 a) $x + 7$ b) $x + 13$ c) $x - 7$ d) $x - 13$

Pause – page 104
1 a) $3n$ d) $\dfrac{n}{2} - 15$ g) $\dfrac{n}{2} - 8$

  b) $\dfrac{n}{2}$ e) $3n - 30$ h) $3n + 15$

  c) $3n + 25$ f) $\dfrac{n}{2} + 12$

2 a) $3 + e$ e) $\dfrac{5}{w}$ i) $\dfrac{t}{2}$

  b) $v - 18$ f) $18 - v$ j) $\dfrac{2}{t}$

  c) $5w$ g) $12d$ k) $m - 2$

  d) $\dfrac{w}{5}$ h) $y + 13$ l) $2 - m$

Pause – page 106
1 a) i) 3   ii) 4   iii) 5   iv) 6
  b) $b = p + 1$
2 a) i) 48   ii) 120   iii) 240   iv) 576
  b) $r = 24b$
3 a) i) 460   ii) 400   iii) 269   iv) 51
  b) $n = 500 - s$
4 a) i) £12   iii) £738·20
     ii) £500   iv) £10 481·92
  b) $s = \dfrac{w}{12}$
5 a) i) £5   iii) £7·50
     ii) £6   iv) £5·45
  b) $C = \dfrac{60}{n}$
6 a) i) £240   iii) £189·50
     ii) £190   iv) £227·86
  b) $l = w - 40$

Pause – page 109
1 a) i) 14 cm   iii) 21 cm
     ii) 36 cm   iv) 62·4 cm
  b) $p = 2l + 2h$
2 a) i) £585   ii) £705   iii) £765
  b) $t = 12n + 225$
3 a) i) £19·50   ii) £17·63   iii) £17
  b) $p = \dfrac{225}{n} + 12$
4 a) i) 12   ii) 36   iii) 40   iv) 40
  b) $p = rc$
5 a) i) 9   ii) 25   iii) 64   iv) 16
  b) $p = r^2$
6 a) i) 27   ii) 64   iii) 125   iv) 512
  b) $b = r^3$

Pause – page 111
1 a) 33 b) 13 c) 65 d) 25 e) 0
2 a) 13 b) 7   c) 1   d) 5   e) 6
3 a) 30 b) 8   c) 50 d) 12 e) 57
4 a) 5   b) 55 c) 17 d) 43 e) 17
5 a) 14 b) 14 c) 21 d) 21 e) 21

Pause – page 112
1 a) 12 b) 6   c) 2   d) 9   e) 2
2 a) 60 b) 61 c) 3   d) 1   e) 10
3 a) 4   b) 60 c) 10 d) 14 e) 27
4 a) 49 b) 51 c) 2   d) 0·5 e) 12
5 a) 66 b) 17 c) 19 d) 33 e) 1

Pause – page 113
1 a) 27 e) 30 i) 45   m) 1   q) 81
  b) 28 f) 15 j) 72   n) 25 r) 150
  c) 42 g) 18 k) 2   o) 75 s) 1080
  d) 30 h) 60 l) 2   p) 216 t) 270
2 a) 9   e) 0   i) 100 m) 28 q) 18
  b) 3   f) 20 j) 320 n) 22 r) 21
  c) 11 g) 40 k) 20   o) 28 s) 41
  d) 23 h) 16 l) 117 p) 21 t) 400

Pause – page 114
1 a) -17 c) 25   e) 50
  b) -37 d) 75
2 a) 7   c) -5 e) -6
  b) 13   d) -1
3 a) -8 c) 12   e) -57
  b) -30 d) 50
4 a) 0   c) -80 e) -40
  b) -80 d) 80
5 a) -96 c) 96   e) 132
  b) 32   d) -24

Pause – page 115
1 a) -12 c) -2 e) -2
  b) -18 d) 5
2 a) -60 c) -3 e) -10
  b) -23 d) -7
3 a) -20 c) -5 e) -2
  b) -29 d) -2
4 a) -200 c) -8 e) -120
  b) 150   d) -5
5 a) -6 c) -12 e) -142
  b) -18 d) -4

Pause – page 116
1 a) 3   f) -15 k) 2   p) -216
  b) -8 g) 18   l) 2   q) 81
  c) 42   h) -60 m) 1   r) -150
  d) 30   i) -45 n) 25   s) -1080
  e) -30 j) 72   o) 75   t) -270

518

2 a) 1   f) −20   k) 20   p) 21
b) 7   g) 40   l) −83   q) 18
c) −1   h) 16   m) −12   r) −11
d) 7   i) −100   n) −22   s) 9
e) −20   j) −320   o) −12   t) 400

Pause – page 117

1

m	Am	I
1	75	0·5
2	150	1
3	225	1·5

2 a)

Time (t sec)	Depth ($5t^2$ metres)
1	5
2	20
3	45
4	80
5	125

b)

Time (t sec)	Velocity ($10t$ metres per sec)
1	10
2	20
3	30
4	40
5	50

c) 5 m   d) 20 metres per second

3 a)

Speed (v)	v^2	$20v$	$v^2 + 20v$	d
10	100	200	300	5
20	400	400	800	13·3
30	900	600	1500	25
40	1600	800	2400	40
50	2500	1000	3500	58·3
60	3600	1200	4800	80
70	4900	1400	6300	105

b) No, (3·2 times as far).

Pause – page 119

1 $3e$
2 $6h$
3 $12y$
4 $2a + 2b$
5 $2m + 10$
6 $2x + 13$
7 $2p + 3q + 8$
8 $2x + 2y + 4$
9 $2a + 2b$

Pause – page 120

1 $7m$
2 $4t$
3 $9e$
4 $2x$
5 $11z$
6 $6r$
7 $6a$
8 $5w$
9 $2x$
10 $6m$
11 $-2d$
12 $-5k$
13 $5x + 5y$
14 $9e + f$
15 $10m + 3n$
16 $4s + 7r$
17 $2r + 4q$
18 $9x + 2y$

Pause – page 120

1 $18x + 17y$
2 $4x + 17y$
3 $18x + 7y$
4 $4x - 17y$
5 $6x + 6y - xy$
6 $2p + 3q + 3pq$
7 $m^2 + 8m + 15$
8 $b^2 + 3b - 10$
9 $2a^2 - 14a + 12$
10 $3v^3 + 5v^2 + 2v$
11 $9ab$
12 $11st$

Pause – page 121

1 $12v$
2 $14m$
3 $40t$
4 $35u$
5 $16c$
6 $18z^2$
7 $12z^3$
8 $15m^4$
9 $24r^2$
10 $21n^3$
11 $-6v$
12 $-42m$
13 $-12t$
14 $-15u$
15 $8c$
16 $4z^2$
17 $-20z^3$
18 $-12m^4$
19 $10r^2$
20 $-36n^3$

Pause – page 121

1 $6w + 21$
2 $20m + 10$
3 $18 + 12v$
4 $15r - 20$
5 $12x - 6$
6 $16 - 8y$
7 $4x^2 + 4$
8 $6w^3 + 30$
9 $15 + 6m^2$
10 $15e + 10f$
11 $21u - 14v$
12 $6s - 9t$
13 $-3w - 21$
14 $-10m - 20$
15 $-12 - 18V$
16 $-24 - 16s$
17 $-12r + 16$
18 $-4x + 10$
19 $-14 + 7m$
20 $-4x^2 - 12$

Pause – page 122

1 $16r + 9$
2 $7x + 17$
3 $21 + 9r$
4 $13y + 3$
5 $10x + 42$
6 $9x + 45$
7 $x^2 + 5x + 6$
8 $5x^2 + 3x - 2$
9 $2x^2 + x - 3$
10 $26g - 73$
11 -72
12 $-72r$
13 $8m - 6n$
14 $14n$
15 $2s^2 + 3s - 5$
16 $-s^2 + 3s - 2$
17 $2 - 3s + s^2$
18 $20n^2 - 10n^3 + 30n - 5$
19 $10n^2 - n - 3$
20 $2x^2 - xy - y^2$

Pause – page 123

1 $x^2 + 8x + 15$
2 $x^2 + 12x + 35$
3 $2x^2 + 7x + 6$
4 $2e^2 + 14e + 20$
5 $2a^2 + 11a + 15$
6 $4w^2 + 12w + 9$
7 $3x^2 + 4x - 7$
8 $2m^2 + 5m - 25$
9 $6n^2 - 17n - 3$
10 $10q^2 - 9q - 9$
11 $2x^2 + x - 1$
12 $3x^2 + 13x - 10$
13 $2e^2 + 10e - 28$
14 $x^2 + 2x - 35$
15 $2x^2 + x - 6$
16 $9 - x^2$
17 $9 - 4x^2$
18 $-w^2 - 4w + 45$
19 $25 - 10w + w^2$
20 $2 - x - 10x^2$

Pause – page 124

1 $5(x + 2)$
2 $3(y + 3)$
3 $2(t + 7)$
4 $7(t + 2)$
5 $3(2x + 5)$
6 $5(x - 3)$
7 $3(g - 7)$
8 $7(2 - x)$
9 $2(8 - z)$
10 $11(w - 2)$
11 $5(2m + 3)$
12 $2(4y + 5)$
13 $5(6 + 5e^2)$
14 $5(2u - 7)$
15 $3(8 - 5y^2)$
16 $x(x + 1)$
17 $x(2x + 3)$
18 $w(1 - 2w)$
19 $y(x + 1)$
20 $z(3y - 1)$

Pause – page 124

1	$10(2x + 1)$	**11**	$2x(x + 5)$
2	$6(2y + 3)$	**12**	$3y(y - 5)$
3	$12(t + 2)$	**13**	$4w(2 - 3w)$
4	$14(3t + 1)$	**14**	$5x(y + 3)$
5	$12(3x + 1)$	**15**	$6y(x + 2)$
6	$15(x^2 - 2)$	**16**	$4f(2e - 5)$
7	$8(2 - z)$	**17**	$3z(2y - 5)$
8	$22(2w^2 - 1)$	**18**	$6y(2z + 3)$
9	$3x(x + 3)$	**19**	$ab(a + 1)$
10	$2y(2y - 3)$	**20**	$a^2(b + 1)$

Pause – page 126

1
a) $(x + 3)(x + 5)$ f) $(x - 2)(x - 5)$
b) $(x + 3)(x + 4)$ g) $(2x + 1)(x + 3)$
c) $(x - 3)(x + 4)$ h) $(x + 9)(2x + 1)$
d) $(x + 5)(x - 2)$ i) $(x + 4)(2x - 3)$
e) $(x - 3)(x - 7)$ j) $(x - 7)(3x - 5)$

2
a) $(x + 3)(x + 6)$ f) $(x + 2)(x + 9)$
b) $(x + 5)(x + 1)$ g) $(x + 18)(x + 1)$
c) $(x + 2)(x + 4)$ h) $(x + 4)(x + 6)$
d) $(x + 3)(x + 5)$ i) $(2x + 1)(x + 3)$
e) $(x + 7)(x + 1)$ j) $(2x + 3)(x + 1)$

3
a) $(x - 3)(x - 2)$ f) $(x - 7)(x - 3)$
b) $(x - 5)(x - 1)$ g) $(x - 4)(x - 5)$
c) $(x - 2)(x - 4)$ h) $(x - 7)(x - 4)$
d) $(x - 3)(x - 5)$ i) $(2x - 1)(x - 1)$
e) $(x - 6)(x - 1)$ j) $(2x - 3)(x - 4)$

4
a) $(x - 3)(x + 2)$ f) $(x - 1)(x + 4)$
b) $(x - 2)(x + 5)$ g) $(x - 6)(x + 5)$
c) $(x + 3)(x - 2)$ h) $(x + 7)(x - 3)$
d) $(x + 2)(x - 5)$ i) $(2x - 1)(x + 3)$
e) $(x + 1)(x - 4)$ j) $(2x - 4)(x + 3)$

SECTION 2
SOLVING EQUATIONS

Pause – page 127

1	$d = 1$	**8**	$s = 2$	**15**	$y = -2$
2	$e = 4$	**9**	$s = 25$	**16**	$m = -1$
3	$x = 2$	**10**	$x = 10$	**17**	$g = -2$
4	$x = 7$	**11**	$d = -6$	**18**	$s = -5$
5	$y = 7$	**12**	$e = -6$	**19**	$s = -6$
6	$m = 7$	**13**	$x = -7$	**20**	$s = -7$
7	$g = 0$	**14**	$x = -10$		

Pause – page 128

1	$d = 21$	**8**	$s = 52$	**15**	$y = 4$
2	$e = 12$	**9**	$s = 29$	**16**	$m = 1$
3	$x = 16$	**10**	$x = 20$	**17**	$g = 0$
4	$x = 19$	**11**	$d = 16$	**18**	$s = 5$
5	$y = 21$	**12**	$e = 10$	**19**	$s = -2$
6	$m = 33$	**13**	$x = 11$	**20**	$s = 3$
7	$g = 2$	**14**	$x = 16$		

Pause – page 128

1	$x = 6$	**8**	$x = 10$	**15**	$z = -11$
2	$m = 4$	**9**	$r = 8$	**16**	$x = 2$
3	$t = 5$	**10**	$z = 2$	**17**	$u = -5$
4	$u = 7$	**11**	$s = -11$	**18**	$t = 10$
5	$e = 2$	**12**	$c = -4$	**19**	$x = -14$
6	$x = 2$	**13**	$m = -10$	**20**	$x = 50$
7	$m = 4$	**14**	$m = 10$		

Pause – page 129

1	$x = 10$	**8**	$x = 49$	**15**	$y = -18$
2	$z = 12$	**9**	$z = 30$	**16**	$y = -18$
3	$m = 100$	**10**	$z = 30$	**17**	$y = 18$
4	$m = 10$	**11**	$y = -12$	**18**	$x = -50$
5	$w = 12$	**12**	$j = -12$	**19**	$x = 50$
6	$f = 25$	**13**	$y = -18$	**20**	$m = -150$
7	$y = 63$	**14**	$y = -18$		

Pause – page 130

1	$x = 2$	**8**	$x = 4$	**15**	$y = 8$
2	$x = 5$	**9**	$x = -2$	**16**	$t = 15$
3	$x = 7$	**10**	$x = -3$	**17**	$y = 4$
4	$x = 2$	**11**	$z = 6$	**18**	$y = -9$
5	$m = 1$	**12**	$x = 8$	**19**	$r = 12$
6	$x = 2$	**13**	$x = 9$	**20**	$x = 20$
7	$x = 5$	**14**	$x = 10$		

21
a) $2e + 10 = 18$ d) $4z + 10 = 90$
  $e = 4$ $z = 20$
b) $4h + 20 = 64$ e) $4a + 12 = 32$
  $h = 11$ $a = 5$
c) $2x + 60 = 84$ f) $3m + 2 = 32$
  $x = 12$ $m = 10$

Pause – page 132

1	$w = 4$	**5**	$r = 2$	**9**	$r = 2$	**13**	$y = 2$
2	$m = 2$	**6**	$x = 5$	**10**	$x = 7$	**14**	$x = 4$
3	$v = 3$	**7**	$y = 2$	**11**	$r = 1$	**15**	$n = -2$
4	$s = 1$	**8**	$m = 3$	**12**	$y = 2$	**16**	$x = -3$

Pause – page 133

1	$p = -7$	**6**	$m = 11$	**11**	$p = 8$
2	$x = 8$	**7**	$x = -4$	**12**	$x = 5$
3	$p = 11$	**8**	$a = 19 \cdot 5$	**13**	$s = 0 \cdot 5$
4	$e = 2$	**9**	$x = 6$	**14**	$x = \frac{5}{18}$
5	$w = 3$	**10**	$x = 4$		

Pause – page 134

1	$x = 1 \cdot 68$	**5**	$m = 1$	**9**	$x = 1 \cdot 46$
2	$x = 0 \cdot 05$	**6**	$x = 2 \cdot 45$	**10**	$x = 0 \cdot 24$
3	$x = 5 \cdot 44$	**7**	$x = 5 \cdot 09$	**11**	$z = 6 \cdot 8$
4	$x = 0 \cdot 33$	**8**	$x = 3 \cdot 71$	**12**	$x = 7 \cdot 56$

Pause – page 135

1	T	**5**	F	**9**	T	**13**	T
2	T	**6**	T	**10**	F	**14**	T
3	F	**7**	F	**11**	T	**15**	F
4	T	**8**	F	**12**	T	**16**	T

Pause – page 136

1
$$-7\ -6\ -5\ -4\ -3\ -2\ -1\ \ 0\ \ 1\ \ 2\ \ 3\ \ 4\ \ 5\ \ 6\ \ 7$$

2
$$-7\ -6\ -5\ -4\ -3\ -2\ -1\ \ 0\ \ 1\ \ 2\ \ 3\ \ 4\ \ 5\ \ 6\ \ 7$$

3
$$-7\ -6\ -5\ -4\ -3\ -2\ -1\ \ 0\ \ 1\ \ 2\ \ 3\ \ 4\ \ 5\ \ 6\ \ 7$$

4
$$-7\ -6\ -5\ -4\ -3\ -2\ -1\ \ 0\ \ 1\ \ 2\ \ 3\ \ 4\ \ 5\ \ 6\ \ 7$$

5
$$-7\ -6\ -5\ -4\ -3\ -2\ -1\ \ 0\ \ 1\ \ 2\ \ 3\ \ 4\ \ 5\ \ 6\ \ 7$$

6
$$-7\ -6\ -5\ -4\ -3\ -2\ -1\ \ 0\ \ 1\ \ 2\ \ 3\ \ 4\ \ 5\ \ 6\ \ 7$$

7
$$-7\ -6\ -5\ -4\ -3\ -2\ -1\ \ 0\ \ 1\ \ 2\ \ 3\ \ 4\ \ 5\ \ 6\ \ 7$$

8
$$-7\ -6\ -5\ -4\ -3\ -2\ -1\ \ 0\ \ 1\ \ 2\ \ 3\ \ 4\ \ 5\ \ 6\ \ 7$$

9
$$-7\ -6\ -5\ -4\ -3\ -2\ -1\ \ 0\ \ 1\ \ 2\ \ 3\ \ 4\ \ 5\ \ 6\ \ 7$$

10
$$-7\ -6\ -5\ -4\ -3\ -2\ -1\ \ 0\ \ 1\ \ 2\ \ 3\ \ 4\ \ 5\ \ 6\ \ 7$$

Pause – page 136

1 a)
b) x values are 3, 4, 5

2 a)
b) x values are 0, 1, 2

3 a)
b) x values are 0, 1, 2, 3, 4, 5, 6

4 a)
b) x values are -2, -1, 0

5 a)
b) x values are 0, 1,

6 a)
b) x values are -4, -3, -2, -1

7 a)
b) x values are -2, -1

8 a)
b) x values are -4, -3, -2, -1, 0, 1, 2

9 a)
b) x values are 2, 3, 4, 5

10 a)
b) x values are -1, 0, 1, 2, 3

Pause – page 137

1 $x < 3$

2 $y \geqslant -3$

3 $e > -7$

4 $y \geqslant 6$

5 $x > 6$

6 $x \leqslant 2$

7 $x < 5$

8 $x \geqslant 4$

9 $x < -2$

10 $x > -3$

11 $z \leqslant 6$

12 $x > 4$

13 $w \geqslant 6$

14 $m < 4$

15 $v \geqslant 0$

16 $p < -5$

17 $x \geqslant -6$

18 $p \leqslant -5$

19 $e > 3$

20 a) $z > 4$ b) $a \leqslant 6$ c) $m \geqslant 2$

Pause – page 139

1 $2 \leqslant x < 5$

$$-7\ -6\ -5\ -4\ -3\ -2\ -1\ \ 0\ \ 1\ \ 2\ \ 3\ \ 4\ \ 5\ \ 6\ \ 7$$

2 $3 \leqslant x \leqslant 6{\cdot}5$

$$-7\ -6\ -5\ -4\ -3\ -2\ -1\ \ 0\ \ 1\ \ 2\ \ 3\ \ 4\ \ 5\ \ 6\ \ 7$$

3 $3 < x \leqslant 7$

$$-7\ -6\ -5\ -4\ -3\ -2\ -1\ \ 0\ \ 1\ \ 2\ \ 3\ \ 4\ \ 5\ \ 6\ \ 7$$

4 $3 < x \leqslant 7$

$$-7\ -6\ -5\ -4\ -3\ -2\ -1\ \ 0\ \ 1\ \ 2\ \ 3\ \ 4\ \ 5\ \ 6\ \ 7$$

5 $-6 \leqslant x \leqslant 3$

$$-7\ -6\ -5\ -4\ -3\ -2\ -1\ \ 0\ \ 1\ \ 2\ \ 3\ \ 4\ \ 5\ \ 6\ \ 7$$

6 $-6 < x \leqslant 3$

$$-7\ -6\ -5\ -4\ -3\ -2\ -1\ \ 0\ \ 1\ \ 2\ \ 3\ \ 4\ \ 5\ \ 6\ \ 7$$

7 $1{\cdot}5 \leqslant x < 2$

$$-7\ -6\ -5\ -4\ -3\ -2\ -1\ \ 0\ \ 1\ \ 2\ \ 3\ \ 4\ \ 5\ \ 6\ \ 7$$

8 $3 < x < 4$

$$-7\ -6\ -5\ -4\ -3\ -2\ -1\ \ 0\ \ 1\ \ 2\ \ 3\ \ 4\ \ 5\ \ 6\ \ 7$$

Pause – page 140

1 $x > -6$

$$-7\ -6\ -5\ -4\ -3\ -2\ -1\ \ 0\ \ 1\ \ 2\ \ 3\ \ 4\ \ 5\ \ 6\ \ 7$$

2 $x \leqslant 2$

$$-7\ -6\ -5\ -4\ -3\ -2\ -1\ \ 0\ \ 1\ \ 2\ \ 3\ \ 4\ \ 5\ \ 6\ \ 7$$

3 $x > -3$

$$-7\ -6\ -5\ -4\ -3\ -2\ -1\ \ 0\ \ 1\ \ 2\ \ 3\ \ 4\ \ 5\ \ 6\ \ 7$$

4 $x \geqslant -2$

$$-7\ -6\ -5\ -4\ -3\ -2\ -1\ \ 0\ \ 1\ \ 2\ \ 3\ \ 4\ \ 5\ \ 6\ \ 7$$

5 $x > 1$

$$-7\ -6\ -5\ -4\ -3\ -2\ -1\ \ 0\ \ 1\ \ 2\ \ 3\ \ 4\ \ 5\ \ 6\ \ 7$$

6 $x \leqslant -1$

$$-7\ -6\ -5\ -4\ -3\ -2\ -1\ \ 0\ \ 1\ \ 2\ \ 3\ \ 4\ \ 5\ \ 6\ \ 7$$

7 $x \geqslant -1$

$$-7\ -6\ -5\ -4\ -3\ -2\ -1\ \ 0\ \ 1\ \ 2\ \ 3\ \ 4\ \ 5\ \ 6\ \ 7$$

8 $x > 4$

$$-7\ -6\ -5\ -4\ -3\ -2\ -1\ \ 0\ \ 1\ \ 2\ \ 3\ \ 4\ \ 5\ \ 6\ \ 7$$

Pause – page 141

1 $e = t - 5$ **3** $d = \dfrac{c}{3}$ **5** $x = \dfrac{y - 3}{2}$

2 $r = m + 4$ **4** $V = IR$ **6** $r = tx - y$

7 $v = \dfrac{s - u}{t}$ **9** $h = \dfrac{2A}{a + b}$

8 $x = \dfrac{2y}{3}$ **10** $v = \dfrac{2s - tu}{t}$

Pause – page 142

1 $x = 5{\cdot}7, y = 2{\cdot}3$ **3** $x = 2, y = 1$

2 $x = 6, y = 5$ **4** $x = 4, y = 3$

5 $x = 1, y = 2$ **8** $p = 4, q = -1$

6 $x = 0, y = 4$ **9** $x = 0, y = -1$

7 $m = 2, n = 2$ **10** $x = 2, y = 5$

Pause – page 143

1 $p = 7, q = 6$ **6** $x = 0, y = -7{\cdot}5$

2 $m = 2, n = -2$ **7** $e = -1, f = 7$

3 $x = 2, y = 0$ **8** $r = 6, s = 9$

4 $a = 2, b = 3$ **9** $m = -3, n = -4$

5 $q = 5, x = 4$ **10** $x = 4, y = -1$

Pause – page 144

1 $m = 3, n = 2$ **6** $x = 4, y = 2{\cdot}5$

2 $m = 5, n = 1$ **7** $m = -1, n = 1$

3 $x = 6, y = -2$ **8** $x = 3, y = 5$

4 $t = -3, s = 4$ **9** $t = 6, s = 5$

5 $e = 12, f = -6$ **10** $a = 0, b = -7$

Pause – page 145

1 $x = 24, y = 16$ **2** fork 80 p, knife 90 p

3 fish & chips £2·50, chicken & chips £2·75

4 length 12 cm, width 6 cm

5 coloured pencil 20 p, felt tip 30 p

Pause – page 146

1 a) $-3, -6$ e) $-1, -7$ i) $-1{\cdot}5, -5$
   b) $-1, -5$ f) $-2, -9$ j) $-2{\cdot}5, -2$
   c) $-2, -4$ g) $-5, -7$
   d) $-3, -5$ h) $-2, -8$

2 a) $3, 2$ e) $1, 6$ i) $3, 5$
   b) $1, 5$ f) $3, 7$ j) $\frac{1}{2}, 4$
   c) $2, 4$ g) $4, 7$
   d) $3, 5$ h) $2, 9$

3 a) $-1, -3$ e) $5, 1$ i) $1{\cdot}5, 1$
   b) $0, 7$ f) $4, -3$ j) $4, -2$
   c) $-3, -6$ g) $2{\cdot}5, -6$
   d) $5, -1$ h) $2, -6$

Pause – page 148

1 0·6 **3** 0·5 **5** 0·4 **7** 1·7 **9** 2·2

2 7·7 **4** 0 **6** 4·7 **8** 2·6 **10** 1·4

SECTION 3
NUMBER PATTERNS AND NUMBER SEQUENCES

Pause – page 149

1 $21, 25$; $+4$ **3** $243, 729$; $\times 3$

2 $15, 13$; -2 **4** $25, 12.5$; $\div 2$

5 $37, 50$; add the next odd number

6 $16, 22$; add the next natural number

7 $26, 31$; $+5$

8 $23, 30$; add the next natural number

9 $364, 1093$; add the next power of 3

10 $7, 2$; subtract the next natural number

11 $13, 21$; add the two previous numbers

12 $625, 3125$; $\times 5$

Pause – page 150

1 a) 20 b) 100 c) 99 d) 17 e) 25

Pause – page 151

1 $2n - 1$ 3 $5n + 1$ 5 $7n - 3$
2 $12n - 7$ 4 $3n + 1$ 6 $2n + 6$

Pause – page 152

1 a)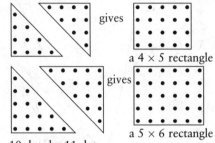

gives

a 4×5 rectangle

gives

a 5×6 rectangle

b) 10 dots by 11 dots
c) 2550 dots, 1275

d) $n(n + 1)$, $\dfrac{n(n + 1)}{2}$

2 a) $(n + 1)^2 + \dfrac{n(n + 1)}{2} = \dfrac{3n^2 + 5n + 2}{2}$

b) $(n + 1)^2 + \dfrac{4n(n + 1)}{2} = 3n^2 + 4n + 1$

3 a) 6   b) 15   c) 3   d) $\dfrac{n(n - 1)}{2}$

4 a) 380   b) 90   c) $n(n - 1)$

Pause – page 154

1	$x \to x + 4$	4	$x \to x^2$
	$-3 \to 1$		$-3 \to 9$
	$-2 \to 2$		$-2 \to 4$
	$-1 \to 3$		$-1 \to 1$
	$0 \to 4$		$0 \to 0$
	$1 \to 5$		$1 \to 1$
	$2 \to 6$		$2 \to 4$
	$3 \to 7$		$3 \to 9$

2	$x \to 2x$	5	$x \to x^2 + 3x - 4$
	$-3 \to -6$		$-3 \to -4$
	$-2 \to -4$		$-2 \to -6$
	$-1 \to -2$		$-1 \to -6$
	$0 \to 0$		$0 \to -4$
	$1 \to 2$		$1 \to 0$
	$2 \to 4$		$2 \to 6$
	$3 \to 6$		$3 \to 14$

3	$x \to \dfrac{x}{2}$	6	$x \to \dfrac{12}{x}$
	$-3 \to -1.5$		$-3 \to -4$
	$-2 \to -1$		$-2 \to -6$
	$-1 \to -0.5$		$-1 \to -12$
	$0 \to 0$		$0 \to$ Not defined
	$1 \to 0.5$		$1 \to 12$
	$2 \to 1$		$2 \to 6$
	$3 \to 1.5$		$3 \to 4$

Pause – page 155

1 a)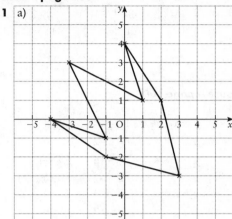

b)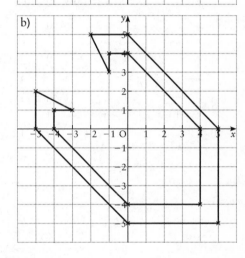

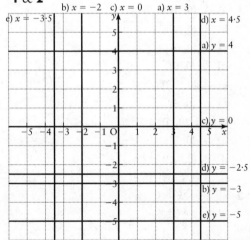

Pause – page 156

1 & 2

b) $x = -2$ c) $x = 0$ a) $x = 3$
e) $x = -3.5$ d) $x = 4.5$
a) $y = 4$
c) $y = 0$
d) $y = -2.5$
b) $y = -3$
e) $y = -5$

3 a) $y = -2$ c) $x = 0$ e) $y = 0$
   b) $x = 2$ d) $y = 7$ f) $x = -5$

1 a) 2   c) 1   e) $\frac{1}{5}$ g) $\frac{3}{2}$ i) −2   k) −5

   b) 3   d) $\frac{1}{2}$ f) $\frac{2}{3}$ h) −1   j) −2   l) $-\frac{1}{4}$

Pause – page 160

1 a)

x	−3	−2	−1	0	1	2	3
$y = x + 4$	1	2	3	4	5	6	7

   b)

x	−3	−2	−1	0	1	2	3
$y = x - 4$	−7	−6	−5	−4	−3	−2	−1

   c)

x	−3	−2	−1	0	1	2	3
$y = 2x$	−6	−4	−2	0	2	4	6

   d)

x	−3	−2	−1	0	1	2	3
$y = \dfrac{x}{2}$	−1·5	−1	−0·5	0	0·5	1	1·5

   e)

x	−3	−2	−1	0	1	2	3
$y = 2x + 4$	−2	0	2	4	6	8	10

   f)

x	−3	−2	−1	0	1	2	3
$y = \dfrac{x}{2} - 4$	−5·5	−5	−4·5	−4	−3·5	−3	−2·5

   g)

x	−3	−2	−1	0	1	2	3
$y = 4 - x$	7	6	5	4	3	2	1

   h)

x	−3	−2	−1	0	1	2	3
$y = x^2$	9	4	1	0	1	4	9

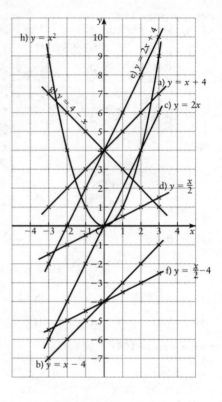

Pause – page 162

1 a) 1, (0, 3)

   b) 1, (0, −3)

   c) 3, (0, 0)

   d) $\frac{1}{4}$, (0, 0)

   e) 3, (0, −3)

   f) $\frac{1}{4}$, (0, 3)

   g) −2, (0, 4)

   h) $\frac{3}{4}$, (0, 3)

   i) 2, (0, −2)

   j) $-\frac{1}{2}$, (0, 8)

2 a) $y = 2x + 3$
   b) $y = 3x + 2$

   c) $y = \dfrac{x}{2} - 1$

   d) $y = -x$
   e) $y = 12 - 2x$

Pause – page 164

No graphs provided. Lines cut axes at:

1 a) $(0, \frac{1}{4})$, (−1, 0)

   b) (0, −4), (2, 0)

   c) (0, 18), (12, 0)

   d) (0, −8), (12, 0)

   e) (0, 4), (−8, 0)

   f) (0, 5), (4, 0)

   g) (0, −7), (7, 0)

   h) (0, 10), (10, 0)

   i) (0, 7), (3, 0)

   j) (0, 4), (6, 0)

Pause – page 165

1 a) $x = 3·75, y = 0·75$

   b) $x = -2, y = -2$

   c) $x = 2, y = 2$

   d) $x = -2, y = 1·5$

   e) $x = 12, y = 2$

1 b)

x	−3	−2	−1	0	1	2	3
y	10	5	2	1	2	5	10

2 a)

x	−3	−2	−1	0	1	2	3
y	8	3	0	−1	0	3	8

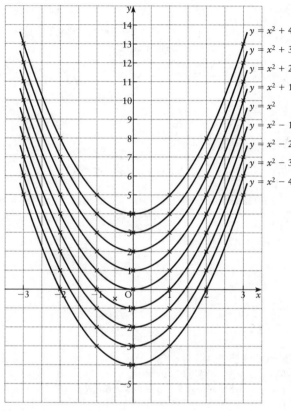

3 a) The curves are all the same shape as x^2. The number added or subtracted at the end gives the intercept on the y-axis, so $y = x^2 - 2$ cuts the y-axis at −2.

b) i) $y = x^2 + 10$ is the same shape as $y = x^2$. It cuts the y-axis at 10.

ii) $y = x^2 - 15$ is the same shape as $y = x^2$. It cuts the y-axis at −15.

Pause – page 168

1 b)

x	−3	−2	−1	0	1	2	3
y	18	8	2	0	2	8	18

d)

x	−3	−2	−1	0	1	2	3
y	4·5	2	0·5	0	0·5	2	4·5

e) i) $y = 5x^2$ goes through the origin. It is 'steeper' than $y = x^2$, but the same general shape.

ii) $y = \frac{1}{4}x^2$ goes through the origin. It is less 'steep' than $y = x^2$, but the same general shape.

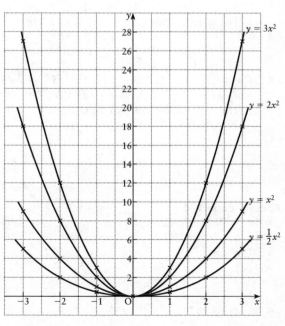

2

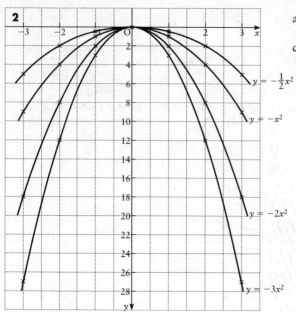

a)

x	-3	-2	-1	0	1	2	3
y	-9	-4	-1	0	-1	-4	-9

c) i) $y = -4x^2$ goes through the origin. It is the same general shape as $y = -x^2$ but goes up and down more steeply.

ii) $y = -\frac{1}{3}x^2$ goes through the origin. It is the same general shape as $y = -\frac{1}{2}x^2$ but is less 'steep'.

Pause – page 170

1 a)

x	-4	-3	-2	-1	0	1	2	3
y	6	0	-4	-6	-6	-4	0	6

2 a)

x	-5	-4	-3	-2	-1	0	1	2	3	4	5
y	7.5	3	-0.5	-3	-4.5	-5	-4.5	-3	-0.5	3	7.5

3 a)

x	0	0.5	1	1.5	2	2.5	3	3.5	4	4.5	5	5.5	6	6.5	7	7.5	8
$16 - 2x$	16	15	14	13	12	11	10	9	8	7	6	5	4	3	2	1	0
Area	0	7.5	14	19.5	24	27.5	30	31.5	32	31.5	30	27.5	24	19.5	14	7.5	0

c) The maximum area is obtained when $x = 4$ metres.

4 a)

t	0	0.5	1	1.5	2	2.5	3	3.5	4
s	0	1.2	4.9	11.0	19.6	30.6	44.1	60	78.4

c) $38.5\,\text{m}$

5 a)

v	10	15	20	25	30	35	40	45	50	55	60	65	70
d	5	8.8	13.3	18.8	25	32.1	40	48.8	58.3	68.8	80	92.4	105

c) The stopping distance of a car going at 50 mph is more than double that of a car going at 30 mph.

Pause – page 174

1 a) $x - 2.8$ or -2.8
 b) $x = 2.2$ or -2.2
 c) $x = 2.6$ or -2.6

2 a) $x = 3$ or -3
 b) 2.7 or -3.7

3 a)

x	-1	0	1	2	3	4	5
y	5	0	-3	-4	-3	0	5

c) $x = 0$ or 4, $x = 3.4$ or 0.6

4 a) $x = 3$ or 1, $x = 0$ or 4
 b) The line $y = -2$ does not cross the graph.

Pause – page 175

1 a)

x	−3	−2	−1	0	1	2	3
y	−27	−8	−1	0	1	8	27

1, 2 and 3

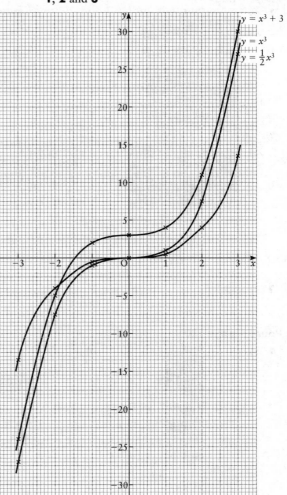

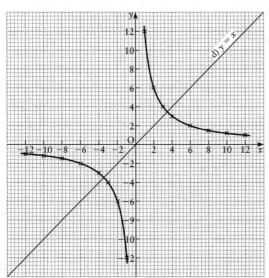

Pause – page 176

1 a)

x	−12	−10	−8	−6	−4	−3	−2	−1
y	−1	−1·2	−1·5	−2	−3	−4	−6	−12

x	1	2	3	4	6	8	10	12
y	12	6	4	3	2	1·5	1·2	1

d) $x = 3·5$ or $−3·5$

2 a)

R	1	2	3	4	5	6	7	8
T	4	2	1·3	1	0·8	0·7	0·6	0·5

d) 2·7

1 a)

c)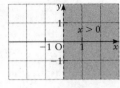

e)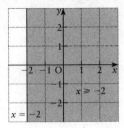

b)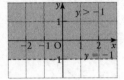

d)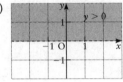

f)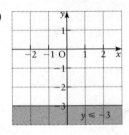

2 a) $x > 3$ b) $y \leqslant 2$ c) $x < -3$ d) $y > 1$

Pause – page 180

1 a)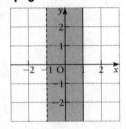

d)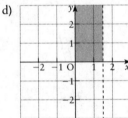

g)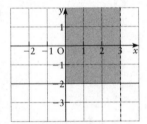

b)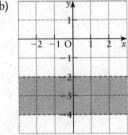

e)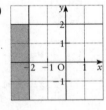

h)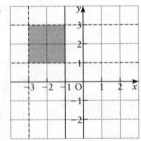

c)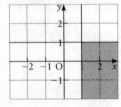

f)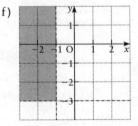

2 a) $-2 < x \leqslant 2$

b) $-2 \leqslant x \leqslant 0, \ 0 \leqslant y \leqslant 1$

c) $-3 \leqslant x \leqslant 0, \ 2 < y < 3$

d) $-1 \leqslant x \leqslant 1, \ -1 \leqslant y < 1$

Pause – page 182

1 a)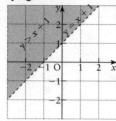

c)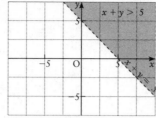

e)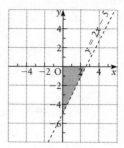

b)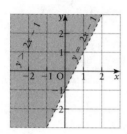

d)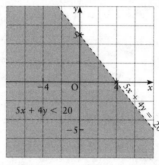

f)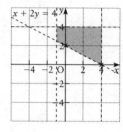

2 a) $16x$ pence   b) $25y$ pence

c) The total she spends is $16x + 25y$ which has to be 200 pence or less.

d)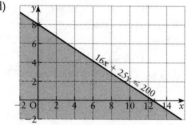

e) Anita can buy 8 skittles at 25 p each.

3 a) $y \leqslant x$, $1 \leqslant x \leqslant 4$ c) $x + y \leqslant 3$, $-1 \leqslant y < 2$

b) $4x + 7y \leqslant 28$, $x > 0$, $y > 0$ d) $y \leqslant 2x$, $-1 \leqslant y < 3$

Pause – page 185

1 a) 2 km   b) 5 minutes   c) 2 km   d) $7\frac{1}{2}$ minutes   e) 5 km

f) i) 24 km per hour   ii) 24 km per hour   iii) 12 km per hour

2 a) 12 noon   c) 16 km   e) 20 km per hour

b) $\frac{3}{4}$ hour   d) 4 km   f) 10 km per hour

5 a) 09 : 36   b) 7 km   c) 96 minutes   d) 4 km/h

6 a) 36 miles   b) 3 hours 48 minutes   d) 4·48 p.m.

7 a) Time $= \dfrac{\text{Distance}}{\text{Speed}} = \frac{7·5}{9}$ hours $= \frac{7·5}{9} \times 60$ minutes $= 50$ minutes

b) and c) (see graph)   d) 28 minutes   e) $\dfrac{42\text{ minutes}}{150\text{ minutes}} = \dfrac{14}{50}$

b) and c)

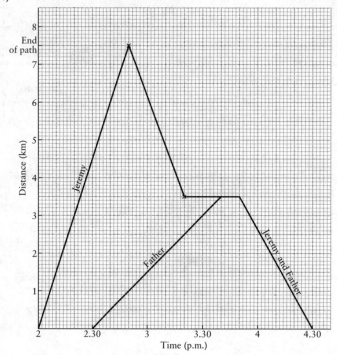

Pause – page 189

1 a) $C = 20d + 80$
 c) For up to three days work the Rick Dastardly Agency is cheaper. For four days work they charge the same price, £160. For more than four days the Purple Panther Agency is cheaper.

2 $c = 5d + 15$ (c is the cost in millions of pounds, d is the length of motorway in km)
 b) £6 250 000 per km
 c) £8 750 000 per km

3 a) 20 bulbs cost £2, 21 cost £1·89 and 22 cost £1·98. Anyone buying 20 bulbs might as well buy 21 or 22. Orders of 41, 42, 43 or 44 bulbs cost less than 40 bulbs (45 bulbs cost the same as 40 bulbs).

4 a) $C = 25 + 0·02m$
 d) A   e) B   f) C

Pause – page 192

1 a) £40
 b) There is a standing charge of £10.
 c) i) 500
   ii) 30
   iii) $\frac{3}{50}$
 d) 50 units cost £3

2 a) 25 cm   b) 270 g
 c) i) 500   ii) 25   iii) $\frac{1}{20}$
 d) 20 g stretches the spring by 1 cm

3 a) 15 cm   b) 28 min
 c) i) 160   ii) −15   iii) $-\frac{1}{4}$
 d) The candle burns down 1 cm in 4 minutes.

4 i) C   ii) B   iii) A

Rewind – page 194

1 a) $y = x + 5$ c) $F = \dfrac{9C}{5} + 32$
 b) $r = 3t - 4s$ d) $v = u + at$

2 a) 9   c) 21   e) 18   g) 54   i) 107
 b) −2   d) 19   f) 18   h) 648   j) 48

3 a) −3   c) −7   e) 2   g) −2   i) 27
 b) −6   d) 29   f) −22 h) 8   j) −8

4 36 p

5 a) $20x + 6y$ e) $2m^2 + 3m + 27$
 b) $x + 8y$ f) $2a^2 - 16a + 14$
 c) $8x - 3y$ g) $9v^3 + 2v^2 + 2v$
 d) $6x + 7y$ h) $7s^2t - st^2$

6 a) $12r + 20$ f) $52r - 2$
 b) $20 - 28r$ g) $8m - 6n$
 c) $5n - 3m$ h) $1 - a$
 d) $ab - a$ i) $12n^2 - 22n - 5$
 e) $12r^2 - 8r^3 + 36r$ j) $3xy - x^2 - 2y^2$

7 a) $x^2 + 5x + 4$ d) $b^2 - 2b + 1$
   b) $x^2 + 6x - 27$ e) $x^2 - y^2$
   c) $5x^2 - 12x - 9$ f) $x^2 + 6xy - 9y^2$
8 Oil left $= 12\,000 - 800t$
9 a) $5(4x + 7)$
   b) $b(1 - 6a)$
   c) $5e(3e - f)$
   d) $xy(y + x + 1)$
   e) $cd(ab + a + b)$
   f) $7mn(2m - 3n + 1)$
10 a) $m = 6$ i) $v = -6$
   b) $t = 9$ j) $x = 2$
   c) $x = -2 \cdot 167$ k) $m = 5$
   d) $t = 59$ l) $m = -5$
   e) $r = 25$ m) $x = 24$
   f) $r = -10$ n) $x = 9$
   g) $x = 50$ o) $x = -0 \cdot 5$
   h) $t = -65$ p) $x = -6$
11 a) $p = -4$ e) $a = 32$
   b) $x = 5$ f) $x = 1$
   c) $m = -36$ g) $p = 0$
   d) $x = 12 \cdot 5$ h) $x = -4$
12 a) $x = 5$ b) $y = 5 \cdot 4$
13 a) $x < 6$ e) $x > 12$
   b) $y > -15$ f) $w < -15$
   c) $e \leqslant -8$ g) $e \geqslant 6 \cdot 6$
   d) $y \geqslant 26$ h) $h \leqslant 3 \cdot 5$
14 a) $x = y - 5$ f) $r = \dfrac{tx - m}{t}$
   b) $t = p + 7$ g) $u = s - vt$
   c) $r = \dfrac{c}{2\pi}$ h) $x = \dfrac{8y}{7}$
   d) $C = \pi d$ i) $a = \dfrac{2A}{b} - b$
   e) $x = \dfrac{y + 1}{4}$ j) $t = \dfrac{2s}{u + v}$
15 a) $x = 5, y = 4$ d) $a = 2, b = -1$
   b) $x = 3, y = 4$ e) $p = 5, q = 7 \cdot 5$
   c) $p = 17, q = 10$
16 a) $a = 3, b = 2$ c) $m = 5, n = 17$
   b) $e = 2, f = -2$ d) $q = 7 \cdot 5, p = 10 \cdot 5$
17 a) $m = 4, n = 1$ d) $x = 5, y = 2 \cdot 5$
   b) $m = -2, n = 3$ e) $t = -2, s = -5$
   c) $x = 3, y = -1$
18 a) $2 \cdot 4$ c) $1 \cdot 1$
   b) $1 \cdot 6$ d) $2 \cdot 3$
19 a) 22 c) 3999
   b) 4000 d) 29
20 a) 2, 5, 8, 11, 14 and 29
   b) 9, 13, 17, 21, 25 and 45
   c) 360, 180, 120, 90, 72 and 36
   d) 1, 8, 27, 64, 125 and 1000
21 a) $4n - 3$
   b) $11n - 5$
   c) $9 \times 2^{n-1}$

22 a) $x \to x + 5$ d) $x \to x(x + 1)$
      $-3 \to 2$ $-3 \to 6$
      $-2 \to 3$ $-2 \to 2$
      $-1 \to 4$ $-1 \to 0$
      $0 \to 5$ $0 \to 0$
      $1 \to 6$ $1 \to 2$
      $2 \to 7$ $2 \to 6$
      $3 \to 8$ $3 \to 12$
   b) $x \to 3x$ e) $x \to x^2 - 3x + 4$
      $-3 \to -9$ $-3 \to 22$
      $-2 \to -6$ $-2 \to 14$
      $-1 \to -3$ $-1 \to 8$
      $0 \to 0$ $0 \to 4$
      $1 \to 3$ $1 \to 2$
      $2 \to 6$ $2 \to 2$
      $3 \to 9$ $3 \to 4$
   c) $x \to \dfrac{x}{2} + 3$ f) $x \to \dfrac{6}{x}$
      $-3 \to 1 \cdot 5$ $-3 \to -2$
      $-2 \to 2$ $-2 \to -3$
      $-1 \to 2 \cdot 5$ $-1 \to -6$
      $0 \to 3$ $0 \to$ not defined
      $1 \to 3 \cdot 5$ $1 \to 6$
      $2 \to 4$ $2 \to 3$
      $3 \to 4 \cdot 5$ $3 \to 2$

23 a)

24 a) $y = -3$ b) $x = 5$
25 a) 2   c) 1   e) $\frac{1}{5}$
   b) $\frac{1}{2}$ d) -1 f) -5
26 a) 1, (0, 5)
   b) 0, (0, -6)
   c) 2·5, (0, 1·5)
   d) $\frac{1}{6}$, (0, 0)
   e) 2, (0, -7)
   f) $\frac{1}{2}$, (0, 1)
   g) 2, (0, 6)
   h) $\frac{1}{2}$, (0, $-\frac{1}{2}$)
27 a) $y = 3x + 2$ c) $y = -2x$
   b) $y = \frac{1}{2}x - 5$ d) $y = \frac{-1}{4}x + 5$

28 a) and b)

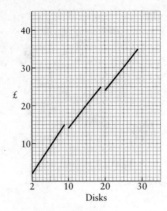

Disks

c) It is cheaper to buy 10 discs (£14) than 9 (£14·60).
It is cheaper to buy 20 discs (£24) than 19 (£24·80).

29 a)

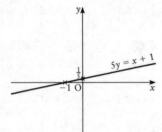

$5y = x + 1$

b)

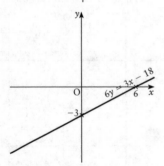

$6y = 3x - 18$

c)

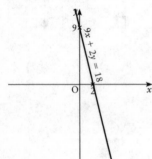

$9x + 2y = 18$

d)

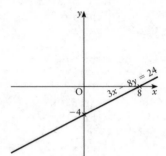

$3x - 8y = 24$

e)

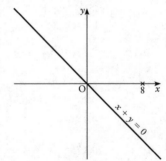

$x + y = 0$

f)

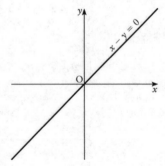

$x - y = 0$

30 a) $x = 1\frac{1}{4}, y = 4\frac{1}{2}$ b) $x = 0, y = 4$

31 a)

x	−5	−4	−3	−2	−1	0	1	2
y	0	−5	−8	−9	−8	−5	0	7

32 a)

Size of rectangle (height and width)	9 × 0	8 × 1	7 × 2	6 × 3	5 × 4	4 × 5	3 × 6	2 × 7	1 × 8	0 × 9
Area (cm²)	0	8	14	18	20	20	18	14	8	0

c) 20·25 cm²

33 a)

x	-3	-2	-1	0	1	2	3	4	5
y	-7	0	5	8	9	8	5	0	-7

c) i) $x = -2, x = 4$ ii) $x = 3\cdot2, x = -1\cdot2$ iii) $x = 3\cdot4, x = -2\cdot4$

35

Number of students on coach	5	10	15	20	25	30	35	40
Cost per student (£)	24	12	8	6	4·80	4	3·43	3

b) 24 students pay exactly £5, 25 or more pay less.

36 a)

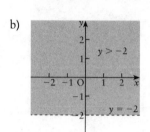

c)

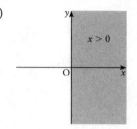

b)

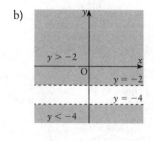

d)

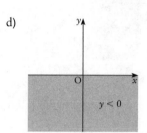

37 a)

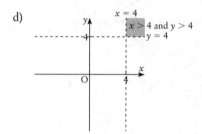

d)

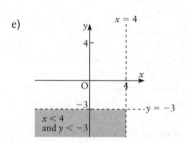

b)

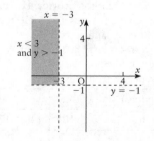

e)

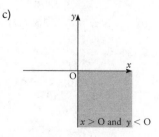

c)

f)

38 a)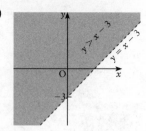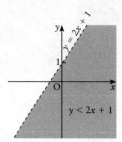

c)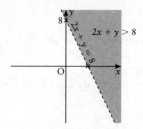

b)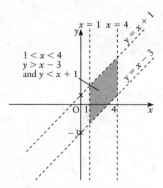

d)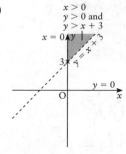

39 a)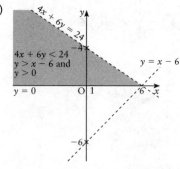

b)

40 a) i) 24   iii) 40   v) $4x + 5y$
ii) $4x$ iv) $5y$

b) $4x + 4y$ is the number of seats. This has to be at least 40.

c) £$30x$ is the cost of hiring x Astras, £$40y$ is the cost of hiring y Carltons, this must not be more than £360.

d) The number of Astras, x, cannot be more than 9.
The number of Carltons, y, cannot be more than 7.

f) 9 Astras and 1 Carlton provide 41 seats and costs £310. 5 Astras and 4 Carltons provide 40 seats and costs £310.

41 Police car overtakes lorry at 12·05 pm, 35 miles from bridge.

42 Graph to be drawn from this table:

Weight (pounds)	1	2	3	4
Time (minutes)	70	105	140	175

Fast Forward – page 202

1 a) $y = 2x + 1$ b) 4

2 a) $C = 7n$ b) $C = 7n - 10$

3 a) 14   b) 7   c) 125   d) 3

4 a) $2x^2 - 5x - 12$ b) $5x(2x - 1)$

5 a) $3pq(4p - 5q)$
b) $2x^2 + 7x - 15$
c) $n = \dfrac{C - 120}{40}$

6 a) 1440   b) 43·5

7 a) 800 watts   b) $I = \sqrt{\dfrac{P}{R}}$

8 6·1

9 a) i) $x + 13$ ii) $x + 13 = 2(x - 7)$
b) 27

10 224 p or £2·24

11 a) £140   b) £3·50

12 a) $x + 50$ b) $2(x + 50)$ c) $4x + 150$
d) $4x + 150 = 670 \Rightarrow x = 130$

13 a) i) $8y$ ii) $6y + 6$
b) i) $8y = 6y + 6$ ii) $y = 3$

14 a) $u = 14$ c) $3ax(x - y)$

   b) $\frac{13}{40}$ or $0{\cdot}325$ d) $t = \sqrt{\dfrac{2s}{a}}$

15 a) $3a + 1$ b) $a = 6$

16 $x = 5$

17 a) $x = 2$ b) $x = 1{\cdot}5$

18 $x = 8{\cdot}5$

19 $x = 2$

20 a) i) $3x$ ii) $x - 9$ iii) $3x - 9$
   b) i) $3x - 9 = 6(x - 9)$ ii) $x = 15$

21 a) $x = 3$ b) $p = 3, q = -3$

22 a) i) $70p - q = 60,\ 56p - q = 39$
   ii) $p = 1{\cdot}5,\ q = 45$
   b) 22

23 a) $x + y = 19$ c) $x = 13{\cdot}5, y = 5{\cdot}5$
   b) $x - y = 8$

24 $x = 7, y = 3$

25 $x = 3{\cdot}3$

26 $x = 2{\cdot}9$

27 $x = 2{\cdot}74$

28 a) 1440 b) 6

29 20

30 a) 86 b) 5

31 $C = \dfrac{V - u}{tu}$

32 a) $a = \dfrac{v^2 - u^2}{2s}$ b) $0{\cdot}5$

33 $-3, -2, -1, 0, 1$

34 a) $n < 3\frac{3}{4}$ b) 3

35 a) $x \leqslant 2$ c) $x < -2$
   b) $-5 < x < 5$

36 a) $-1, 0, 1, 2, 3$ b) $17, 18, 19, 20, 21$

37 a) $110, 135$ b) $1, -6$

38 a) 26
   b) The rule is $+6$
   c) Line 20 $100 \rightarrow 106$

39 a) $35, 42$
   b) Add 7 to the previous number.

40 $4n + 3$

41 a) $323, 971$ b) $3x + 2$

42 a) x^9 b) 1

43 a)

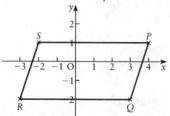

   b) parallelogram

44 a)

x	y
0	1
1	3
3	7

   c) $y = 2x + 1$
   d) $x = 4$

45 a)

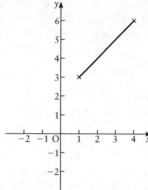

   b) $a = 3$ c) $\begin{array}{l}1 \rightarrow 3 \\ 5 \rightarrow 7 \\ 10 \rightarrow 12 \\ x \rightarrow x + 2\end{array}$

46 $\frac{-1}{2}$

47 b) $(0, 4)$

48 a) $8\,\text{m}$ c) $d = 5\frac{1}{2} t$
   b) $5\frac{1}{2}$ metres per second

49 a)

x	0	20	40	60
y	0	32	64	96

   c) i) $83{\cdot}2\,\text{km}$ ii) $26{\cdot}9\,\text{km}$
   d) $320\,\text{km}$

50 a) $y = 3x + 2,\ y = 5x + 2$
   b) $y = 3x + 2,\ y = 3x - 3$

51 c) $x - y = 3$
   d)

x	3	7	10
y	0	4	7

   e) long life battery (x) $8{\cdot}5$ hours
   standard life battery (y) $5{\cdot}5$ hours

52 b) $x = 2, y = 5$

53 $2 \leqslant x \leqslant 4$

54 a)

x	-2	-1	0	1	2	3
y	-5	-3	-1	1	3	5

   b) & d)
   c) $x = \frac{3}{4}$

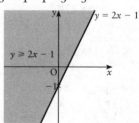

55 a)

V	0	1	2	3	4	5
h	0	5	20	45	80	125

   c) i) $61\,\text{cm}$ ii) $2{\cdot}4$ metres per sec

56 a)

x	-3	$-2{\cdot}5$	-2	$-1{\cdot}5$	-1	$-0{\cdot}5$	0
y	9	$6{\cdot}25$	4	$2{\cdot}25$	1	$0{\cdot}25$	0

x	$0{\cdot}5$	1	$1{\cdot}5$	2	$2{\cdot}5$	3
y	$0{\cdot}25$	1	$2{\cdot}25$	4	$6{\cdot}25$	9

   c) ii) -1
   d) $x = 1{\cdot}3,\ x = -2{\cdot}3$

57 a)

t	0	1	2	3	4	5	6
h	24	30	32	30	24	14	0

    i) 24 m   iii) 6 secs

    ii) 8 m   iv) 12 metres per second

58 a) $h = \dfrac{36 - x^2}{2x}$

  b) Volume $= hx^2$

$$= \frac{(36 - x^2)x^2}{2x} = \frac{36x - x^3}{2}$$

$$= 18x - \frac{x^3}{2}$$

  c)

x	0	1	2	3	4	5	6
V	0	17·5	32	40·5	40	27·5	0

  e) 1·85 cm by 1·85 cm by 8·77 cm

59 a)

v	10	30	40	60	80
s	15	75	120	240	400

  c) 54 miles per hour   d) 400 feet

60 a)

x	1	1·5	2	3	4	5	6
y	5	2·5	2	3	5	7·4	10

  c) ii) $x = 5·4$, $x = 1·1$

  d) $2x^2 - 13x + 12 = 0$

61 a)

x	−2	−1	0	1	2	3	4	5
$x + 1$	−1	0	1	2	3	4	5	6
$x - 4$	−6	−5	−4	−3	−2	−1	0	1
y	6	0	−4	−6	−6	−4	0	6

  c) −6·25

62 a) $36 \div 5$ is not a whole number

  b) i) (2, 18), (3, 12), (4, 9), (6, 6), (9, 4),

     (12, 3)

    iii) 7·2

63 i) c)   ii) d)

SHAPE AND SPACE

SECTION 1
GEOMETRY

Pause – page 228

1 5·8 m

2 22·5 km

4 b) i) 115°   ii) 0·9 m

5 a) Equilateral

Pause – page 231

1 23 km	**6** 112 km	**11** 8·9 cm
2 77 km	**7** 28 km	**12** 14·3 cm
3 65 km	**8** 88 km	**13** 1·7 km
4 196 km	**9** 152 km	**14** 2·9 cm
5 126 km	**10** 102 km	**15** 12·1 cm

Pause – page 232

1 100 m	**6** 410 m	**11** 1·47 cm
2 400 m	**7** 300 m	**12** 8·31 cm
3 180 m	**8** 500 m	**13** 10 cm
4 220 m	**9** 130 m	**14** 13 cm
5 810 m	**10** 5·97 cm	

64 a) vi) $y = \dfrac{3}{x}$ c) iv) $y = 3x^2$

  b) i) $y = x + 3$

65 a) i) 51FF   ii) 25FF

  b) i) £6.50   ii) £4.70

66 b) i) 2·6 gallons   ii) 20·4 litres

67 b) i) £296   iii) £8

    ii) $12\frac{1}{2}$ hours   iv) £240

68 a) 10·00

  b) 15 miles

  c) 30 min

  d) The first part.

  e) 10 mph

  f) 12·30

  g) 6 mph

  i) 10·50

  j) 5·5 miles

69 a) The motorist leaves Bridgend at 09·00 and travels for 50 miles until 10·00. He then rests for half an hour. The motorist starts again at 10·30 and travels a further 50 miles in 1 hour. He arrives in Aberystwyth at 11.30.

70 a) 1.3 m/s   b) 17 r.p.m.

  c) i) 4·7 m/s ii) 28 m

  d) i) 570 m ii) 274 revolutions

71 a) The sound level increased a lot at first then the rate of increase slowed until the sound level was constant.

  b) The electricity was switched off.

72 a) i) 4 m   ii) 2 pm   iii) $11\frac{1}{2}$ hours

  b) 5 hours

Pause – page 234

1 a) 036°   b) 216°

2 a) 315°   b) 135°

3 a) 345°   b) 165°

4 a) i) 70°   ii) 070°

  b) i) 9 cm   ii) 180 m

  c) i) 160 m   ii) 138°

Pause – page 237

6 A → E,   B → F,   C → D

7 a) square based pyramid

  b) triangular prism

  c) triangular-based pyramid

Pause – page 250

1 a) 5°   g) 80°, 60°, 40°

  b) 47°   h) 25·5°

  c) 30°   i) 30°

  d) 40°, 50°   j) 340°, 20°

  e) 67°   k) 135°

  f) 145°

1 $a = 140°$ $b = 40°$ $c = 140°$
2 $d = 85°$ $e = 95°$ $f = 85°$
3 $g = 55°$ $h = 125°$ $i = 55°$
4 $j = 162°$ $k = 18°$ $l = 162°$
5 $m = 34°$ $n = 146°$ $p = 34°$
6 $q = 60°$ $t = 30°$ $r = 90°$ $s = 60°$

Pause – page 253
1 a) $b = f = d = 123°$, $a = c = e = g = 57°$
   b) $a = d = f = i = k = n = l = 63°$,
      $b = c = e = g = h = j = m = p = 117°$
   c) $160°$
   d) $a = 70°$, $b = 80°$, $c = 70°$, $d = 30°$,
      $e = 80°$
   e) $210°$

Pause – page 256
1 a) $30°$ d) $57°$ g) $30°$ j) $20°$
   b) $61°$ e) $84°$ h) $18°$ k) $74°$
   c) $49°$ f) $x = 15°$ i) $42°$ l) $115°$
2 a) equilateral c) ACD
   b) i) $60°$ ii) $120°$
3 i) $56°$ ii) $84°$
4 a) $40°$ b) $40°$
5 $x = 79°$, $y = 22°$, $z = 248°$
6 a) $48°$ b) $84°$ c) $84°$

Pause – page 262
1 a) $360°$ b) $90°$ c) $90°$ d) $50°$
2 a) $1440°$ b) $144°$ c) $36°$
3 a) $720°$ b) $120°$ c) $60°$ d) $95°$
4 a) $1080°$ b) $135°$ c) $45°$
5 a) 24 b) 30 c) 12 d) 36
6 20
7 a) i) $70°$ ii) $265°$
   b) i) $100°$ ii) $35°$ iii) $65°$
8 a) $110°$ b) i) $125°$ ii) $70°$ iii) $165°$
9 a) i) $65°$ ii) $25°$ iii) $40°$ iv) $40°$
   b) i) $48°$ ii) $48°$ iii) $66°$
10 20

Pause – page 266
4 $36°$

SECTION 2
TRANSFORMATION GEOMETRY

Pause – page 268
1 b) A $\begin{pmatrix} +1 \\ +3 \end{pmatrix}$ B $\begin{pmatrix} +4 \\ +2 \end{pmatrix}$ C $\begin{pmatrix} +4 \\ -1 \end{pmatrix}$ D $\begin{pmatrix} +4 \\ -3 \end{pmatrix}$
     E $\begin{pmatrix} 0 \\ -4 \end{pmatrix}$ F $\begin{pmatrix} -5 \\ -4 \end{pmatrix}$ G $\begin{pmatrix} -5 \\ 0 \end{pmatrix}$ H $\begin{pmatrix} -3 \\ +3 \end{pmatrix}$

Pause – page 281
1 a) $3 : 6 = 1 : 2$, scale factor $= 2$
   b) BC $= 4.5$ cm PR $= 24$ cm
2 a) $2 : 4 = 1 : 2$, scale factor $= 2$
   b) MP $= 10$ cm FE $= 3$ cm NO $= 14$ cm

3 LM $= 8$ cm NJ $= 6$ cm
   BC $= 9$ cm DE $= 12$ cm
4 a) $x = 1.2$ m $y = 2.4$ m $z = 2$ m
   b) 0.96 m^2, 2.16 m^2, 3.84 m^2, 6 m^2
   c) $0.96 : 6 = 1 : 6.25$

Pause – page 283
1 a) translation $\begin{pmatrix} +1.5 \\ 0 \end{pmatrix}$ c) rotation of $180°$
   b) translation $\begin{pmatrix} +3 \\ -2 \end{pmatrix}$
2 c) reflection in $y = x$
4 a) i) reflection in y-axis ($x = 0$)
      ii) reflection in $y = x$
      iii) rotation, $90°$ clockwise, centre $(0, 0)$
   b) rotation, $180°$, centre $(0, 0)$;
      translation $\begin{pmatrix} -4 \\ -2 \end{pmatrix}$
   c) reflection in y-axis ($x = 0$) followed by
      reflection in x-axis ($y = 0$)
   d) ii) $9 : 4$

Pause – page 288

4	Name	Diagonal	Axes	Order
a)	rectangle	no	2	2
b)	rhombus	yes	2	2
c)	parallelogram	no	0	2
d)	kite	yes	1	1
e)	square	yes	4	4

SECTION 3
PERIMETER, AREA AND VOLUME

Pause – page 291
1 4 cm^2 **3** 12 cm^2 **5** 4 cm^2 **7** 12 cm^2
2 16 cm^2 **4** 10 cm^2 **6** 8 cm^2 **8** 7.5 cm^2

Pause – page 293
1 a) 36 cm and 32 cm^2
   b) 166 m and 1380 m^2
   c) 556 mm and $18\,645$ mm^2
   d) 25.8 cm and 35.6 cm^2
   e) 13.2 m and 9.45 m^2
   f) 21.4 cm and 24.82 cm^2
   g) 7.9 m and 3.375 m^2
   h) 226 cm and 2136 cm^2
2 a) 25 cm and 30 cm^2
   b) 16 m and 9.25 m^2
   c) 28.2 m and 28.72 m^2
   d) 50 m and 57 m^2. This question raises
      interesting points about the minimum
      data needed to solve a problem.
      We cannot, for example be certain from
      the diagram that the two keys are each
      8 m by 3 m. We can however be certain
      that put together they would form a
      rectangle 8 m by 6 m and hence we can
      calculate the area.

3 a) 6 m   c) £239·76   e) £39·96
    b) 24 m^2 d) 4 m^2

4 a) 36 cm^2 b) i) 6 cm

5 a) 400 m^2 b) 104

6 a) 1·8 m long, 1·2 m wide   c) £31·32
    b) 2·16 m^2

Pause – page 296

1 a) 40 cm^2 d) 5·076 m^2
    b) 50·76 m^2 e) 4058·4 cm^2
    c) 32·93 mm^2

2 a) 15·4 cm^2 b) 48 cm^2

Pause – page 298

1 a) 20 cm^2 d) 2·338 m^2
    b) 23·38 m^2 e) 2029·2 cm^2
    c) 16·465 mm^2

2 a) 55 cm^2 b) 14·5 m^2

3 a) 156 cm   b) 432 cm^2 c) 792 cm^2

Pause – page 299

1 a) 12·5 cm^2 d) 14·25 cm^2
    b) 55·935 m^2 e) 99·715 m^2
    c) 25·2 mm^2

2 a) 34·5 cm^2 b) 544 mm^2

3 a) 118 m   d) 51·43 kg
    b) £177   e) 3
    c) 720 m^2

Pause – page 301

1 16 cm^2 **4** 1950 mm^2

2 147 m^2 **5** 26 m^2

3 75 m^2

6 a) 2500 cm^2 b) 1500 cm^2 c) 3 : 5

Pause – page 303

1 a) 25·1 cm   c) 11·0 m   e) 753·6 m
    b) 50·2 cm   d) 67·8 mm   f) 270·0 mm

2 (using $\pi = 3.14$)
    a) $d = 100$ cm $r = 50$ cm
    b) $d = 31.8$ mm $r = 15.9$ mm
    c) $d = 18.9$ cm $r = 9.45$ cm
    d) $d = 11.8$ mm $r = 5.9$ mm

3 a) 204·1 cm   b) 2041 cm

4 254·96 cm

5 a) 113·04 cm   b) 37·68 cm

6 a) 188·4 cm   b) 3

7 469·9 cm

Pause – page 306

1 a) 12·56 cm^2 f) 3·14 cm^2
    b) 38·465 cm^2 g) 7·065 cm^2
    c) 94·985 cm^2 h) 63·585 cm^2
    d) 572·265 m^2 i) 28·26 m^2
    e) 9498·5 cm^2 j) 176·625 mm^2

2 a) 10 cm   c) 2·00 m
    b) 2·24 m   d) 5·28 mm

3 10·8 kg

4 a) 589·3466 cm^2 b) £206·27

5 a) 24 m by 13$\frac{1}{2}$ m   c) 285·535 m^2
    b) 38·465 m^2 d) 12

6 a) 3·82 m   c) 4·5216 cm^2

Pause – page 310

1 a) 9 cm^3 e) 384·65 cm^3
    b) 336 000 cm^3 f) 0·4823 m^3
    c) 4500 mm^3 g) 588·75 cm^3
    d) 870 cm^3 h) 128·52 m^3

2 a) 17·0 cm   b) 198·0 cm^3 c) 5

3 2·19 cm

4 a) 461·58 cm^3 b) 3 cm

Pause – page 314

1 a) 31 cm^2 e) 296·73 cm^2
    b) 46 900 cm^2 f) 3·6575 m^2
    c) 2100 mm^2 g) 464 cm^2
    d) 687 cm^2 h) 149·94 m^2

2 a) 6450 cm^2 b) 33 750 cm^3

3 a) 36 cm^3 c) 72 cm^2

Pause – page 316

1 a) V   c) L   e) A   g) A
    b) A   d) V   f) V   h) L

2 a) V   c) A   e) L   g) A
    b) A   d) V   f) A   h) A

3 *abc* gives a volume
    2(*a* + *b*) gives the perimeter of a face
    bd gives the area of a face

4 a) *ab*, 4π*rl*

SECTION 4
PYTHAGORAS AND TRIGONOMETRY

Pause – page 318

1 a) 4225   b) 166·41   c) 1·69   d) 53 361

2 2862·25 cm^2

Pause – page 318

1 8   **3** 9   **5** 1   **7** 7 cm

2 2   **4** 3   **6** 6 cm   **8** 80 cm

9 a) 7·75   b) 27·09   c) 2·39   d) 30.60

10 6·71 cm

Pause – page 320

1 10·30 cm   **4** 10·44 cm   **7** 4·72 m

2 13 cm   **5** 7·81 cm   **8** 8·5 m

3 6·1 cm   **6** 260·77 km   **9** 12·28 m

10 a) 3·16 units   c) 3·16 units
     b) 5.66 units   d) isosceles

Pause – page 324

1 12 m   **3** 110 m   **5** 1·53 m   **7** 2·45 m

2 3 m   **4** 7·14 m   **6** 2·38 m

8 a) 4·47 cm   b) 17·88 cm^2

9 31·22 cm^2 **10** 7·90 cm

Pause – page 326

1 a) 26 cm   b) 32 cm^2 c) 3 cm
2 a) 3 cm   b) 5·20 cm   c) 78 cm^2
3 215 m
4 a) AB = 7 m,
      loop = 5·6 + 4·2 + 7 = 16·8 m
   b) i) 4·9 m   ii) 3·43 m   iii) 12·01 m^2
5 10 cm
6 a) 104 m   b) i) 5th   ii) C   c) 21 m

Pause – page 332

1 9 m   2 18 m   3 360 m

Pause – page 335

1 a) 2·887 cm   c) 14·301 mm
   b) 43·301 m
2 4·12 m   3 1·820 m   4 0·839 km

Pause – page 337

1 a) 0·2679491   c) 0·7673269
   b) 4·0107809   d) 1·7710985
2 a) 7·369 m   c) 211·523 mm
   b) 24·185 cm
3 81·906 m   4 5·485 m   5 9·696 m

Pause – page 340

1 a) 14·036°   c) 56·310°   e) 0·997°
   b) 36·686°   d) 85·156°
2 a) 57·995°   b) 50·194°   c) 15·297°
3 61·928°
4 44·29° and 135·71°
5 a) 37·569°   b) 115·11 cm c) 23·479°
6 a) 035·5°   b) 305·5°
7 a) 315°   b) 135°
8 a) 344°   b) 164°

Pause – page 344

1 a) 6·180 m   c) 22·717 mm
   b) 12·379 cm
2 4·698 m   4 14·501 m   6 70 km
3 578·509 m   5 93·2 km   7 144·5 miles

Pause – page 347

1 a) 14·478°   c) 30°   e) 0·859°
   b) 48·159°   d) 59·997°
2 a) 36·870°   b) 50·805°   c) 54·903°
3 The roof has a slope of about 11°, so the
   tiles cannot be used.
4 The incline has a slope of about 6·9° so a
   rack and pinion could be used.
5 50·3°   6 61·9°

Pause – page 350

1 a) 8·480 cm   c) 21·155 m
   b) 10·554 mm
2 38·302 km   5 8·362 cm
3 2·785 m   6 2·9 m
4 8·989 m

Pause – page 353

1 a) 61·315°   b) 29·893°   c) 43·090°
2 043·8°   4 31·1°   6 78·5°
3 69·19°   5 342°

Pause – page 357

1 1·212 m   4 10·06 cm
2 14·522 m   5 6·25 m
3 2·38 m   6 1·57 m

Rewind – page 359

3 b) 14·5 m
5 cuboid, triangular prism
7 a) 18°, acute
   b) 55°, acute; 35°, acute
   c) 148·5° obtuse
   d) 338°, reflex
   e) 20°, acute
8 $a = c = d = f = 125°$ $b = e = g = 55°$
   $j = h = k = m = 60°$ $i = n = l = 120°$
   $o = y = v = 63°$ $p = q = 117°$ $s = 109°$
   $r = t = u = x = 71°$ $v = z = 46°$
9 $a = 40°$ $b = 29°$ $c = 70°$ $d = 75°$
   $e = 30°$ $f = 69°$ $g = 69°$ $h = 70°$
   $i = 50°$ $j = 50°$ $k = 80°$ $l = 98°$
10 a) square   d) parallelogram
    b) trapezium   e) rectangle
    c) kite   f) rhombus
11 a) 900°   c) 128·57°
    b) 360°   d) 51·43°
12 a) 1440°, 360°, 144°, 36°
    b) 1800°, 360°, 150°, 30°
13 18
14 equilateral triangle, square, hexagon
23 $x = 1·8$ m, $y = 2·4$ m, $z = 1·4$ m
26 b) i) 4   ii) 3
27 a) i) 5   b) 108°   c) no
28 12 m and 5·24 m^2
29 10·7 m and 5·24 m^2
30 39·75 cm^2
31 22 cm and 26 cm^2
32 30 m^2
33 18 cm and 12 cm^2
34 41·075 cm^2
35 87·08 cm^2 and 32·86 cm
36 14·18 m^2 and 13·35 m
37 13 mm and 26 mm
38 12 cm and 6 cm
39 a) 64 cm^3 b) 31·4 cm^3 c) 17·3 m^3
40 $R = \dfrac{mnt}{5}$ represents a volume
    $T = 4\pi r^2$ represents an area
    $S = 12l + 3m + 4n$ represents a length
41 a) 10 m   b) 2·6 cm
42 a) 36 cm   b) 0·7 cm
43 5·657 cm   44 2·828 cm

6 a) Type Y
c) i) Type X
ii) The range of weights for Type X is from 0·5 to 3·0 kg and for Type Y from 1·0 to 3·0 kg.

SECTION 3
MAKING COMPARISONS BETWEEN SETS OF DATA

Pause – page 421

1 a) m = 8   r = 6
b) m = 3   r = 5
c) m = £2   r = £1·50
d) m = 1   r = 8
e) m = 5·77   r = 6·2
f) m = 15·3 cm   r = 17 cm
g) m = 495·7 g   r = 791 g
h) m = 197·8 cm   r = 270 cm
i) m = 950g   r = 800 g
j) m = 1744·5   r = 85

2 a) m = 56·2   r = 85
    m = 51·1   r = 83
b) The mean of the Year 10 students is lower. The ranges are very similar.

3 a) m = 25·2   r = 59
    m = 20·7   r = 60
b) Maureen's mean score is higher than Eric's. The ranges are very similar.

4 a) 22·5 metres   b) 4·1 metres

5 a) 225   b) 583   c) 50·5

Pause – page 423

1 a) Estate A   m = 1·78   r = 6
    Estate B   m = 2·82   r = 6
b) The mean number of children per family is much higher for Estate B. The range is the same.

2 a) City A   m = 51·66   r = 5
    City B   m = 52·22   r = 6
b) The mean price of petrol in City B is higher than City A. The range in City B is also slightly greater.

3 a) School A   m = 4·4   (D–E)
    School B   m = 3·64 (C–D)
b) School A   r = 7 (A–U)
    School B   r = 7 (A–U)
c) The average grade for School B is higher than the average grade for School A. The ranges are the same. If the grades are taken to the nearest grade, both schools have an average grade of D.

Pause – page 426

1 a) Rooksend   m = 68·2 cm   r = 160 cm
    Streamside   m = 99 cm   r = 160 cm

b) The mean for Rooksend is much less than the mean for Streamside. Rooksend may contain younger trees or trees of a different species. The range for the two woods is the same.

2 a) Maureen   m = 91·7   r = 179
    Eric   m = 76·7   r = 119
b) Maureen has a much higher mean score than Eric and is probably the better player. Eric has a lower range than Maureen and is more consistent, with fewer really high or really low scores.

3 a) Diet A Sample   m = 28·9 kg   r = 10 kg
    Diet B Sample   m = 29·32 kg   r = 12 kg
b) The pigs on Diet B have a higher mean weight than the pigs on Diet A. Diet B is the more successful. The range for Diet B is greater because two pigs reached the 34·5–36·5 kg class interval.

4 a) Before   m = 82·125 kg   r = 69 kg
    After   m = 74·525 kg   r = 69 kg
The mean weight after the sponsored slim is nearly 8 kg less than before the sponsored slim. The diets have been successful overall. The ranges are the same because people dropped out of the highest class interval and into the lowest class interval.

Pause – page 429

1 a) med = 7   r = 6
b) med = 3   r = 4
c) med = £4·50   r = £4·50
d) med = £4·25   r = £4·50
e) med = 0   r = 6
f) med = 6·1   r = 6·2
g) med = 14·5 cm   r = 17 cm
h) med = 452·5 g   r = 791 g
i) med = 190 cm   r = 306 cm
j) med = 950 g   r = 800 g
k) med = 1745   r = 85

2 a) Year 11   med = 51·5   r = 85
    Year 10   med = 50·5   r = 83
b) The median of the Year 10 students is lower. The ranges are very similar.

3 a) Maureen   med = 20   r = 59
    Eric   med = 20   r = 60
b) The median for both players is the same. The ranges are very similar.

Pause – page 431

1 a) med = 3   r = 5
b) med = 2   r = 4

2 a) med = 39·5   r = 7
b) med = 41   r = 7

c) Flares can only just claim to have 'average contents 40 matches' when the median is corrected to the nearest whole number. Squibs claim is sensible and perhaps even an underestimate. They could claim 'average contents 41 matches' on the basis of this survey.

3 a) med = 1   r = 3
   b) med = 2   r = 2
   c) Line B is producing more televisions with faults than Line A. This is shown by the higher median for Line B. The range for Line A is higher but this is caused by a single television with 3 faults.

4 a) med = 2   r = 5
   b) med = 3   r = 5
   c) The median for boys is higher and we conclude that boys are more likely than girls to eat chips with their school meal.

Pause – page 434

1 b) **Rooksend**   median = 63 cm
      Streamside   median = 103 cm
   c) **Rooksend** $Q_1 = 42$ cm $Q_2 = 84$ cm
      IQR = 42 cm
      Streamside $Q_1 = 80$ cm $Q_2 = 123$ cm
      IQR = 43 cm
   d) 30 trees   e) 62 trees

2 b) **Maureen**   median = 90
      Eric   median = 76
   c) **Maureen** $Q_1 = 50$ $Q_2 = 132$ IQR = 82
      Eric $Q_1 = 54$ $Q_2 = 93$ IQR = 39
   d) 28 times   e) 9 times

3 b) **Before**   median = 83 kg
      After   median = 74.5 kg
   c) **Before** $Q_1 = 72$ kg $Q_2 = 93$ kg
      IQR = 21 kg
      After $Q_1 = 66.5$ kg $Q_2 = 80$ kg
      IQR = 13.5 kg
   d) 45   e) 18

4 b) **Group A**   range = 25 − 18 = 7 min
      IQR = 21.9 − 20.2 = 1.7 min
      Group B   range = 24 − 19 = 5 min
      IQR = 21.4 − 19.8 = 1.6 min
   c) Group B is a 'tighter' distribution with values less spread out than Group A.

5 b) **Group A**   median = 31 range = 60
      IQR = 35 − 29 = 6
      Group B   median = 35 range = 60
      IQR = 41 − 30 = 10
   c) Group B has a higher median and more students with higher marks. Group A is a 'tighter' distribution than Group B with the marks grouped more closely about the median.

6 b) **Ms Salt** $Q1 = 3.5$ mins
      median = 4.4 mins
      $Q2 = 5.4$ mins
      Ms Pepper $Q1 = 4.95$ mins
      median = 5.65 mins
      $Q2 = 6.5$ mins
   c) For Ms Salt: range = 8 − 1 = 7 mins
      IQR = 5.4 − 3.5 = 1.9 mins
      For Ms Pepper:
      range = 8 − 3 = 5 mins
      IQR = 6.5 − 4.95 = 1.55 mins
      These figures show the distribution of times for Ms Pepper is 'tighter'. There is less variation in Ms Pepper's times than Ms Salt's.

Pause – page 439

1 a) 6   f) no sensible mode
   b) 3   g) 12 cm and 15 cm
   c) £1.50   h) 12 g
   d) 0   i) 340 cm
   e) no mode   j) no mode

2 City A = 52
   City B = 53

3 Rooksend: $60 \leqslant c < 80$
   Streamside: $120 \leqslant c < 140$

Pause – page 440

		Mean	Median	Mode	Range
1	a)	3.63	3	3	7
	b)	14.09	12	12	18
	c)	37.93	38	37	5
	d)	100.77	101	100	3
	e)	3	2	1	9

		Burners	Flames
2	a)	40.34	39.87
	b)	40	40
	c)	41	40
	d)	8	6

   e) Burners have a slightly higher 'average' contents than Flames. Flames have a 'tighter' distribution with less variation in the number of matches per box.

3 a)–d)	Mean	Median	Mode	Range
Year 1	4.783	6	6	8
Year 2	3.767	5	5	7

   e) The 'average' number of eggs per nest dropped in Year 2. The range was also less in Year 2.

4 a)–d)

	Mean	Median	Q1	Q2	Mode	Range	IQR
Before	354.7	352	334	376	340–360	140	42
After	315.7	316	296	334	320–340	120	38

Pause – page 472

1 $\frac{1}{1296} = 0.000\,772 = 0.0772\%$

2 $\frac{1}{256} = 0.0039 = 0.39\%$

3 $\frac{1}{8} = 0.125 = 12.5\%$

4 a) $\frac{27}{343} = 0.0787 = 7.87\%$

   b) $\frac{64}{343} = 0.187 = 18.7\%$

   c) $\frac{36}{343} = 0.105 = 10.5\%$

   d) $\frac{36}{343} = 0.105 = 10.5\%$

5 a) $\frac{27}{1728} = 0.0156 = 1.56\%$

   b) $\frac{64}{1728} = 0.037 = 3.7\%$

   c) $\frac{125}{1728} = 0.0723 = 7.23\%$

   d) $\frac{60}{1728} = 0.0347 = 3.47\%$

   e) $\frac{60}{1728} = 0.0347 = 3.47\%$

Pause – page 475

1 a) $\frac{21}{25} = 0.84 = 84\%$

   b) $\frac{13}{25} = 0.52 = 52\%$

2 a) $\frac{17}{125} = 0.136 = 13.6\%$

   b) $\frac{35}{125} = 0.28 = 28\%$

3 $\frac{2}{27} = 0.0741 = 7.41\%$

4 $\frac{648}{1000} = \frac{81}{125} = 0.648 = 64.8\%$

5 $\frac{271}{1000} = 0.271 = 27.1\%$

6 a) $\frac{5}{10} = \frac{1}{2} = 0.5 = 50\%$

   b) $\frac{7}{10} = 0.7 = 70\%$

   c) $\frac{3}{10} = 0.3 = 30\%$

7 a) $\frac{398}{1000} = 0.398 = 39.8\%$

   b) $\frac{496}{1000} = 0.496 = 49.6\%$

   c) $\frac{6}{1000} = 0.006 = 0.6\%$

   d) $\frac{504}{1000} = 0.504 = 50.4\%$

Rewind – page 477

1 See section at start of Data Handling.

2

Mark	Tally	Frequency
1	|	1
2	||||	5
3	|||	3
4	||||	5
5	||||	5
6	|||| ||	7
7	|||| |||	8
8	|||| |	6
9	|||| ||	7
10	|||	3

4 77, 4100   78, 4150   79, 3700   80, 3800
   81, 4450   82, 5050   83, 4500   84, 4250
   85, 7200   86, 5800   87, 6500

		Mean	Median	Q1	Q2	Mode	Range	IQR
7	a)	4·105	4			2	7	
7	b)	22·33	22			21	4	
7	c)	0·545	1			1	6	
7	d)	330·77	300			200	400	
8	a)	39·11	39		40	39	7	
8	b)	1·847	2		2	2	6	
9	a)	15·438	15·5	11·5	19·5	15–20	30	8
9	b)	4·515	4·4	4·1	4·8	4·0–4·5	9	0·4

10 b) 5

11 a) $\frac{1}{2} = 0.5 = 50\%$

    b) $\frac{5}{6} = 0.833 = 83.3\%$

    c) $\frac{2}{3} = 0.667 = 66.7\%$

    d) $\frac{1}{6} = 0.167 = 16.7\%$

12 a) $\frac{1}{4} = 0.25 = 25\%$

    b) $\frac{1}{2} = 0.5 = 50\%$

    c) $\frac{1}{52} = 0.0192 = 1.92\%$

    d) $\frac{3}{13} = 0.231 = 23.1\%$

    e) $\frac{3}{4} = 0.75 = 75\%$

    f) $\frac{7}{13} = 0.538 = 53.8\%$

    g) $\frac{1}{26} = 0.0385 = 3.85\%$

    h) $\frac{4}{13} = 0.308 = 30.8\%$

13 a)

Females	11	25	36	46	57	69	85	97	107	122
Total	20	40	60	80	100	120	140	160	180	200
Relative frequency	$\frac{11}{20}$	$\frac{25}{40}$	$\frac{36}{60}$	$\frac{46}{80}$	$\frac{57}{100}$	$\frac{69}{120}$	$\frac{85}{140}$	$\frac{97}{160}$	$\frac{107}{180}$	$\frac{122}{200}$

    b) $\frac{122}{200}$ c) 732

14 a) $\frac{7}{18} = 0.389 = 38.9\%$

    b) $\frac{1}{36} = 0.028 = 2.8\%$

    c) $\frac{5}{18} = 0.278 = 27.8\%$

    d) $\frac{1}{6} = 0.167 = 16.7\%$

    e) $\frac{1}{3} = 0.333 = 33.3\%$

    f) 0   g) 0   h) $\frac{1}{18} = 0.0556 = 55.6\%$

15 a) $\frac{1}{16} = 0.0625 = 6.25\%$

    b) $\frac{1}{32\,768} = 0.000\,030\,5 = 0.003\,05\%$

    c) $\frac{1}{60\,466\,176} = 0.000\,000\,016\,5$
                $= 0.000\,001\,65\%$

16 a) $\frac{19}{25} = 0.76 = 76\%$

    b) $\frac{6}{25} = 0.24 = 24\%$

17 $\frac{79}{100} = 0.79 = 79\%$

Fast Forward – page 482

2 a) Monday

   b) i) 2500   ii) 3200

   c) $6000 \times \pounds14.20 = \pounds85\,200$

3 a)

Colour	Tally	Frequency
White	ℍℍ ℍℍ	10
Blue	|||	3
Red	ℍℍ ℍℍ ||||	14
Green	|||	3

4 a) i) 6 ii) 3

b)

Mushrooms	Days	Sub total	Cumulative frequency
0	3	0	3
1	3	3	6
2	6	12	12
3	4	12	16
4	4	16	20
5	4	20	24
6	3	18	27
7	1	7	28
8	1	8	29
9	1	9	30

Total = 105

c) 2 d) 4 e) 3·5

f) $\frac{20}{30} = \frac{2}{3} = 0·667 = 66·7\%$

5 a) 80 and 50

7 a) & b)

Type of crisp	Frequency	Angle
Plain	9	72°
Salt and Vinegar	16	128°
Cheese and Onion	11	88°
Beef	6	48°
Crispy bacon	3	24°

8 a) 46%, 28%, 16%, 10%

9 a) 150° b) $\frac{120}{360} = \frac{1}{3}$

10 a) 10% c) £126 000 000

b) 63 : 9 = 7 : 1

11 a) 28 − 3 = 25

b)

Class interval	Frequency
1–5	2
6–10	4
11–15	5
16–20	7
21–25	13
26–30	4

12 a) i) $\frac{32}{200} = 0·16 = 16\%$

ii) $\frac{106}{200} = 0·53 = 53\%$

13 a) 6 b) 2 c) 3 d) 6

14 a) 7·2 and 6·8

b) 8 and 3

c) Samantha — because she has a higher mean
Teresa — because she is more consistent

15 a) 20 d) £1·40

b) £29·80 e) £1·30

c) £1·49

16 a) 12 b) 29 cm

c) The mean is the same but the range this year is greater. This means there is wider variation in the lengths than in last year's competition.

17 a) 61 mm

b) i) 352 ii) 55

c) The Gambia

d) Although the Gambia has a higher mean rainfall, the range is far higher. This means the variation between months is greater, with no rainfall at all from December to April. This is more likely to lead to shortages.

18 a) 30

b) Ian's range is 80 and David's range is 38. They have the same mean but David is the more consistent batsman, with fewer really high scores or very low scores.

19 a) 4°C b) 1°C c) 1·4°C

20 a)

Class Interval	Frequency f	Mid-interval valve x	fx
1–5	16	3	48
6–10	28	8	224
11–15	26	13	338
16–20	14	18	252
21–25	10	23	230
26–30	3	28	84
31–35	1	33	33
36–40	0	38	0
41–45	2	43	86
	Total 100		Total 1295

b) i) 6–10 ii) 11–15

c) 12·95

21 b) £18·90

c) 6·4 miles

22 c) i) 179·75 cm ii) 8·5 cm iii) 37

23 a) i) 222 g ii) 48 g

b) 110

24 b) i) 43·5 mph ii) 10·5

25 a) 56·3

c) i) 84 ii) 48

26 b) Strong positive correlation.

27 b) There is a negative correlation between value and age, as the cars get older they are worth less.

c) The car which is $5\frac{1}{2}$ years old is worth only £2200, which is less than the cars which are 6 and 7 years old. It may have a very high mileage or have been in an accident or had some other factor to decrease its value.

e) £6900